MATHEMATICS
AND MUSIC

COMPOSITION, PERCEPTION, AND PERFORMANCE

MATHEMATICS
AND MUSIC

COMPOSITION, PERCEPTION, AND PERFORMANCE

JAMES S. WALKER • GARY W. DON

UNIVERSITY OF WISCONSIN
EAU CLAIRE, USA

CRC Press
Taylor & Francis Group
Boca Raton London New York

CRC Press is an imprint of the
Taylor & Francis Group, an **informa** business

A CHAPMAN & HALL BOOK

CRC Press
Taylor & Francis Group
6000 Broken Sound Parkway NW, Suite 300
Boca Raton, FL 33487-2742

Printed on acid-free paper
Version Date: 20130129

International Standard Book Number-13: 978-1-4398-6709-9 (Paperback)

Library of Congress Cataloging-in-Publication Data

Walker, James S.
   Mathematics and music : composition, perception, and performance / James S. Walker, Gary W. Don.
     pages cm
     Includes bibliographical references and index.
     ISBN 978-1-4398-6709-9 (pbk. : alk. paper)
     1. Music--Mathematics. 2. Music and geometry. 3. Music theory. I. Don, Gary W. II. Title.

ML3805.W25 2013
781--dc23
                                                2013002169

Visit the Taylor & Francis Web site at
http://www.taylorandfrancis.com

and the CRC Press Web site at
http://www.crcpress.com

To Angela

Contents

Preface

No school would eliminate the study of language, mathematics, or history from its curriculum, yet the study of music, which encompasses so many aspects of these fields and can even contribute to a better understanding of them, is entirely ignored.

—Daniel Barenboim

The purpose of this book is to explore the connections between mathematics and music. This may seem to be a curious task. Aren't mathematics and music from separate worlds, mathematics from the world of science and music from the world of art? While mathematics does belong to the world of science, one of the goals of science is to understand everything that we experience, and music is no doubt an essential part of human experience. Mathematics has been described as the science of patterns, and we shall see that there are many patterns in music that can be described with mathematics. Mathematics has also been described as the language of the universe, and music itself has been described in such a poetic way. In fact, connections between these two subjects go back thousands of years. For example, the classical Greek mathematician, Pythagoras, contributed the essential ideas for how we quantify changes of pitch for musical tones (musical intervals). The connections between mathematics and music have grown enormously since those ancient days. We will try to explore as many of these connections as possible, in a way that presents both the mathematics and the music to as wide an audience as possible.

Summary of Chapters

Here is a brief description of the main topics covered in the book. For more details, please consult the Table of Contents.

Chapter 1 describes the scientific approach to musical pitch, first worked out by Helmholtz in the 19th century. Helmholtz's theory, which relates pitch to frequency, provides a foundation for understanding different musical scales. One very distinctive aspect of our treatment of this material, is that we use the method of *spectrograms*. A spectrogram is a graphical portrait of the tones within a musical passage, plotting these tones in terms of their frequencies and the time during which they are sounding. We believe that spectrograms are an important tool for understanding and appreciating music, and that they are not difficult to interpret correctly. So we introduce them before we describe the mathematics used to create them; we postpone that discussion to Chapter 4. Although some might object to using a mathematical technique before describing the details underlying it, we believe that the spectrogram examples described here are so compelling, and so dramatically illustrate this material, that we simply had to include them. In any case, they should provide a strong motivation for learning the mathematics of spectrograms described in Chapter 4. Chapter 2 provides a brief introduction to musical notation. It describes just enough notation so that all readers, even those who are not musicians, should be able to read the brief score excerpts that we include in the book. There are a number of such score excerpts in Chapter 3, where we provide some background in basic music theory. This basic music theory is surprisingly mathematical. We emphasize the different musical transformations—scale shiftings, transpositions, inversions—that composers have employed for centuries. These transformations do have a clear mathematical interpretation.

As described in the last paragraph, in Chapter 4 we discuss the mathematics of *spectrograms*. In addition to the mathematics, we also provide some interesting musical illustrations, such as the phenomenon known as ***beating*** and its relation to musical consonance and dissonance. In Chapter 5 we demonstrate how spectrograms provide revealing insights into musical structure. These insights would be difficult if not impossible to obtain through listening alone, because listening involves mostly short-term memory, while spectrograms can display an analysis of several minutes of music. Furthermore, when videos of spectrograms are traced out as the music is played they allow us to see ahead what tones are to be played, thereby enhancing our anticipation of the music's development. Spectrograms also allow us to detect, and more deeply appreciate, subtle aspects of musical sound quality such as vibrato, dynamic emphasis, and percussion. All of these insights would be difficult, if not impossible, to gain if one only analyzed scores. Spectrograms provide a powerful tool for analyzing the music that we hear, rather than the notes prescribed for musicians to play. Having another tool for analyzing music, in addition to musical scores, is very valuable. One way that spectrograms and scores work together is that spectrograms reveal the overtone structure of the notes played from a musical score. This overtone structure is very important for understanding musical intervals, which are the building blocks of melody and harmony.

We have described some of the many valuable contributions that spectrograms make to the study and appreciation of music. Our students generally consider the material on spectrograms in Chapters 4 and 5 to be the highlight of the book. Following these chapters, we incorporate rhythm into our study of the mathematical aspects of music. In Chapter 6 we describe how pitch and rhythm share many of the same mathematical features. Most books on music, both in music theory and in mathematical treatments, focus exclusively on pitch and harmony. We believe our treatment of rhythm provides our book with a more complete description of music.

The six chapters just described form the core material of the book. The two chapters that follow them describe more advanced mathematical aspects of music. Throughout the book, we make use of geometrical diagrams to aid us in understanding the basic logic of pitch organization and harmony. Chapter 7 explores this connection of geometry with music theory more deeply. Chapter 8 describes some of the ways that computers can be used for synthesizing music. Electronically synthesized music is widely used, and we have tried to explain how it works without getting overwhelmed by technicalities.

Web site

To aid in the study of this book, there is an accompanying web site. To access this site, go to the CRC web site:

www.crcpress.com

and do a search using the book title:

Mathematics and Music: Composition, Perception, and Performance.

You will then arrive at the web page for our book, where you can click on a link to access the supporting web site. There are links at the book's web site for videos of many of the spectrograms we discuss in the book. You can also download the musical scores we examine in the book, playable with the free music software MUSESCORE. We have supplied an online bibliography with many links to free downloadable articles on math and music. Finally, there are links to other web sites related to math and music, including all the ones mentioned in the book.

Prerequisites

To read this book, one needs to have a good background in high school mathematics. We will not assume, however, the ability to read music. The book aims to teach some mathematics, so there are exercises at the end of each section. It also aims to teach how the mathematics relates to music, so many of the exercises involve musical examples. At the end we hope the reader will have a greater mastery of some fundamental mathematics, and a deeper appreciation of music. An appreciation of music made deeper because it is informed by both its mathematical and aesthetic structure.

Music Software

The world of recorded music has been enormously changed in the last three decades or so with the introduction of computer technology. In this book, we use computers to aid in applying mathematics to the analysis of music, and also to the creation of new music. Mostly, we use two *free* software programs. These two free programs are

1. AUDACITY. An audio editor. We have used it for creating and playing spectrograms.

2. MUSESCORE. A musical scoring program. We have used it to create brief passages of musical scores, which you can play on MUSESCORE when studying these passages in the text.

The book can be studied without working with these programs, although we encourage you to try them. We provide some tutorials on using these programs in Appendix B.

Order of Chapters

Chapters are mostly organized sequentially. Each chapter uses, to a degree, material from preceding chapters. Chapter 8 is an exception, as it can be read immediately following Chapter 4. Although chapters proceed sequentially, there is some flexibility in how they can be covered in a classroom setting. For example, in our Mathematics and Music course at UW-Eau Claire, we have successfully taught the material using the following sequence:

Chapter 1, Chapter 4, Chapter 5, Chapter 3, Chapter 6, Chapter 8, Chapter 7.

Since typically at least half of the class can play and read music, Chapter 2 is given as optional reading at the start of the class for those students who need to learn basic music notation. Having students work in groups on material, such as Chapter 3 with its emphasis on music theory, can be very helpful for those students who have a great interest in understanding music but lack performance ability. We have found, however, that even students who are not musicians can master the elementary material in Chapter 2 on their own, and then read the music theory in Chapter 3 with understanding.

Acknowledgments

It is a pleasure to acknowledge as many people as I can, who have helped me with this project. Gary Don, associate professor of music at UWEC, has been a constant supportive colleague from the world of music. Simply listing him as musical consultant for this book does not really do justice to the enjoyable interactions and collaborations that we have been engaged in for over a decade. My Mathematics Department Chair at UWEC, Alex Smith, has done everything in his power to help me teach my Mathematics and Music course. Without his hard work on my behalf, this book would simply not exist. Steven Krantz, Professor of Mathematics at Washington University-St. Louis, has given me a lot of encouragement and help in publishing my papers on this subject. The scholarly support programs at UWEC—the Center for Excellence in Teaching and Learning and the Office of Research and Sponsored Programs—have generously provided me with grants for pursuing the research and writing activities needed for producing this book. One extremely important grant was for funding my sabbatical leave at Macalester College in the academic year 2011-12. While I was at Macalester, I was able to teach a course in Mathematics and Music. I particularly want to thank Karen Saxe, Chair of the Department of Mathematics, Statistics, and Computer Science, and Mark Mazullo, Chair of the Department of Music, at Macalester for arranging my position as Visiting Professor in those departments. My students have given me a lot of help as well. I would especially like to thank Lara Conrad, Hannah Stoelze, Michael Jacobs, Stewart Wallace, Jeanne Knauf, Andrew Jannsen, Gary Baier, Andrew Detra, Kaitlyn Johnstone, Joshua Fuchs, Abigail Doering, Thomas Kokemoor, Carmen Whitehead, Andrew Hanson, Xiaowen Cheng, Jarod Hart, Karyn Muir, Brent McKain, Yeng Chang, and Marisa Berseth. Finally, a heartfelt thanks to my wife, Angela Huang. I am very grateful for her patient support, and I dedicate this book to her.

About the authors

Principal Author

James S. Walker received his doctorate from the University of Illinois, Chicago, in 1982. He has been a professor of Mathematics at the University of Wisconsin-Eau Claire since 1982. His publications include papers on Fourier analysis, wavelet analysis, complex variables, logic, and mathematics and music. He is also the author of five books on Fourier analysis, FFTs, and wavelet analysis.

Musical Consultant

Gary W. Don received his doctorate in music theory from the University of Washington in 1991. He teaches theory and aural skills, 20th-century techniques, counterpoint, and form and analysis as an associate professor at the University of Wisconsin-Eau Claire. The topics of his published articles include Goethe's influence on music theorists of the 19th and 20th centuries, overtone structures in the music of Debussy, and theory pedagogy.

Chapter 1

Pitch, Frequency, and Musical Scales

...in nature itself, a single note sets up a harmony of its own; and this harmonic series has been the (unconscious) basis of Western European harmony, and the tonal system.

—Deryk Cooke

The scientific study of music was put on a firm footing with the seminal work of Hermann von Helmholtz in the middle of the 19th century. His masterpiece, *On the Sensations of Tone,* is still worth studying today. In this chapter we describe Helmholtz's ideas and show how they provide a rationale for musical scales.

1.1   Pitch and Frequency

There is a close connection between pitches in musical tones and the mathematical concept of *frequency*. In the 19th century, Helmholtz did an experiment with tuning forks. He attached a pen to one of the tines of a tuning fork and drew the fork across a piece of paper while it was sounding a specific pitch. The vibration of the pen traced out a simple waveform. We will refer to it as a ***pure tone waveform***. See Figure 1.1.

The most fundamental aspect of a pure tone waveform is that it repeats itself periodically. In physics, the distance from one peak of the wave to the next is called its ***wavelength***. Another term for wavelength is ***cycle***. We have marked one cycle for the pure tone waveform in Figure 1.1. The number of cycles in the pure tone waveform that occur in 1 second is called its ***frequency***. We have this formula for frequency:

$$\text{frequency} = \frac{\text{number of cycles}}{\text{second}}.$$

The unit of cycles/sec for measuring frequency is also called Hz, which is short for Hertz (another German physicist who did fundamental work in the study of frequency). For example, if the cycle shown in Figure 1.1 has a time duration of 0.025 seconds, then the frequency of the pure tone waveform is $1/0.025 = 40$ Hz.

Nowadays, with digital technology, we can record a representation of the sound wave from a tuning fork as a further demonstration of Helmholtz's idea. In Figure 1.2 we show the plot of such a digital waveform recorded from a tuning fork. This tuning fork was designed to match the pitch for the key of middle C on a piano.[1] As described in the caption of Figure 1.2, the frequency for the pure tone waveform produced by the tuning fork is 262 Hz. The maximum y-values of the waveform are all approximately 5000 (and the minimum values are all approximately -5000). This number, 5000, is called the ***amplitude*** of the pure tone. The larger a pure tone's amplitude, the louder the volume of its sound.

To conclude our analysis of pure tones from tuning forks, we look at an extremely important method for displaying the single, constant frequency of the pure tone over time. This method of

---

[1]See Figure 1.12 on p. 15.

1

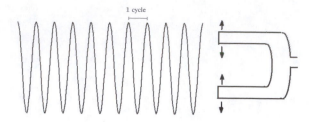

Figure 1.1. Illustration of the famous experiment of Helmholtz. A pen is attached to a tine of tuning fork. As tuning fork is struck and drawn across a piece of paper at a uniform speed, the pen traces out a pure tone waveform. Distance marked between two peaks of waveform is called a *cycle*.

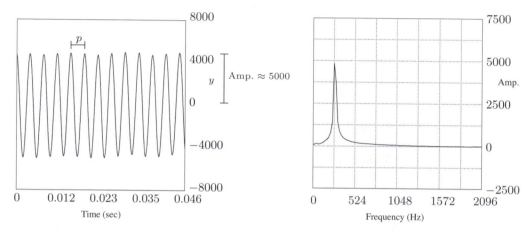

Figure 1.2. Left: Waveform from recording of tuning fork with one cycle marked, $p = 0.00382$ seconds. Frequency is about 1 cycle/0.00382 sec $= 262$ Hz, so note from tuning fork is middle C (c.f. Table 1.2, p. 13). Right: Amplitude for waveform. The height of the spike at frequency 262 Hz is approximately 5000. In Chapter 4 we describe how this amplitude plot was obtained.

display is called a *spectrogram*. In Figure 1.3 we show a spectrogram of a recording of the sound from the tuning fork discussed above. We will describe how this spectrogram was produced in Chapter 4. We are using it now because it provides such an easily interpretable and compelling picture of the frequency content of this pure tone. The single bright, horizontal band in the spectrogram is centered on the single frequency of 262 Hz that we found for this tuning fork's pure tone.

We have now shown that there is a definite connection between frequency and pitch. For a pure tone from a tuning fork, there will be a single precise frequency for its waveform. Music, of course, is played on musical instruments rather than tuning forks. So we now turn to the question of frequency and pitch for musical instruments.

1.1.1 Instrumental Tones

There is a huge variety of musical instruments, including human voices, violins, pianos, clarinets, and many others. The tones produced by these instruments are far more complex, and musically interesting, than the tones produced by tuning forks. See the left sides of Figures 1.4 and 1.5 for examples of portions of waveforms from a flute and piano playing the same note. These waveforms have a fundamental cycle, at least approximately. We show this approximate fundamental cycle on the left side of Figure 1.4 for the flute tone, where it is easier to see in the graph. This fundamental cycle determines the frequency of approximately 329 Hz for the note being played. We can see, however, that these waveforms are not cycling in nearly so uniform a manner as the tuning fork waveform

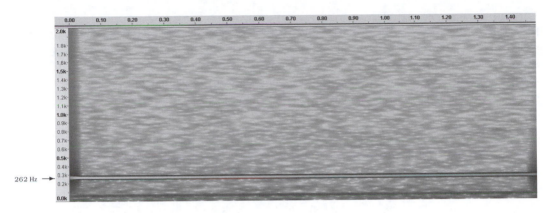

Figure 1.3. Spectrogram of recording of tone from tuning fork. Horizontal axis is time axis, labeled in units of seconds along top. Vertical axis is frequency axis, labeled in units of kHz (1 kHz = 1000 Hz) along left side. Bright white band is centered on frequency 262 Hz, the frequency for tuning fork tone (c.f. Figure 1.2). Background features of spectrogram come from faint background noise, largely inaudible, present during recording process. To watch a video of this spectrogram tracing out in time, as the sound from the tuning fork is played, please visit the book's web site and click on the link *Videos*.

shown in Figure 1.2. In fact, as we shall examine more closely in Chapter 4, they are *combinations* of several pure tone waveforms of differing frequency and loudness. For both of these instruments, these combined pure tone waveforms have frequencies that are positive integer multiples of 329 Hz:

$$329 \text{ Hz}, \quad 2 \cdot 329 \text{ Hz}, \quad 3 \cdot 329 \text{ Hz}, \quad 4 \cdot 329 \text{ Hz}, \quad 5 \cdot 329 \text{ Hz}, \quad 6 \cdot 329 \text{ Hz}, \ldots \quad (1.1)$$

These frequencies are called the ***harmonics*** for these instrumental tones.

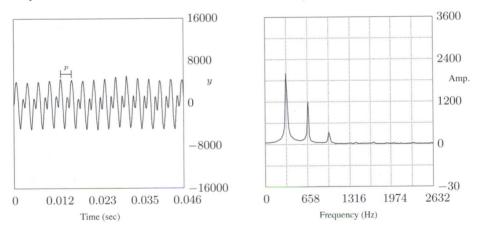

Figure 1.4. Left: Waveform for a recording of flute playing single note. Time p for an approximate cycle is $p = 0.00304$ seconds. Fundamental frequency is approximately 329 Hz. Right: Amplitudes of harmonics for waveform. Harmonics of 329 Hz, $2 \cdot 329$ Hz, and $3 \cdot 329$ Hz, are clearly marked by spikes. Heights of spikes correspond with amplitudes of each pure tone within the complete tone. In Chapter 4 we describe how this plot of amplitudes for harmonics was obtained.

For the flute note, the first two harmonics of 329 Hz and $2 \cdot 329 = 658$ Hz are the loudest, the third harmonic $3 \cdot 329 = 987$ Hz is fainter, and the higher multiples of 329 Hz are fainter still. See the right side of Figure 1.4. In the graph shown there, the heights of the peaks at 329 Hz, 658 Hz, and 987 Hz, correspond to how loud those pitches would be *if those pitches were heard separately*. They are not heard separately, however. It is their combination that produces the complex sound of the flute's tone. We will discuss this point in more detail at the end of this section.

Similarly, the note played by the piano is a combination of harmonics. For the piano, however, the graph on the right of Figure 1.5 shows that the amplitudes of the harmonics are much more equally distributed in size. An interesting feature of this graph is that the magnitude for the harmonic $2 \cdot 329$ Hz is actually larger than the magnitude for 329 Hz. Nevertheless, we refer to 329 Hz as the *fundamental* for this piano note since it corresponds to the pitch that the note is sounding at (which is the same pitch as the flute note). We shall emphasize this point again later, as it is a subtle one: ***The fundamental harmonic for a note is the frequency that determines the note's pitch, and that may or may not be the loudest harmonic produced when the note is played.***

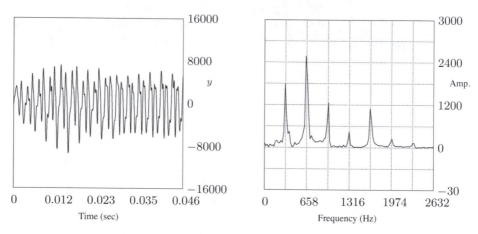

Figure 1.5. Left: Waveform for recording of piano playing single note. Right: Amplitudes of harmonics within piano note, harmonics marked by spikes at multiples of 329 Hz. First spike is at 329 Hz, second spike at $2 \cdot 329$ Hz, third spike at $3 \cdot 329$ Hz, up to seventh spike at $7 \cdot 329$ Hz.

The graphs of amplitudes of harmonics of flute and piano tones shown on the right of Figures 1.4 and 1.5 were obtained by computer processing of recordings of these tones. We shall explain in Chapter 4 how this processing is done.

We summarize this discussion with a definition of these harmonics in instrumental tones.

Definition 1.1.1. *For an instrumental tone that contains frequencies of the form:*

$$\nu_o, \quad 2\nu_o, \quad 3\nu_o, \quad 4\nu_o, \quad 5\nu_o, \quad 6\nu_o, \ \ldots$$

*The smallest frequency, ν_o, is called the **fundamental**. The other frequencies are called **overtones**. All of the frequencies are called **harmonics**. The **first harmonic** is ν_o, the **second harmonic** is $2\nu_o$, the **third harmonic** is $3\nu_o$, and so on.*

Remark 1.1.1. The physical explanation for why tones from musical instruments contain multiple harmonics is beyond the scope of this book. See Chapter 3 of *Fourier Analysis*, by James S. Walker (Oxford University Press, 1988) for a discussion of stringed instruments. For other instruments, consult the book *Music, Physics and Engineering*, by Harry F. Olson (Dover, 1967).

1.1.2 Pure Tones Combining to Create an Instrumental Tone

We have described how instrumental tones are combinations of pure tones. We shall now demonstrate this for a single trumpet tone. In Figure 1.6, we show a spectrogram of a recording of a trumpet playing the note of middle C, with fundamental $\nu_o = 262$ Hz, and of the individual pure tones that combine to create this instrumental tone. As a static picture, this spectrogram shows single bright bands corresponding to the individual harmonics of the trumpet tone, and how they combine to produce the complete tone. However, to fully appreciate this spectrogram, it is absolutely necessary to watch a video of it. Please visit the book web page listed in the Preface, and click on the link, Videos. As the spectrogram traces out, you will hear the individual harmonics sounding just like individual tuning forks, as they should because they correspond to the pure tones making up the trumpet tone. At the end of the spectrogram, you will hear all of these pure tones playing together, creating the complete trumpet sound.

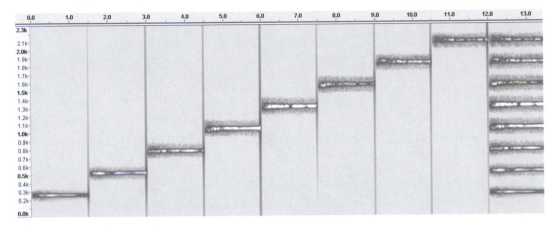

Figure 1.6. Spectrogram illustrating combination of harmonics in trumpet tone for single note. From 0.00 to 1.5 seconds: tone's fundamental of 262 Hz displayed as bright band. From 1.5 to 3.0 seconds: the tone's second harmonic of 524 Hz ($= 2 \cdot 262$ Hz) displayed as another bright band. From 3.0 to 4.5 seconds, tone's third harmonic of 786 Hz ($= 3 \cdot 262$ Hz) displayed as third bright band. Each of first 8 harmonics are displayed in ascending order from left to right, finishing at 12.0 seconds. Sounds from these harmonics are indistinguishable from tuning fork tones. (Thin vertical bars at start and end of individual harmonics — heard as clicking noises in the playback — are artifacts of process of clipping these harmonics out from original trumpet note recording.) Variations in brightness in these harmonic bands correspond to variations in loudness of sound from the harmonics: brighter parts of bands corresponding to louder parts of sound from the harmonics. From 12.0 seconds onward, these 8 harmonics combine to create one tone, the tone of a trumpet.

The reader may have noticed that the separate eight pure tones displayed in this last example sound like a portion of an ascending scale. In the next section, we make this idea precise by discussing the connection between harmonics of instrumental tones and the notes used on musical scales.

Exercises

1.1.1. A pure tone has duration $p = 0.02$ seconds for one cycle. What is its frequency?

1.1.2. A pure tone has duration $p = 0.004$ seconds for one cycle. What is its frequency?

1.1.3. For the graph of amplitudes of harmonics shown on the left of Figure 1.7, estimate the frequencies of the harmonics and find the fundamental frequency.

1.1.4. For the graph of amplitudes of harmonics shown on the right of Figure 1.7, estimate the frequencies of the harmonics and find the fundamental frequency.

1.1.5. On the left of Figure 1.8 there is a graph of the magnitudes of the harmonics from a female pronouncing the vowel ā (long a). Estimate the frequencies of the harmonics and find the fundamental frequency.

1.1.6. On the right of Figure 1.8 there is a graph of the amplitudes of the harmonics from a male pronouncing the vowel ā (long a). Estimate the frequencies of the harmonics and find the fundamental frequency. What difference do you observe compared to the previous exercise with the female speaker, and how do you explain this difference?

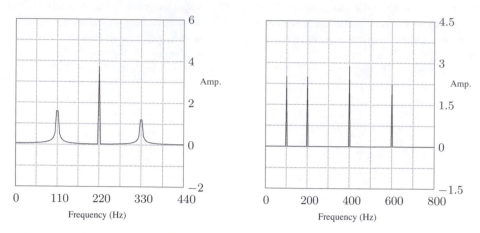

Figure 1.7. Left: Graph of amplitudes of harmonics for Exercise 1.1.3. Right: Graph of amplitudes of harmonics for Exercise 1.1.4.

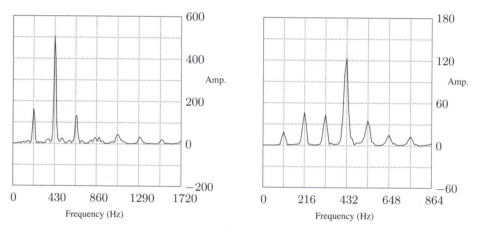

Figure 1.8. Left: Graph of amplitudes of harmonics for Exercise 1.1.5, female pronouncing vowel $\bar{a}$ (long a). Right: Graph of amplitudes of harmonics for Exercise 1.1.6, male pronouncing same vowel.

1.1.7. Why does the tone from the flute (with magnitude of its harmonics graphed in Figure 1.4), sound more pure than the tone from the piano (with magnitude of its harmonics graphed in Figure 1.5)? On the other hand, why does the tone from the piano sound more rich (more complex) than the tone from the flute?

1.1.8. On the left of Figure 1.9 there is a graph of the amplitudes of the harmonics from a male pronouncing the vowel $\bar{e}$ (long e). Estimate the frequencies of the harmonics and find the fundamental frequency.

1.1.9. On the right of Figure 1.9 there is a graph of the amplitudes of the harmonics from a trumpet playing one note. Estimate the frequencies of the harmonics and find the fundamental frequency. Using Table 1.2 on page 13, determine which note is being played.

1.2 Overtones, Pitch Equivalence, and Musical Scales

We have seen that the tones from musical instruments contain harmonics that are all multiples of a fundamental frequency. This fact provides a basis for an equivalence of two tones, when the frequency for one tone is twice the frequency of the other tone. For example, suppose one tone has fundamental

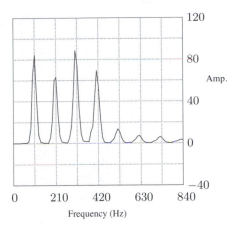
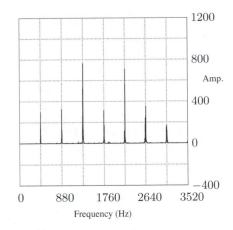

Figure 1.9. Left: Graph of amplitudes of harmonics for Exercise 1.1.8, male pronouncing vowel ē (long e). Right: Graph of amplitudes of harmonics for Exercise 1.1.9, trumpet playing one note.

110 Hz, and a second tone has fundamental 220 Hz. Then the first tone's harmonics are (in Hz):

$$110, \quad 220, \quad 330, \quad 440, \quad 550, \quad 660, \quad 770, \quad 880, \quad 990, \quad 1100, \quad 1210, \dots$$

and the second tone's harmonics are

$$220, \qquad 440, \qquad 660, \qquad 880, \qquad 1100, \qquad \dots$$

All of the harmonics for the tone with fundamental 220 Hz are also harmonics for the tone with 110 Hz. Clearly, this will happen whenever two tones have fundamentals of ν_o and $2\nu_o$. We will say that the tone with frequency $2\nu_o$ is ***an octave higher in pitch*** than the tone with frequency ν_o. The term octave comes from the use of 8 notes on the scales commonly used in Western music. On such scales, the eighth tone has double the fundamental of the first tone.

1.2.1 Pitch Equivalence

When one tone is an octave higher in pitch, then the two tones will be nearly indistinguishable ***when they are sounded together***. They are ***harmonically equivalent***. This harmonic equivalence is usually referred to in music theory as ***octave equivalence***. Harmonic equivalence (or octave equivalence) can be shown by playing two notes that are an octave apart, first separately and then together. In Figure 1.10 we show a spectrogram of the resulting sound. This spectrogram shows the harmonics from the tones traced out over time, as bright horizontal bands. One can see how the second tone, with pitch an octave higher, has all of its harmonics contained within those of the first tone. So when the two tones are sounded together, the horizontal bands in the spectrogram appear almost indistinguishable from those of the first tone. The sound of the two tones together sounds much like the first tone, almost as if that first tone was played by striking its key on the piano in a slightly different manner, rather than what was actually done (striking two keys together).

Since the term *octave equivalence* is standard in music theory we shall employ it from now on. It should be remembered, however, that octave equivalence refers exclusively to notes played in harmony. When notes an octave apart are played separately, they are easily distinguishable by their differences in pitch. It is only when they are played together in harmony that they become equivalent. On the other hand, even with notes played separately in melody, there is often an underlying harmonic scheme for which octave equivalence does play a role in analyzing and appreciating the music. For this reason, musicians train themselves to hear the equivalency of separate notes an octave apart in pitch. We will discuss the idea of an underlying harmonic scheme for a melody in Chapter 3.

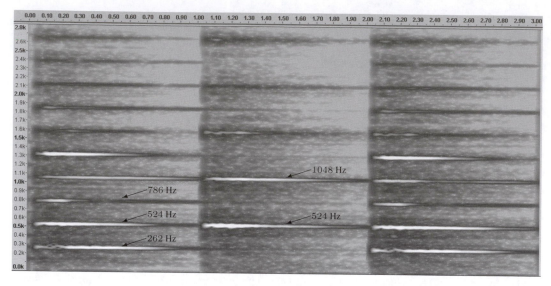

Figure 1.10. Spectrogram of piano tones, illustrating harmonic equivalence (octave equivalence). From 0.00 to 1.00 seconds: piano tone with fundamental frequency 262 Hz. From 1.00 to 2.00 seconds: piano tone with fundamental of twice that frequency. From 2.00 to 3.00 seconds: these two tones sounding together. Tracing out of harmonics from each tone is apparent as horizontal bands, with brightest white bands being the most intense (loudest). The two tones sounding together have an almost identical graph as first tone.

1.2.2 Musical Scales

There are a wide variety of musical scales used throughout the world. The harmonics of instrumental tones can be used to explain some features of the most commonly employed musical scales. We shall describe two types of scales. First, a pentatonic scale, which has 5 distinct notes. Second, an octave scale, which has 8 notes. ***Our aim here is not to trace the historical development of these scales***. Instead, we aim to show how they have a mathematical structure related to the harmonics of their notes. Similar discussion could be given for many different musical scales. We chose these two because of their wide use throughout the world.

Suppose we begin with a specific note, whose tone has a fundamental frequency ν_o. We shall refer to this note as C. The frequencies of the harmonics for C are

$$\nu_o, \quad 2\nu_o, \quad 3\nu_o, \quad 4\nu_o, \quad 5\nu_o, \quad 6\nu_o, \quad \ldots \tag{1.2}$$

The 2nd harmonic, $2\nu_o$, is the fundamental for a tone that we have shown is octave-equivalent to the tone for C. Therefore, we shall also call this note C, and it will be the ending note of our scale. All subsequent notes on our scale will lie between these two C notes, *hence they will have fundamentals that lie between ν_o and $2\nu_o$.*

The third harmonic, $3\nu_o$, is the fundamental for a higher pitched tone than the ending tone of our scale. If we divide $3\nu_o$ by 2, we obtain an octave-equivalent pitch with fundamental $\frac{3}{2}\nu_o$. That frequency lies halfway between ν_o and $2\nu_o$. We shall use G to designate the note with fundamental frequency $\frac{3}{2}\nu_o$. In Figure 1.11, we show a spectrogram of two tones with fundamentals ν_o and $\frac{3}{2}\nu_o$. The figure shows that if these two tones are played in immediate succession (or even together), then their harmonics are either perfectly matched or well-separated from one another. This phenomenon is called ***acoustic consonance.*** When two distinct harmonics are not well-separated, then the sound waves from these harmonics will clash (interfere). This interference creates a roughness in the sound, which is called ***acoustic dissonance.*** We will explore these notions of acoustic consonance and dissonance more thoroughly in Chapter 4.

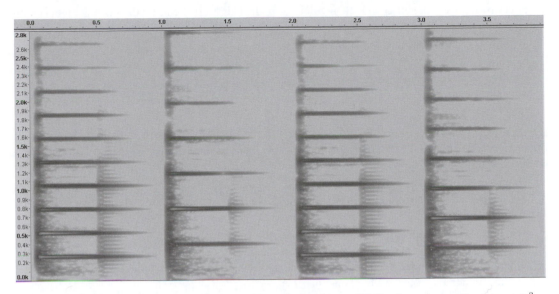

Figure 1.11. Spectrogram of four tones. First tone: Fundamental $\nu_o = 262$ Hz. Second tone: Fundamental $\frac{3}{2}\nu_o$. Third tone: Fundamental ν_o again. Fourth tone: Fundamental $\frac{5}{4}\nu_o$.

Returning to Figure 1.11, we see that the odd-numbered harmonics of the tone for G fit halfway between successive harmonics of the tone for C, and its even-numbered harmonics match with harmonics of the tone for C. In fact, if the note C is lowered in pitch by one octave, then the harmonics for G are a subset of the harmonics of this lower pitch C. To see this, we note that $\frac{1}{2}\nu_o$ is the fundamental for the octave lower C note. Therefore, the harmonics for this C and for G satisfy

$$\text{octave lower C:} \quad \tfrac{1}{2}\nu_o \quad \nu_o \quad \tfrac{3}{2}\nu_o \quad 2\nu_o \quad \tfrac{5}{2}\nu_o \quad 3\nu_o \quad \tfrac{7}{2}\nu_o \quad 4\nu_o \quad \tfrac{9}{2}\nu_o \quad 5\nu_o \quad \cdots$$
$$\text{G:} \qquad\qquad\qquad\qquad \tfrac{3}{2}\nu_o \qquad\qquad\qquad 3\nu_o \qquad\qquad \tfrac{9}{2}\nu_o \qquad\qquad \cdots$$

The octave lower C provides a perfectly acoustically consonant *harmonic foundation* for the pitch G. This use of a lower pitch, bass note as a harmonic foundation for other notes is a very common practice in music.[2]

We now have three notes on our scale:

$$
\begin{array}{ccc}
\text{C} & \text{G} & \text{C} \\
\nu_o & \tfrac{3}{2}\nu_o & 2\nu_o
\end{array}
$$

Throughout the world, almost all musical scales contain two notes whose fundamentals are in the ratio 2 to 1, and a third note whose fundamental lies halfway between them. Musical scales, however, contain more than three notes. To get another note for our scale, we continue looking at the harmonics in (1.2). The next harmonic would be $4\nu_o$, but this corresponds to a note octave-equivalent to C. So we go to the next harmonic $5\nu_o$. The frequency $5\nu_o$ does not lie between ν_o and $2\nu_o$. To get an octave-equivalent tone that does lie between ν_o and $2\nu_o$, we divide $5\nu_o$ by 4. We shall use E to designate the note with fundamental frequency $\frac{5}{4}\nu_o$. In Figure 1.11, we show a spectrogram of two tones with fundamentals ν_o and $\frac{5}{4}\nu_o$. The figure shows that if these two notes are played in immediate succession (or even together), then they have a high degree of acoustic consonance. Every fourth harmonic for E matches a harmonic for C. This is because $4 \cdot \frac{5}{4}\nu_o = 5\nu_o$. The other harmonics for E also fit nicely in between harmonics for C and G. In fact, if the note C was lowered in pitch by two octaves, then the harmonics for E and G would all match perfectly with subsets of the harmonics of this lower pitch

---

[2]For example, see Exercises 1.4.10 and 1.4.11.

C. This two-octave lowered pitch C then would provide a perfectly acoustically consonant harmonic foundation for the other two notes, E and G.

We now have four notes on our scale:

C	E	G	C
ν_o	$\frac{5}{4}\nu_o$	$\frac{3}{2}\nu_o$	$2\nu_o$

It is interesting to note that $\frac{5}{4}\nu_o = \frac{1}{2}\left(\nu_o + \frac{3}{2}\nu_o\right)$, so the fundamental for E is halfway between the fundamentals for C and G. A large percentage of the musical scales in the world have a note in this position on the scale.

To get additional notes on our scale we could continue with the process of looking at harmonics of C. We shall take a different route, however, which we need to do in order to obtain two of the most commonly used scales.

The note G has a fundamental that is $\frac{3}{2}$ times the fundamental for C. If we also multiply the fundamental for G by $\frac{3}{2}$, then we get $\frac{9}{4}\nu_o$. That frequency is greater than $2\nu_o$. To obtain an octave-equivalent tone with fundamental between ν_o and $2\nu_o$, we divide $\frac{9}{4}\nu_o$ by 2, obtaining $\frac{9}{8}\nu_o$. We shall use D to designate the note with fundamental $\frac{9}{8}\nu_o$.

To get the fundamental for D we multiplied the fundamental for G by $\frac{3}{2}$, and applied octave equivalency. For consistency, we would also like E to be related to a note on our scale by multiplying the fundamental of that note by $\frac{3}{2}$ (and employing octave equivalency if needed). We can find that note by dividing the fundamental for E by $\frac{3}{2}$, which produces the frequency $\frac{5}{4}\nu_o / \frac{3}{2} = \frac{5}{6}\nu_o$. The frequency $\frac{5}{6}\nu_o$ does not lie between ν_o and $2\nu_o$. To get an octave-equivalent tone that does lie between ν_o and $2\nu_o$, we multiply $\frac{5}{6}\nu_o$ by 2, obtaining the frequency $\frac{5}{3}\nu_o$. We shall use A to designate the tone with fundamental frequency $\frac{5}{3}\nu_o$. We now have the following scale:

C	D	E	G	A	C	
ν_o	$\frac{9}{8}\nu_o$	$\frac{5}{4}\nu_o$	$\frac{3}{2}\nu_o$	$\frac{5}{3}\nu_o$	$2\nu_o$	(1.3)

This is a *pentatonic major scale*. Many cultures throughout the world use a pentatonic major scale. Chinese folk music is a prime example. Most Western music, however, is based on an octave scale.

Octave scale

We will now derive one type of octave scale by adding more notes to the pentatonic major scale. To obtain the fundamental for G we multiplied ν_o by $\frac{3}{2}$. If we look for a note whose fundamental, when multiplied by $\frac{3}{2}$, is equal to ν_o (the fundamental for C), we get a fundamental of $\frac{2}{3}\nu_o$. The frequency $\frac{2}{3}\nu_o$ does not lie between ν_o and $2\nu_o$. To get an octave-equivalent tone, we multiply by 2, obtaining the frequency $\frac{4}{3}\nu_o$. We shall use F to designate the tone with fundamental frequency $\frac{4}{3}\nu_o$.

To complete our octave scale, we multiply the fundamental $\frac{5}{4}\nu_o$ for note E by $\frac{3}{2}$, obtaining the frequency $\frac{15}{8}\nu_o$. We use B to designate the tone with fundamental frequency $\frac{15}{8}\nu_o$.

Our octave scale is now complete. Here are its notes and frequencies:

C	D	E	F	G	A	B	C	
ν_o	$\frac{9}{8}\nu_o$	$\frac{5}{4}\nu_o$	$\frac{4}{3}\nu_o$	$\frac{3}{2}\nu_o$	$\frac{5}{3}\nu_o$	$\frac{15}{8}\nu_o$	$2\nu_o$	(1.4)

This scale is called a ***just major scale***. In Table 1.1 we show a standard tuning system for this type of scale. It is also called an ***eight-tone just scale***. The term "just" refers to the fact that the harmonics for various combinations of notes—such as C, E, and G—have perfect acoustic consonance. Several other combinations of notes also have perfect acoustic consonance; we describe some of them in the

Table 1.1 PITCH AND FREQUENCY FOR EIGHT-TONE JUST SCALE*

Frequency Ratio	Octave 1		Octave 2		Octave 3		Octave 4		Octave 5		Octave 6		Octave 7	
1	C	32	C	64	C	128	C	256	C	512	C	1024	C	2048
9/8	D	36	D	72	D	144	D	288	D	576	D	1152	D	2304
5/4	E	40	E	80	E	160	E	320	E	640	E	1280	E	2560
4/3	F	43	F	85	F	171	F	341	F	683	F	1365	F	2731
3/2	G	48	G	96	G	192	G	384	G	768	G	1536	G	3072
5/3	A	53	A	107	A	213	A	427	A	853	A	1707	A	3413
15/8	B	60	B	120	B	240	B	480	B	960	B	1920	B	3840
2	C	64	C	128	C	256	C	512	C	1024	C	2048	C	4096

*Fundamentals for C notes are exact. If needed, other fundamentals are rounded to nearest Hz.

exercises.

The frequencies shown in (1.4) have an interesting pattern of ratios. Here we show these ratios, written as multiplying factors of the fundamentals for successive notes:

$$\nu_o \xrightarrow{\cdot\frac{9}{8}} \frac{9}{8}\nu_o \xrightarrow{\cdot\frac{10}{9}} \frac{5}{4}\nu_o \xrightarrow{\cdot\frac{16}{15}} \frac{4}{3}\nu_o \xrightarrow{\cdot\frac{9}{8}} \frac{3}{2}\nu_o \xrightarrow{\cdot\frac{10}{9}} \frac{5}{3}\nu_o \xrightarrow{\cdot\frac{9}{8}} \frac{15}{8}\nu_o \xrightarrow{\cdot\frac{16}{15}} 2\nu_o. \qquad (1.5)$$

$$\text{C} \qquad \text{D} \qquad \text{E} \qquad \text{F} \qquad \text{G} \qquad \text{A} \qquad \text{B} \qquad \text{C}$$

Our brains interpret equal frequency ratios as equivalent changes in pitch. For example, going from C to D is heard as an equivalent change in pitch as going from F to G. In both cases, the frequency ratios of the pitch changes are $\frac{9}{8}$. Similarly, going from E to F is heard as an equivalent change in pitch as going from B to C. In both cases, the frequency ratios of the pitch changes are $\frac{16}{15}$. We will explore this point further later in the chapter, as it is a crucial fact about pitch change.

The ratios in (1.5) are not all the same, however. The fractions $\frac{9}{8}$ and $\frac{10}{9}$ differ by about 1% in magnitude. Consequently there is a slight variation in the pitch change in going from C to D, compared to going from D to E. Likewise there is a slight variation in pitch change in going from F to G, versus going from G to A. These variations in pitch change are the cause of different major scales, using our just tuning, to be inconsistent with each other. For example, if one builds a major scale starting with the note G having frequency $\frac{3}{2}\nu_o$, then the second note in that scale is $\frac{9}{8} \cdot \frac{3}{2}\nu_o = \frac{27}{16}\nu_o$. This note is very close in pitch to A in the scale shown in (1.4), but not exactly the same. For instance, if $\nu_o = 256$ Hz, the fundamental for C in the fourth octave in Table 1.1, then $\frac{27}{16}\nu_o = 432$ Hz. The frequency for A in the fourth octave in Table 1.1 is 427 Hz. So the two notes will have a noticeable difference in pitch, the tone for A at 427 Hz will sound slightly lower in pitch (slightly flatter) than the tone for A at 432 Hz. These two major scales will sound out of tune if played together. Different musical instruments are often tuned to major scales with different starting notes, so this inconsistency is a real problem with just tuning.

There is one further problem with the 8-tone just scale that we described above: It is incomplete. For example, the note D has fundamental $\frac{9}{8}$ times the fundamental ν_o for the starting note C. For reasons of symmetry, or completeness, we look for a note whose fundamental times $\frac{9}{8}$ will equal a fundamental for C. In order for this fundamental to lie between ν_o and $2\nu_o$, we divide $\frac{9}{8}$ into $2\nu_o$ to get $\frac{16}{9}\nu_o$. The note with fundamental $\frac{16}{9}\nu_o$ lies between A and B. Depending on the musical context, it is either called A$^\sharp$ (A-sharp) or B$^\flat$ (B-flat). We will not pursue this question further, since introducing new notes into the just scale only increases the inconsistencies in frequency ratios that we have already observed.

By using a different scale, an equal-tempered scale, these inconsistencies in frequency ratios simply disappear. We discuss the 12-tone equal-tempered scale in the next section.

Exercises

1.2.1. Show that if the note G (with fundamental $\frac{3}{2}\nu_o$) is lowered in pitch by two octaves, then the harmonics for D (with fundamental $\frac{9}{8}\nu_o$) all match perfectly with a subset of the harmonics for this lower pitch G.

1.2.2. In the text it was stated that *if the note C (with fundamental ν_o) is lowered in pitch by two octaves, then the harmonics for G (with fundamental $\frac{3}{2}\nu_o$) and the harmonics for E (with fundamental $\frac{5}{4}\nu_o$) all match perfectly with a subset of the harmonics for this lower pitch C.* Verify that this statement is correct.

1.2.3. Show that if the note F (with fundamental $\frac{4}{3}\nu_o$) is lowered in pitch by two octaves, then the harmonics for A (with fundamental $\frac{5}{3}\nu_o$) and the harmonics for C (with fundamental $2\nu_o$) all match perfectly with subsets of the harmonics for this lower pitch F.

1.2.4. Show that if the note G (with fundamental $\frac{3}{2}\nu_o$) was lowered in pitch by two octaves, then the harmonics for B (with fundamental $\frac{15}{8}\nu_o$) and the harmonics for D (with fundamental $\frac{9}{8}\nu_o$) would all match perfectly with subsets of the harmonics for this lower pitch G.

1.2.5. Explain why the sixth harmonic $6\nu_o$ for the note C does not introduce any new note into the just scale we have described.

1.2.6. Suppose the seventh harmonic $7\nu_o$ for the note C is used to introduce a new note into the just scale. Where would this note lie in relation to the scale in (1.4)? What if the eleventh harmonic is used to introduce a new note, where would this note lie on the scale in (1.4)? Why did we skip over the ninth and tenth harmonics?

1.3   The 12-Tone Equal-Tempered Scale

In this section we describe a different musical scale that eliminates the inconsistencies in frequency ratios in the just scale, and has additional notes that make it a relatively complete scale. This musical scale is called the 12-tone equal-tempered scale.

To explain how the 12-tone equal-tempered scale is defined, we need to closely examine the inconsistencies in frequency ratios in the eight-tone just scale. These frequency ratios are shown in (1.5). We notice that these frequency ratios split into two categories, large and small (major and minor):

$$\text{Major:} \quad \tfrac{9}{8} = 1.125, \quad \tfrac{10}{9} = 1.111. \qquad \text{Minor:} \quad \tfrac{16}{15} = 1.0667.$$

The major frequency ratios are used five times in (1.5), while the minor frequency ratio is used twice.

We also observe that the square of the minor frequency ratio $\frac{16}{15}$ is nearly equal to both of the major frequency ratios:

$$\left(\frac{16}{15}\right)^2 = 1.13778$$

$$\approx 1.125 = \frac{9}{8}$$

$$\approx 1.11111 = \frac{10}{9}.$$

Suppose for a moment that these approximations are close enough to be regarded as equalities. Then, denoting the minor frequency ratio $\frac{16}{15}$ by $\mathbf{r}$, we will suppose that $\frac{9}{8} = \mathbf{r}^2$ and $\frac{10}{9} = \mathbf{r}^2$. This fiction allows us to rewrite (1.5) as follows:

$$\nu_o \xrightarrow{\cdot r^2} \mathbf{r}^2\nu_o \xrightarrow{\cdot r^2} \mathbf{r}^4\nu_o \xrightarrow{\cdot r} \mathbf{r}^5\nu_o \xrightarrow{\cdot r^2} \mathbf{r}^7\nu_o \xrightarrow{\cdot r^2} \mathbf{r}^9\nu_o \xrightarrow{\cdot r^2} \mathbf{r}^{11}\nu_o \xrightarrow{\cdot r} \mathbf{r}^{12}\nu_o = 2\nu_o.$$
$$\text{C} \qquad \text{D} \qquad \text{E} \qquad \text{F} \qquad \text{G} \qquad \text{A} \qquad \text{B} \qquad \text{C}$$

(1.6)

From this calculation we have derived the equation $r^{12}\nu_o = 2\nu_o$. Dividing out ν_o, we obtain $r^{12} = 2$. Therefore, r is the twelfth root of 2:

$$r = 2^{1/12}. \tag{1.7}$$

Using $r = 2^{1/12}$ for each minor frequency ratio, and $r^2 = 2^{2/12}$ for each major frequency ratio, completely removes the inconsistency that we described above for the eight-tone just scale.

Using this frequency ratio r, we can also add more notes to our major scale in (1.6). For example, if we divide the highest frequency $2\nu_o = r^{12}\nu_o$ by r^2, then we get $r^{10}\nu_o$. The note with fundamental $r^{10}\nu_o$ lies equally between—in terms of frequency ratios—the fundamentals $r^9\nu_o$ for note A and $r^{11}\nu_o$ for note B. This new note can be written as either A$^\sharp$ or B$^\flat$. The symbol $\sharp$ (a sharp) indicates that the note A has been raised in pitch by the frequency ratio r, while the symbol $\flat$ (a flat) indicates that the note B has been lowered in pitch by the frequency ratio r. What this leads to is that multiplying ν_o successively by r produces a new scale, the ***12-tone equal-tempered scale***. To save verbiage, we shall also refer to it as the ***chromatic scale***.[3] In Table 1.2 we show a standard tuning system for this type of scale. For convenience, we have used only sharps in this table. In each octave, the frequencies

Table 1.2   PITCH AND FREQUENCY FOR 12-TONE EQUAL-TEMPERED SCALE[*]

Frequency Ratio	Octave 1		Octave 2		Octave 3		Octave 4		Octave 5		Octave 6		Octave 7	
1	C	33	C	65	C	131	C	262	C	523	C	1047	C	2093
$2^{1/12}$	C$^\sharp$	35	C$^\sharp$	69	C$^\sharp$	139	C$^\sharp$	277	C$^\sharp$	554	C$^\sharp$	1109	C$^\sharp$	2218
$2^{2/12}$	D	37	D	73	D	147	D	294	D	587	D	1175	D	2349
$2^{3/12}$	D$^\sharp$	39	D$^\sharp$	78	D$^\sharp$	156	D$^\sharp$	311	D$^\sharp$	622	D$^\sharp$	1245	D$^\sharp$	2489
$2^{4/12}$	E	41	E	82	E	165	E	330	E	659	E	1319	E	2637
$2^{5/12}$	F	44	F	87	F	175	F	349	F	699	F	1397	F	2794
$2^{6/12}$	F$^\sharp$	46	F$^\sharp$	93	F$^\sharp$	185	F$^\sharp$	370	F$^\sharp$	740	F$^\sharp$	1475	F$^\sharp$	2960
$2^{7/12}$	G	49	G	98	G	196	G	392	G	784	G	1568	G	3136
$2^{8/12}$	G$^\sharp$	52	G$^\sharp$	104	G$^\sharp$	208	G$^\sharp$	415	G$^\sharp$	831	G$^\sharp$	1661	G$^\sharp$	3322
$2^{9/12}$	A	55	A	110	A	220	A	440	A	880	A	1760	A	3520
$2^{10/12}$	A$^\sharp$	58	A$^\sharp$	117	A$^\sharp$	233	A$^\sharp$	466	A$^\sharp$	932	A$^\sharp$	1865	A$^\sharp$	3729
$2^{11/12}$	B	62	B	124	B	247	B	494	B	988	B	1976	B	3951
2	C	65	C	131	C	262	C	523	C	1047	C	2093	C	4186

[*]Frequencies for fundamentals are rounded to nearest Hz, except for the A notes which are exact.

are the same for the five pairs: C$^\sharp$, D$^\flat$, and D$^\sharp$, E$^\flat$, and F$^\sharp$, G$^\flat$, and G$^\sharp$, A$^\flat$, and A$^\sharp$, B$^\flat$.

The tuning system for the chromatic scale shown in Table 1.2 is now established as an International Standard. It is used, for example, in tuning many orchestral instruments, including pianos and harps, and also guitars.[4] In the next section, we will describe how the chromatic scale provides the basis for the most commonly used major scales in Western music. In contrast to the just scale, major scales beginning with different notes are perfectly consistent when the 12-tone equal-tempered system is used.

The first column in Table 1.2 is included to emphasize the fact that the notes in each octave are all generated by the same frequency ratios, relative to the initial note C. Using $r = 2^{1/12}$, and ν_o for the frequency of the C note that begins an octave, the frequencies of each note in that octave have this

---

[3]There are other chromatic scales besides the 12-tone equal-tempered scale, but we will not discuss them in this book.

[4]An interesting exception is that stringed instruments in an orchestra are typically tuned with their fundamentals in ratios of 3 : 2. See Section 1.4.4 on page 20 for further discussion.

pattern:

$$\nu_o \cdot \mathbf{r}^0 \qquad \nu_o \cdot \mathbf{r}^1 \qquad \nu_o \cdot \mathbf{r}^2 \qquad \nu_o \cdot \mathbf{r}^3 \qquad \nu_o \cdot \mathbf{r}^4 \qquad \nu_o \cdot \mathbf{r}^5$$

$$\text{C} \qquad\quad \text{C}^\sharp \qquad\quad \text{D} \qquad\quad \text{D}^\sharp \qquad\quad \text{E} \qquad\quad \text{F}$$

$$\nu_o \cdot \mathbf{r}^6 \qquad \nu_o \cdot \mathbf{r}^7 \qquad \nu_o \cdot \mathbf{r}^8 \qquad \nu_o \cdot \mathbf{r}^9 \qquad \nu_o \cdot \mathbf{r}^{10} \qquad \nu_o \cdot \mathbf{r}^{11} \qquad \nu_o \cdot \mathbf{r}^{12}$$

$$\text{F}^\sharp \qquad\quad \text{G} \qquad\quad \text{G}^\sharp \qquad\quad \text{A} \qquad\quad \text{A}^\sharp \qquad\quad \text{B} \qquad\quad \text{C}$$

(1.8)

The final note C, with fundamental $\nu_o \cdot \mathbf{r}^{12} = \nu_o \cdot 2$, is an octave above the initial note C.

Example 1.3.1. Tuning note for the chromatic scale. The fundamentals shown for the notes in Table 1.2 are based on A in the 4th octave as reference note. We will denote this note by A_4. The fundamental of 440 Hz for A_4 is exact, as are the other fundamentals for all of the A notes. The frequencies in the table for other notes, however, are only approximations to their exact values. For instance, the exact value for the fundamental of C_4, C in the 4th octave, is not 262 Hz (as shown in Table 1.2). Its exact value can be found from the fact that the fundamental for A_4 is $\nu_o \cdot \mathbf{r}^9$, where ν_o is the fundamental for C_4. We calculate as follows:

$$\begin{aligned}
\nu_o &= (\nu_o \cdot \mathbf{r}^9) \cdot \mathbf{r}^{-9} \\
&= (440) \cdot \frac{1}{\mathbf{r}^9} \\
&= \frac{440}{2^{9/12}} \\
&= 261.625565300599.
\end{aligned}$$

Therefore, the fundamental for C_4 is 261.625565300599 Hz. The value of 262 Hz shown in Table 1.2 is an approximation to this more exact value.

Example 1.3.2. Finding fundamentals of notes using C as reference. As shown in the previous example, the note C_4 has fundamental

$$261.625565300599 \text{ Hz} \approx 262 \text{ Hz}.$$

Therefore $C_4^\sharp$ has fundamental

$$\begin{aligned}
261.625565300599 \cdot \mathbf{r}^1 &= 261.625565300599 \cdot 2^{1/12} \\
&= 277.182630976873 \\
&\approx 277 \text{ Hz}
\end{aligned}$$

as shown in the table. Moreover, G_4 has fundamental

$$\begin{aligned}
261.625565300599 \cdot \mathbf{r}^7 &= 261.625565300599 \cdot 2^{7/12} \\
&= 391.99543598175 \\
&\approx 392 \text{ Hz}
\end{aligned}$$

as shown in the table.

The chromatic scale provides the basis for most of the scales used throughout the Western world. We look at these scales in the next section.

Exercises

1.3.1. On the right of Figure 1.4, page 3, there is a graph of the amplitudes for harmonics of a flute tone. Using Table 1.2, estimate what note is being played.

1.3.2. On the right of Figure 1.5, page 4, there is a graph of the amplitudes for harmonics of a piano tone. Using Table 1.2, estimate what note is being played. Why is this the same note as for the flute in the previous exercise?

1.3.3. On the right of Figure 1.9, page 7, there is a graph of the amplitudes for harmonics of a trumpet tone. Using Table 1.2, estimate what note is being played.

1.3.4. On the left of Figure 1.9, page 7, there is a graph of the amplitudes for harmonics for a male speaking the vowel ē (long e). Setting aside the fact that this male is not singing, find the notes in Table 1.1 and in Table 1.2 that are the best estimates to the note for the tone of this person's vowel.

1.3.5. On the left of Figure 1.8, page 6, there is a graph of the amplitudes for harmonics for a female speaking the vowel ā (long a). Setting aside the fact that this female is not singing, find the notes in Table 1.1 and in Table 1.2 that are the best estimates to the note for the tone of this person's vowel.

1.3.6. On the right of Figure 1.8, page 6, there is a graph of the amplitudes for harmonics for a male speaking the vowel ā (long a). Setting aside the fact that this male is not singing, find the notes in Table 1.1 and in Table 1.2 that are the best estimates to the tone of this person's vowel.

1.4 Musical Scales within the Chromatic Scale

In this section we will describe the standard Western musical scales. These are most easily explained using a piano keyboard. Almost all pianos are tuned to play the pitches described in Table 1.2. The layout for a piano keyboard is shown in Figure 1.12. The notes on this piano keyboard span a pitch range of 7 octaves,[5] a huge pitch range which contains the pitch ranges of just about every instrument used in music.

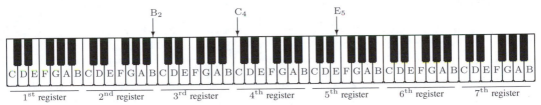

Figure 1.12. Piano keyboard spanning 7 octave registers. The 1st octave register contains notes with lowest pitch. Each successive register contains notes with higher pitch than preceding one, until notes with highest pitch in 7th register. In each register, white keys are labeled by notes from C-major scale. If it is important to state what register a specific note belongs to, then a subscript is used. For example, the note C in 4th register is C_4, also known as middle C. Note B in second register is B_2, and so forth.

1.4.1 The C-Major Scale

The most straightforward musical scale is the C-major scale, since it consists entirely of white notes on the keyboard. It corresponds to 8 adjacent white keys, an *octave,* on the keyboard shown in Figure 1.12. In Figure 1.13 we show one section of the keyboard with these 8 adjacent white keys. These white keys are labeled C, D, E, F, G, A, B, C. The last C is octave-equivalent to the first C, just an octave higher in pitch.

These 8 white keys on the piano will play a C-major scale:

$$C \quad D \quad E \quad F \quad G \quad A \quad B \quad C. \tag{1.9}$$

If you play this scale, you will hear the familiar sound of "Do-Re-Mi-Fa-So-La-Ti-Do" that school children learn to sing in music classes.

---

[5]A typical piano keyboard has four additional keys, three to the left and one to the right of the keys shown in Figure 1.12. To explain the basic principles more clearly, we have omitted these keys, which are rarely played.

The black keys are for notes whose pitches lie above or below the pitches of the notes on adjacent white keys. For example, $C^\sharp$ has a pitch above C but below D. On the other hand, $D^\flat$ has a pitch below D but above C. On the piano, which is tuned to the chromatic scale, the two notes $C^\sharp$ and $D^\flat$ *have the same pitch.* They are called **enharmonic.** We have indicated this by writing both $C^\sharp$ and $D^\flat$ on the first black key on the left side of Figure 1.13. Similar remarks apply to the rest of the black keys. If we were to strike both white and black keys in order from left to right, beginning with the note C on the left, then we would play this scale:

$$C \quad C^\sharp \quad D \quad D^\sharp \quad E \quad F \quad F^\sharp \quad G \quad G^\sharp \quad A \quad A^\sharp \quad B \quad C \qquad (1.10)$$

which is the chromatic scale discussed in the previous section. A C-major scale is called a *diatonic scale.* A chromatic scale and a C-major scale are shown in Figure 1.15, along with two other scales

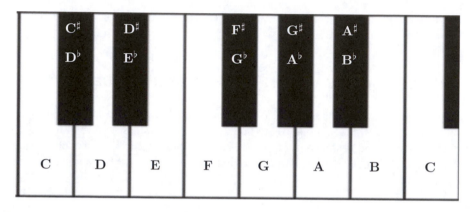

Figure 1.13. Keys on piano keyboard spanning one octave, starting at C. White keys are C-major scale. All black and white keys are chromatic scale.

that we shall discuss soon.

In Figure 1.13 there are black keys in between some white keys but not others. For example, the key for D is 2 keys over to the right from the key for C. Likewise, the key for E is 2 keys over from the key for D. However, the key for F is just 1 key over from the key for E. We have this pattern of how many keys over to the right each white key is to the next on the piano:

$$C \xrightarrow{+2\ keys} D \xrightarrow{+2\ keys} E \xrightarrow{+1\ key} F \xrightarrow{+2\ keys} G \xrightarrow{+2\ keys} A \xrightarrow{+2\ keys} B \xrightarrow{+1\ key} C.$$

Because a move of +2 keys occurs much more often than a move of +1 key, musical terminology refers to a move of +2 keys as a **whole step** and a move of +1 key as a **half step.** A whole step equals two half steps.[6] Using this terminology we can write instead

$$C \xrightarrow{2\ half\ steps} D \xrightarrow{2\ half\ steps} E \xrightarrow{1\ half\ step} F \xrightarrow{2\ half\ steps} G \xrightarrow{2\ half\ steps} A \xrightarrow{2\ half\ steps} B \xrightarrow{1\ half\ step} C.$$

This pattern of half steps exactly matches with the powers of the frequency factor **r** used in (1.6) on page 12:

$$\nu_o \xrightarrow{\cdot r^2} r^2 \nu_o \xrightarrow{\cdot r^2} r^4 \nu_o \xrightarrow{\cdot r} r^5 \nu_o \xrightarrow{\cdot r^2} r^7 \nu_o \xrightarrow{\cdot r^2} r^9 \nu_o \xrightarrow{\cdot r^2} r^{11} \nu_o \xrightarrow{\cdot r} r^{12}\nu_o = 2\nu_o.$$

$$C \qquad\quad D \qquad\quad E \qquad\quad F \qquad\quad G \qquad\quad A \qquad\quad B \qquad\quad C$$

Consequently, we can use this pattern of key distances or half steps:

$$2 \quad\ 2 \quad\ 1 \quad\ 2 \quad\ 2 \quad\ 2 \quad\ 1 \qquad\qquad (1.11)$$

---

[6]Another terminology is *semitone* for half step, and *whole tone* for whole step. We shall use the terms half step and whole step, since they might be familiar already to many readers.

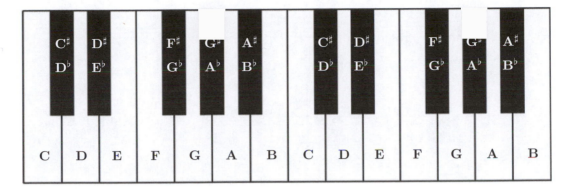

Figure 1.14. Keys on piano keyboard spanning two registers.

to create major scales with any starting note.

1.4.2 Other Major Scales

To create another major scale, we use the pattern of half steps in (1.11), starting with a different note than C. Based on our previous discussion of harmony, we will choose the fifth note in a given scale to begin a new scale. For the C-major scale, the fifth note is G. So we shall start with the note G and use the half step numbers in (1.11) as key distances on the piano keyboard. This will produce the G-major scale. To perform our work, we will start with the first key marked G on the left side of the piano keyboard section shown in Figure 1.14. Referring to this figure, and using our pattern of half steps, we get the following results:

$$G \xrightarrow{\text{2 half steps}} A \xrightarrow{\text{2 half steps}} B \xrightarrow{\text{1 half step}} C \xrightarrow{\text{2 half steps}} D \xrightarrow{\text{2 half steps}} E \xrightarrow{\text{2 half steps}} F^\sharp \xrightarrow{\text{1 half step}} G.$$

Notice that going two half steps to the right from the key for E gets us to the key marked either $F^\sharp$ or $G^\flat$. We chose $F^\sharp$ as the note for the black key, rather than $G^\flat$. This was done to avoid repeating a note letter. If we had chosen $G^\flat$, then we would have had $G^\flat$ and G as the last two notes for our scale. Clearly, that would lead to some confusion. We shall always follow the standard musical convention of avoiding using the same note letter twice in our scales.

To summarize, we have found that the G-major scale is

$$G \qquad A \qquad B \qquad C \qquad D \qquad E \qquad F^\sharp \qquad G. \tag{1.12}$$

See Figure 1.15.

Remark 1.4.1. Before we do another example, we should point out the meaning of sharps and flats in terms of half steps. If a note has a sharp on it, then that means the note is raised in pitch by one half step. For example, $C^\sharp$ is one half step above C. Or, $E^\sharp$ is one half step above E. This last example is interesting because the note F is also one half step above E. The notes $E^\sharp$ and F are enharmonic (have the same pitch). If a flat is applied to a note, then the note is lowered one half step. So $D^\flat$ is one half step below D. There are enharmonic cases here as well. For instance, the notes $C^\flat$ and B are enharmonic.

Returning to our discussion of major scales, suppose we start on the fifth note D of the G-major scale. We will produce the scale known as D-major. Using our pattern of half steps, starting with the leftmost key for D in Figure 1.14, we get the following results:

$$D \xrightarrow{\text{2 half steps}} E \xrightarrow{\text{2 half steps}} F^\sharp \xrightarrow{\text{1 half step}} G \xrightarrow{\text{2 half steps}} A \xrightarrow{\text{2 half steps}} B \xrightarrow{\text{2 half steps}} C^\sharp \xrightarrow{\text{1 half step}} D.$$

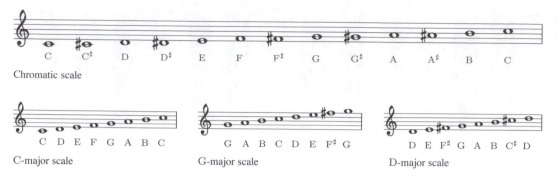

Figure 1.15. Some musical scales. Try playing these scales with a real instrument, or a virtual one (using software such as MUSESCORE). For diatonic scales, shown below chromatic scale, you will hear essentially the same scale but with different pitch ranges.

So the D-major scale is

$$\text{D} \quad \text{E} \quad \text{F}^\sharp \quad \text{G} \quad \text{A} \quad \text{B} \quad \text{C}^\sharp \quad \text{D}. \tag{1.13}$$

See Figure 1.15.

Try playing these three scales, C-major, G-major, and D-major, on a real or virtual instrument to hear the similarity in the pitch changes. In each case, you should hear the familiar scale of "Do-Re-Mi-Fa-So-La-Ti-Do," just over different pitch ranges. The frequency ratios for successive pitches are the same in each scale, and our brains detect this equivalence.

1.4.3 Scales and Clock Arithmetic

The method of using the piano keyboard for describing scales is workable, but a bit cumbersome. There is an elegant mathematical way of creating scales using the notion of *clock arithmetic*. We will introduce this method in an informal way in this section.

Notice that in the piano keyboard shown in Figure 1.12 the sequence of keys in each register just repeats again and again (it is said to be a *periodic* sequence). We can also see that there is a single half step used to go from one key to the next on the chromatic scale formed by the 12 keys in each register. The easiest way to express this periodic repetition is to pair these notes with the 12 hours on a clock face as shown in Figure 1.16. We could imagine physically wrapping the piano keyboard around the clock: Starting with the leftmost key for the note C_1 at hour 0, and pairing each successive key with each successive hour. Each C-key in each register will then pair up with hour 0, and the keyboard will wrap around the clock 7 times. We shall refer to this clock as the ***chromatic clock.***

The hour values on the chromatic clock are the powers of the frequency factor **r** that is used for creating the chromatic scale. The chromatic clock has the top hour start at 0 rather than 12, which will facilitate the arithmetic we shall be using. With the chromatic clock, rather than counting half steps using a piano keyboard, we can instead count hour distances. So, for instance, going from C to D is a 2-hour change, which corresponds precisely to the 2 half steps we used previously. Or, going from E to F is a 1-hour change, corresponding precisely to the 1 half step we used before. When hour 12 is reached, then the equality $\mathbf{r}^{12} = 2$ corresponds to the octave-equivalence of two C-notes differing by one octave.

Remark 1.4.2. On the chromatic clock we have written only the letter names for notes, disregarding their registers. Since we identified the chromatic scale in each register with the hours on the clock, we can regard each note on the clock as specifying a set of pitches from all of the registers. For example, the note **C** on the clock denotes the set of pitches $\{C_1, C_2, \ldots, C_7\}$. So we can write

$$\mathbf{C} = \{C_1, C_2, C_3, C_4, C_5, C_6, C_7\}.$$

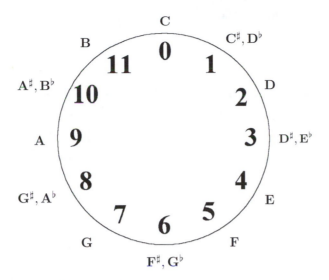

Figure 1.16. Chromatic clock. The chromatic scale arranged on clock face (starting at hour **0**).

Likewise, we have

$$\mathbf{C}^\sharp = \{C_1^\sharp, C_2^\sharp, C_3^\sharp, C_4^\sharp, C_5^\sharp, C_6^\sharp, C_7^\sharp\}$$

and so on. In the examples that follow we will keep track of hour changes around the clock, and whenever we pass the top hour 0 going clockwise it will mean we have passed into a higher register. We will not usually write register subscripts, however, as they can almost always be inferred from context.

Using the chromatic clock, we can create major scales using the sequence

$$2 \quad 2 \quad 1 \quad 2 \quad 2 \quad 2 \quad 1 \tag{1.14}$$

as hours to add in moving around the hour positions on the clock. We agree, however, that when we circle around the top of the clock (the number 0) we will always reset to 0 by subtracting 12 (just as in telling time with this type of clock). This convention will allow us to correctly identify the note names, which is all we need for this application of chromatic clock arithmetic.

Example 1.4.1. C-major scale. To make the C-major scale, we start with the note C at hour 0, and then apply the sequence in (1.14) as hour moves around the chromatic clock. We perform these calculations:

$$0 \xrightarrow{+2} 2 \xrightarrow{+2} 4 \xrightarrow{+1} 5 \xrightarrow{+2} 7 \xrightarrow{+2} 9 \xrightarrow{+2} 11 \xrightarrow[-12 \text{ at top}]{+1} 0 \tag{1.15}$$
$$C \qquad D \qquad E \qquad F \qquad G \qquad A \qquad B \qquad C.$$

Notice that we subtracted 12 when reaching the top of the clock (so hour 0, not 12). Thus we have obtained the C-major scale:

$$C \quad D \quad E \quad F \quad G \quad A \quad B \quad C. \tag{1.16}$$

Example 1.4.2. G-major scale. To make the G-major scale, we start at G which has hour 7. We then make the following calculations:

$$7 \xrightarrow{+2} 9 \xrightarrow{+2} 11 \xrightarrow[-12 \text{ at top}]{+1} 0 \xrightarrow{+2} 2 \xrightarrow{+2} 4 \xrightarrow{+2} 6 \xrightarrow{+1} 7$$
$$G \qquad A \qquad B \qquad C \qquad D \qquad E \qquad F^\sharp \qquad G.$$

Thus, we have obtained the G-major scale:

$$G \quad A \quad B \quad C \quad D \quad E \quad F^{\sharp} \quad G. \tag{1.17}$$

Example 1.4.3. D-major scale. We can also use addition on the chromatic clock to produce the D-major scale. The note D has hour 2, and the calculations go as follows:

$$2 \xrightarrow{+2} 4 \xrightarrow{+2} 6 \xrightarrow{+1} 7 \xrightarrow{+2} 9 \xrightarrow{+2} 11 \xrightarrow[-12 \text{ at top}]{+2} 1 \xrightarrow{+1} 2$$
$$D \qquad E \qquad F^{\sharp} \qquad G \qquad A \qquad B \qquad C^{\sharp} \qquad D$$

to get the D-major scale:

$$D \quad E \quad F^{\sharp} \quad G \quad A \quad B \quad C^{\sharp} \quad D. \tag{1.18}$$

Example 1.4.4. A-major scale. Finally, let's go up a fifth from the D-major scale. We will start from its fifth note, A, and create the A-major scale. The hour number for A is 9 and the calculations go as follows:

$$9 \xrightarrow{+2} 11 \xrightarrow[-12 \text{ at top}]{+2} 1 \xrightarrow{+1} 2 \xrightarrow{+2} 4 \xrightarrow{+2} 6 \xrightarrow{+2} 8 \xrightarrow{+1} 9$$
$$A \qquad B \qquad C^{\sharp} \qquad D \qquad E \qquad F^{\sharp} \qquad G^{\sharp} \qquad A$$

to obtain the A-major scale:

$$A \quad B \quad C^{\sharp} \quad D \quad E \quad F^{\sharp} \quad G^{\sharp} \quad A. \tag{1.19}$$

Remark 1.4.3. Looking at these successive scales, we see that every time we went up a fifth we found the same notes as on the previous scale. Except for the next to the last note, where we obtained a sharped note from the previous scale's note. For example, here are the G-major and D-major scales:

$$G \quad A \quad B \quad C \quad D \quad E \quad F^{\sharp} \quad G$$

$$D \quad E \quad F^{\sharp} \quad G \quad A \quad B \quad C^{\sharp} \quad D.$$

The D-major scale can be rapidly obtained by just writing down the notes of the G-major scale, starting with D, and simply remembering to sharp the C that comes right before the ending D. This phenomenon is summarized in the famous "Circle of Fifths," which we describe in Chapter 2.

1.4.4   Relation between Just and Equal-Tempered Tunings

Although we have treated just and equal tempered tunings separately, we did note the close connection between them. Here is a listing of the frequency ratios for a major scale in equal temperament and the ratios that they approximate in the just scale:

$$\mathbf{r}^0 = 1 \text{ (exact)}$$
$$\mathbf{r}^2 = 1.122462\cdots \approx 1.125 = \frac{9}{8}$$
$$\mathbf{r}^4 = 1.259921\cdots \approx 1.25 = \frac{5}{4}$$
$$\mathbf{r}^5 = 1.334839\cdots \approx 1.3333\cdots = \frac{4}{3}$$
$$\mathbf{r}^7 = 1.498307\cdots \approx 1.5 = \frac{3}{2}$$
$$\mathbf{r}^9 = 1.681792\cdots \approx 1.6666\cdots = \frac{5}{3}$$
$$\mathbf{r}^{11} = 1.887748\cdots \approx 1.875 = \frac{15}{8}$$
$$\mathbf{r}^{12} = 2 \text{ (exact)}$$

(1.20)

Musicians make use of these close approximations. The composer and physicist, John Powell, has concisely described how they do it:

> The "just" system can only be used to good effect by instruments that don't have fixed notes—like violins, violas, cellos, trombones and, most important, the human voice. On these instruments it is possible for good players to adjust their notes so that the combinations of notes always have lovely, simple relationships. (J. Powell, *How Music Works,* Little, Brown and Company, 2010, p. 131.)

By "simple relationships," Powell means that the fundamentals for the combined notes have the simple fractional relationships found in just tuning. These simple fractional relationships provide the perfect matches of different harmonics from the combined notes, a property that equal-tempered tuning does not have. Nevertheless, since the vast majority of Western music is written with the equal-tempered system in mind, we shall concentrate on that system from here on.

Exercises

1.4.1. Use the clock arithmetic method to go up a fifth from the A-major scale, shown in (1.19), to find the E-major scale.

1.4.2. Explain why going up a fifth note in a scale is always done by adding 7 hours on the chromatic clock.

1.4.3. Going down a fifth. Here is the chromatic scale written using flats:

$$\text{C} \quad \text{D}^\flat \quad \text{D} \quad \text{E}^\flat \quad \text{E} \quad \text{F} \quad \text{G}^\flat \quad \text{G} \quad \text{A}^\flat \quad \text{A} \quad \text{B}^\flat \quad \text{B} \quad \text{C}$$

(1.21)

Now, suppose you have the C-major scale:

$$\text{C} \quad \text{D} \quad \text{E} \quad \text{F} \quad \text{G} \quad \text{A} \quad \text{B} \quad \text{C}$$

and you want to make a new major scale that begins with the fifth note *down* from the note C at the right end of this scale. That would be the note F. **(a)** Use the clock arithmetic method to find the F-major scale, which starts with this note F. [Note: Make sure you do not use a note letter more than once, resulting in one flatted note.] **(b)** From the F-major scale that you just found, make a new major scale that begins with the fifth note down from the highest pitch note (the second F) in the F-major scale.

1.4.4. Starting with the note F$^\sharp$, use this sequence of additions on the chromatic clock:

$$+2 \quad +2 \quad +3 \quad +2 \quad +3$$

to generate a pentatonic major scale. (Notice that its five distinct notes are the five black keys on a piano.)

1.4.5. Harmonic minor scales. A harmonic minor scale is obtained from a given starting note by using the following sequence of additions in our clock arithmetic method:

$$+2 \quad +1 \quad +2 \quad +2 \quad +1 \quad +3 \quad +1.$$

Use these additions to generate a harmonic minor scale that begins with C (the harmonic C-minor scale). (Remember to not repeat note letters, this will determine how you employ flats or sharps.)

1.4.6. Natural minor scales. A natural minor scale is obtained from a major scale by starting from its sixth note. For example, if the scale is C-major, then its sixth note is A. The natural A-minor scale is

$$A \quad B \quad C \quad D \quad E \quad F \quad G \quad A.$$

(a) Verify that the natural A-minor scale, given above, corresponds to the following sequence of additions on the chromatic clock:

$$+2 \quad +1 \quad +2 \quad +2 \quad +1 \quad +2 \quad +2.$$

(b) Use these additions to generate a natural minor scale beginning with G (natural G-minor scale).
(c) Explain why this natural G-minor scale has the same notes as the $B^\flat$-major scale.

Modes for Major Scales

A major scale can be played in various modes. Each mode is obtained by using a different starting note for the scale. For example, if the scale is C-major:

$$C \quad D \quad E \quad F \quad G \quad A \quad B \quad C$$

then its ***Dorian mode*** is obtained by starting on its second note:

$$D \quad E \quad F \quad G \quad A \quad B \quad C \quad D.$$

Although this scale was obtained from the C-major scale, it is usually designated by listing its starting note, D, and stating that it is Dorian mode. Thus, it is referred to as D-Dorian.
  If one starts on the fourth note of the C-major scale, then the notes are in ***Lydian mode:***

$$F \quad G \quad A \quad B \quad C \quad D \quad E \quad F$$

This scale is called F-Lydian. If one starts on the fifth note of the C-major scale, then the notes are in ***Mixolydian mode:***

$$G \quad A \quad B \quad C \quad D \quad E \quad F \quad G.$$

This scale is called G-Mixolydian.
  We will not be dealing extensively with modes in this book. These three modes should suffice to illustrate the idea. Their significance will be clearer once we describe chords and chord progressions in Chapter 3. Here are a few exercises that relate these modes to our clock arithmetic method.

1.4.7. (a) Verify that the D-Dorian mode, given above, corresponds to the following sequence of hour additions on the chromatic clock:

$$+2 \quad +1 \quad +2 \quad +2 \quad +2 \quad +1 \quad +2. \tag{1.22}$$

(b) Use this sequence of hour additions to find the scale for the G-Dorian mode. **(c)** Use this sequence of hour additions to find the scale for the C-Dorian mode.

1.4.8. (a) Verify that the G-Mixolydian mode, given above, corresponds to the following sequence of hour additions on the chromatic clock:

$$+2 \quad +2 \quad +1 \quad +2 \quad +2 \quad +1 \quad +2. \tag{1.23}$$

(b) Use this sequence of hour additions to find the scale for the E-Mixolydian mode. **(c)** Use this sequence of hour additions to find the scale for the C-Mixolydian mode.

1.4.9. (a) Verify that the F-Lydian mode, given above, corresponds to the following sequence of hour additions on the chromatic clock:

$$+2 \quad +2 \quad +2 \quad +1 \quad +2 \quad +2 \quad +1. \tag{1.24}$$

(b) Use this sequence of hour additions to find the scale for the D-Lydian mode. **(c)** Use this sequence of hour additions to find the scale for the C-Lydian mode.

1.4.10. In the song *Think of Me* (music by Andrew Loyld Webber), the first piano chord used consists of the simultaneous notes D_2, A_3, D_4, and $F^\sharp_4$. Assuming that D_2 has fundamental ν_o, find the powers of $\mathbf{r}$ that multiply ν_o to obtain the fundamentals for these four notes. What simple fractions are closely approximated by these powers of $\mathbf{r}$, and what do these fractions have to do with the acoustic consonance discussed in Section 1.2?

1.4.11. In Beethoven's *Moonlight Sonata,* the following five notes are used in an (arpeggiated) chord: A_1, A_2, A_3, $C^\sharp_4$, and E_4. Assuming that A_1 has fundamental ν_o, find the powers of $\mathbf{r}$ that multiply ν_o to obtain the fundamentals for these five notes. What simple fractions are closely approximated by these powers of $\mathbf{r}$, and what do these fractions have to do with the acoustic consonance discussed in Section 1.2?

1.4.12. Harmonic major scales. A harmonic major scale is obtained from a given starting note by using the following sequence of additions in our clock arithmetic method:

$$+2 \quad +2 \quad +1 \quad +2 \quad +1 \quad +3 \quad +1.$$

Use these additions to generate a harmonic major scale that begins with C (the harmonic C-major scale). Also use these additions to generate a harmonic major scale that begins with A (the harmonic A-major scale).

1.4.13. Melodic minor scales. A melodic minor scale is obtained from a given starting note by using the following sequence of additions in our clock arithmetic method:

$$+2 \quad +1 \quad +2 \quad +2 \quad +2 \quad +2 \quad +1.$$

Use these additions to find the melodic minor scale that begins with A (the melodic A-minor scale). (The melodic minor scale is used in ascending melodies in minor keys, while descending melodies use the natural minor scale.)

1.5 Logarithms

By working with half steps to represent changes in pitch from one note to the next we have, in effect, done an operation with logarithms. In this section we shall define logarithms in a precise way, and show exactly how logarithms describe the half step changes that we have been working with. We shall also discuss a few other ways they occur in music. Logarithms often occur in mathematics related to human perception, such as the musically important perceptions of pitch (frequency), loudness (sound volume), and rhythm (note durations).

1.5.1 Half Steps and Logarithms

We begin our discussion of logarithms by showing how they relate to the half step changes in pitch that we used for constructing musical scales. In (1.6) we described the C-major scale using these multiplications by powers of $\mathbf{r} = 2^{1/12}$:

$$\nu_o \xrightarrow{\cdot r^2} r^2\nu_o \xrightarrow{\cdot r^2} r^4\nu_o \xrightarrow{\cdot r} r^5\nu_o \xrightarrow{\cdot r^2} r^7\nu_o \xrightarrow{\cdot r^2} r^9\nu_o \xrightarrow{\cdot r^2} r^{11}\nu_o \xrightarrow{\cdot r} r^{12}\nu_o = 2\nu_o. \tag{1.25}$$

$$\text{C} \qquad \text{D} \qquad \text{E} \qquad \text{F} \qquad \text{G} \qquad \text{A} \qquad \text{B} \qquad \text{C}$$

Later, in (1.15), we described this same scale using half step changes:

$$0 \xrightarrow{+2} 2 \xrightarrow{+2} 4 \xrightarrow{+1} 5 \xrightarrow{+2} 7 \xrightarrow{+2} 9 \xrightarrow{+2} 11 \xrightarrow[-12 \text{ at top}]{+1} 0 \tag{1.26}$$

$$\text{C} \qquad \text{D} \qquad \text{E} \qquad \text{F} \qquad \text{G} \qquad \text{A} \qquad \text{B} \qquad \text{C.}$$

We can see that, except for the change to 0 when 12 is reached, this latter calculation is *using addition of the powers of* **r** *rather than multiplication by* **r** *raised to those powers*. These powers of **r** are logarithms. To be more precise, we have the following definition.

Definition 1.5.1. *The* **logarithm**, *base* **r**, *of a positive number* c *is the power of* **r** *that is needed to produce* c. *We write* $\log_{\mathbf{r}} c$ *to express this power. In other words,* $\log_{\mathbf{r}} c$ *satisfies:*

$$\mathbf{r}^{\log_{\mathbf{r}} c} = c. \tag{1.27}$$

At first sight, this definition does not appear to relate to our calculations with the C-major scale. However, if we have $c = \mathbf{r}^p$, then the power needed for obtaining c on the base **r** is p. Therefore, we have the following identity:

$$\log_{\mathbf{r}} \left(\mathbf{r}^p \right) = p. \tag{1.28}$$

Identity (1.28) implies that the hour changes used in (1.26) are the logarithms, base **r**, of the frequency factors used in (1.25).

The logarithm, base **r**, of a given positive number is unique. The reason for the uniqueness of the logarithm is shown in Figure 1.17.

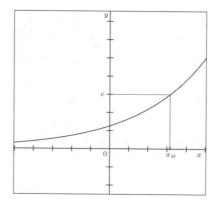

Figure 1.17. Graph of $y = \mathbf{r}^x$. Steadily increasing y-values on the graph imply that the logarithm of any given positive number c is unique. As indicated on the graph, for the value of c shown, there is a unique x-value x_o satisfying $c = \mathbf{r}^{x_o}$. This x-value is the logarithm of c, base **r**. We have $c = \mathbf{r}^{x_o}$, so $x_o = \log_{\mathbf{r}} c$.

We have introduced logarithms by relating them to constructing the C-major scale. However, they also can be used to describe any frequency changes occurring from one pitch to another. Here are some basic illustrations of how this is done.

Example 1.5.1. Frequency change by half steps. (a) If one pitch has fundamental 220 Hz, and another pitch has fundamental 440 Hz, then we have

$$\frac{440}{220} = 2.$$

However, since $\mathbf{r} = 2^{1/12}$, we also know that

$$\mathbf{r}^{12} = 2.$$

Therefore,

$$\log_{\mathbf{r}} \left(\frac{440}{220} \right) = \log_{\mathbf{r}} \left(\mathbf{r}^{12} \right)$$

$$= 12.$$

This result shows that the 12 half steps needed to go up one octave, from 220 Hz to 440 Hz, is the logarithm, base $\mathbf{r}$, of their ratio. **(b)** If we go down in pitch one octave, say from 294 Hz to 147 Hz, then we have

$$\frac{147}{294} = \frac{1}{2}$$

$$= \frac{1}{\mathbf{r}^{12}}$$

$$= \mathbf{r}^{-12}.$$

Therefore,

$$\log_{\mathbf{r}}\left(\frac{147}{294}\right) = \log_{\mathbf{r}}\left(\mathbf{r}^{-12}\right) = -12.$$

This result shows that the -12 half steps needed to go down one octave, from 294 Hz to 147 Hz, is the logarithm, base $\mathbf{r}$, of their ratio. **(c)** If the pitch C_4 has fundamental $\nu_o \approx 262$ Hz, then the pitch G_4 has fundamental $\mathbf{r}^7\nu_o$. We then have

$$\log_{\mathbf{r}}\left(\frac{\mathbf{r}^7\nu_o}{\nu_o}\right) = \log_{\mathbf{r}}\left(\mathbf{r}^7\right)$$

$$= 7.$$

This result shows that the 7 half steps needed to go from C_4 to G_4 is the logarithm, base $\mathbf{r}$, of the ratio of their fundamentals. **(d)** The lowest pitch note in Table 1.2 is C_1, while the highest pitch note is C_8. The pitch C_8 is 7 octaves above the pitch C_1. Therefore, if ν_o is the fundamental for C_1, then $2^7\nu_o$ is the fundamental for C_8. We then have

$$\log_{\mathbf{r}}\left(\frac{2^7\nu_o}{\nu_o}\right) = \log_{\mathbf{r}}\left(2^7\right)$$

$$= \log_{\mathbf{r}}\left(\left(\mathbf{r}^{12}\right)^7\right)$$

$$= \log_{\mathbf{r}}\left(\mathbf{r}^{12\cdot7}\right)$$

$$= \log_{\mathbf{r}}\left(\mathbf{r}^{84}\right)$$

$$= 84.$$

This result shows that 84 half steps are needed to go from C_1 to C_8. An increase by 84 half steps is much easier to comprehend than the increase of over 4100 Hz shown in Table 1.2.

To extend the calculations in this last example to other pitches, we need a method for calculating logarithms, base $\mathbf{r}$, of any frequency ratio. For example, if one pitch has fundamental 200 Hz and another pitch has fundamental 250 Hz, then we would like to calculate

$$\log_{\mathbf{r}}\left(\frac{250}{200}\right) = \log_{\mathbf{r}}(1.25).$$

The method we use for calculating such a logarithm is the *change of base formula*. The change of base formula makes use of a logarithm in another, more convenient, base. So, first we will define logarithms to other bases, and then describe the most convenient base for a logarithm, base 10. Using base 10 logarithms, which all calculators can compute, we will be able to apply the change of base formula to find any desired logarithm.

Definition 1.5.2. *Let b stand for a positive number, greater than 1. The **logarithm**, base b, of a positive number x is the power of b that is needed to produce x. We write $\log_b x$ to express this power. In other words, $\log_b x$ satisfies:*

$$b^{\log_b x} = x. \tag{1.29}$$

As with $\log_r$, it follows from this definition that we also have this identity:

$$\log_b (b^p) = p. \tag{1.30}$$

Example 1.5.2. (a) We have $1000 = 10^3$, so $\log_{10} 1000 = 3$. (b) We have $16 = 2^4$, so $\log_2 16 = 4$. (c) We have $0.01 = 10^{-2}$, so $\log_{10} 0.01 = -2$. (d) We have $243 = 3^5$, so $\log_3 243 = 5$.

Remark 1.5.1. Base 10 logarithms. The logarithm, base 10, is known as the **common logarithm**. It is also denoted by log, **without the base subscript**. Using this notation, we have for example, $\log 1000 = 3$ and $\log 0.01 = -2$.

We are now prepared to state the change of base formula referred to above. We will state it in the form that we will need in our applications, rather than in its most general form.

Theorem 1.5.1. Change of Base Formula. *For every positive number x, the quantity $\log_b x$ satisfies*

$$\log_b x = \frac{\log x}{\log b}. \tag{1.31}$$

Proof. First, we use a basic fact about exponents along with the definition of logarithm, to find that

$$10^{\log b \cdot \log_b x} = \left(10^{\log b}\right)^{\log_b x}$$
$$= b^{\log_b x}$$
$$= x.$$

Thus,

$$10^{\log b \cdot \log_b x} = x$$

which tells us that

$$\log b \cdot \log_b x = \log x.$$

Dividing both sides of this equation by $\log b$ gives us (1.31), and that completes the proof. $\square$

If we use $b = r$, then the change of base formula becomes

$$\log_r x = \frac{\log x}{\log r}. \tag{1.32}$$

Equation (1.32) can be used for finding changes in half steps between any two frequencies.

Example 1.5.3. We can now finish the calculation for finding $\log_r \left(\frac{250}{200}\right)$. Using the change of base formula, we have

$$\log_r \left(\frac{250}{200}\right) = \log_r(1.25)$$

$$= \frac{\log(1.25)}{\log(r)}$$

$$= \log(1.25) \, / \, \log\big(2 \wedge (1/12)\big)$$

$$= 3.86313713864835.$$

The calculation tells us that 250 Hz is about 3.86 half steps above 200 Hz.

Example 1.5.4. A quarter step. Suppose we have the pitch A_4, which has a fundamental frequency of 440 Hz. If we were to raise this pitch by 1 half step, then that would involve multiplying 440 by $\mathbf{r}^1$, while raising it by 2 half steps would involve multiplying 440 by $\mathbf{r}^2$. So, again we note that the number of half steps by which we raise the pitch is the same as the power used on $\mathbf{r}$. Based on this idea, we will say that multiplying 440 by $\mathbf{r}^{1/2}$ is raising the pitch of A_4 by a **quarter step** (half of a half step). In this case, we get

$$440\,\mathbf{r}^{1/2} = 440\,(2^{1/12})^{1/2}$$

$$= 440\,(2 \wedge (1/12)) \wedge (1/2)$$

$$= 452.892984123137.$$

Therefore, a pitch with fundamental 452.892984123137 Hz is a quarter step above the pitch of A_4. Moreover, by the change of base formula, we find that

$$\log_{\mathbf{r}}\left(\frac{452.892984123137}{440}\right) = \log\left(\frac{452.892984123137}{440}\right) / \log \mathbf{r}$$

$$= \log\left(452.892984123137 / 440\right) / \log\left(2 \wedge (1/12)\right)$$

$$= 0.500000000000019.$$

So we have found by our logarithm method that 452.89 Hz is about 0.5 half steps above 440 Hz, i.e., about a quarter step above 440 Hz.

1.5.2 Cents

The logarithm method of comparing two frequencies provides a very fine discrimination, one that accords well with human perception. For example, we have found that 250 Hz is about 3.86 half steps above 200 Hz, and 452.89 Hz is about 0.5 half steps above 440 Hz. Both of these results can be expressed in units of **cents**, a unit for measuring frequency difference that is widely employed in music. The cents measure is analogous to the cents used in monetary calculations. For example, \$3.86 is equal to 3 dollars and 86 cents. In our case, 3.86 half steps is equal to 3 half steps plus 86 cents. Similarly, 0.50 half steps is equal to 50 cents. The unit of cents divides each hour of the chromatic clock, each half step, into 100 equal parts. The cents units thereby provide quite a fine measure for distinguishing frequency. It has been found to provide sufficient accuracy for distinguishing those pitches that the human ear is able to distinguish, which accounts for its wide use in music.

We shall now show that each cent accounts for a multiplication by a frequency factor of $\mathbf{r}^{1/100} = \mathbf{r}^{0.01}$. This factor $\mathbf{r}^{0.01}$ satisfies

$$\mathbf{r}^{0.01} = (2^{1/12})^{0.01}$$

$$= (2 \wedge (1/12)) \wedge (0.01)$$

$$= 1.00057778950655$$

so it represents a very slight change in frequency.

Example 1.5.5. Suppose that $\nu_o' = \mathbf{r}^{0.01}\nu_o$, then ν_o' is equal to multiplying ν_o by the frequency factor $\mathbf{r}^{0.01}$. In this case, we find that

$$\log_{\mathbf{r}}\left(\frac{\nu_o'}{\nu_o}\right) = \log_{\mathbf{r}}\left(\frac{\mathbf{r}^{0.01}\nu_o}{\nu_o}\right)$$

$$= \log_{\mathbf{r}}\left(\mathbf{r}^{0.01}\right)$$

$$= 0.01$$

which shows that multiplying ν_o by $\mathbf{r}^{0.01}$ corresponds to a 1 cent increase in frequency.

As another instance, suppose we multiply ν_o by $\mathbf{r}^{0.01}$ a number of times, say 19 times. This gives $\nu_o' = (\mathbf{r}^{0.01})^{19}\nu_o$. By a basic property of exponents, we have $(\mathbf{r}^{0.01})^{19} = \mathbf{r}^{0.01*19}$, and therefore

$$\log_{\mathbf{r}}\left(\frac{\nu_o'}{\nu_o}\right) = \log_{\mathbf{r}}\left(\frac{(\mathbf{r}^{0.01})^{19}\nu_o}{\nu_o}\right)$$

$$= \log_{\mathbf{r}}\left(\mathbf{r}^{0.01*19}\right)$$

$$= \log_{\mathbf{r}}\left(\mathbf{r}^{0.19}\right)$$

$$= 0.19.$$

This shows that multiplying ν_o by $\mathbf{r}^{0.01}$ a total of 19 times corresponds to a 19 cent increase in frequency.

Looking at these examples, we can see that the number of cents for a change of frequency from ν_o to ν_o' can be computed by the following formula:

$$(\text{cents}) = \frac{100\log(\nu_o'/\nu_o)}{\log(\mathbf{r})}. \tag{1.33}$$

For example, this formula would yield 1 cent for the frequencies $\nu_o' = \mathbf{r}^{0.01}\nu_o$ and ν_o. It would yield 19 cents for the frequencies $\nu_o' = (\mathbf{r}^{0.01})^{19}\nu_o$ and ν_o.

Here are a couple examples of how this formula works.

Example 1.5.6. For the frequencies $452.892984123137\,\text{Hz}$ and $440\,\text{Hz}$, we have

$$(\text{cents}) = \frac{100\log(452.892984123137/440)}{\log(\mathbf{r})}$$

$$= 100\log(452.892984123137/440)\,/\,\log\big(2 \wedge (1/12)\big)$$

$$= 50.0000000000019.$$

This shows that 452.892984123137 is 50 cents greater in frequency than $440\,\text{Hz}$, as we found before. However, we have now calculated it from formula (1.33). Moreover, as explained above, these 50 cents represent a multiplying of $440\,\text{Hz}$ a total of 50 times by the frequency factor $\mathbf{r}^{0.01} = 1.00057778950655$.

Example 1.5.7. For the frequencies $261.63\,\text{Hz}$ and $262\,\text{Hz}$, we have

$$(\text{cents}) = \frac{100\log(262/261.63)}{\log(\mathbf{r})}$$

$$= 100\log(262/261.63)\,/\,\log\big(2 \wedge (1/12)\big)$$

$$= 2.4466004803503.$$

This shows that 262 is about 2.45 cents greater in frequency than $261.63\,\text{Hz}$.

Another formula for cents

There is an alternative formula that can be used to measure cents, which is much easier to enter into a calculator. For measuring the amount of cents between two frequencies, ν_o and ν_o', we compute

$$(\text{cents}) = 1200\log(\nu_o'/\nu_o)/\log(2). \tag{1.34}$$

To see why formula (1.34) is correct, we need the following identity for logarithms.

Theorem 1.5.2. Power Identity. *For each positive number x and real number p, the logarithm $\log_b$ satisfies*

$$\log_b\left(x^p\right) = p \log_b x.$$

Proof. We find that b raised to the power $p \log_b x$ satisfies

$$b^{p \log_b x} = \left(b^{\log_b x}\right)^p$$
$$= (x)^p.$$

Thus, $b^{p \log_b x} = x^p$, so we conclude that $\log_b\left(x^p\right) = p \log_b x$, and that completes the proof. $\square$

We can now establish formula (1.34). Since $\mathbf{r} = 2^{1/12}$, formula (1.33) becomes

$$(\text{cents}) = \frac{100 \log(\nu_o'/\nu_o)}{\log(\mathbf{r})}$$

$$= \frac{100 \log(\nu_o'/\nu_o)}{\log\left(2^{1/12}\right)}.$$

Using Theorem 1.5.2, we have $\log\left(2^{1/12}\right) = \frac{1}{12} \log(2)$. Therefore,

$$(\text{cents}) = \frac{100 \log(\nu_o'/\nu_o)}{\log\left(2^{1/12}\right)}$$

$$= \frac{100 \log(\nu_o'/\nu_o)}{\frac{1}{12} \log(2)}$$

$$= \frac{1200 \log(\nu_o'/\nu_o)}{\log(2)}$$

and that establishes formula (1.34). Here are a couple of examples of using formula (1.34).

Example 1.5.8. **(a)** For the frequencies 261.63 Hz and 262 Hz, we have

$$(\text{cents}) = \frac{1200 \log(262/261.63)}{\log(2)}$$
$$= 2.4466004803503$$

as calculated previously. **(b)** For the frequencies 440 Hz and 448 Hz, we have

$$(\text{cents}) = \frac{1200 \log(448/440)}{\log(2)}$$
$$= 31.1942502395332$$

which shows that 448 Hz is about 31.19 cents greater than 440 Hz.

In this section we have seen that logarithms provide a useful measure for quantifying the changes in frequencies commonly encountered in music. Using half steps, which are logarithms of frequency factors, reduces the size range of the numbers used to describe pitches in music. For example, the change of the fundamental for C_1, which is about 33 Hz, going to the fundamental for C_8, which is about 4186 Hz, is described by just 84 half steps. This change by 84 half steps is much easier to grasp than the frequency change of 4186 Hz. In the exercises for this section, we shall see that logarithms provide us with a similar approach to quantifying sound intensity, which correlates with the loudness we hear. In the next chapter, we shall find that logarithms are also related to the relative durations of the notes typically played in Western music.

Exercises

1.5.1. Find the following logarithms:

(a) $\log_2 32$

(b) $\log_\mathbf{r} \left(\mathbf{r}^6 \right)$

(c) $\log 100$

(d) $\log_4 64$

1.5.2. Find the following logarithms:

(a) $\log_2 128$

(b) $\log_\mathbf{r} \left(\mathbf{r}^9 \right)$

(c) $\log 10000$

(d) $\log_4 256$

1.5.3. Use the graph of $y = 2^x$ on the left of Figure 1.18 to estimate the following logarithms:

(a) $\log_2 5$

(b) $\log_2 10$

(c) $\log_2 15$

(d) $\log_2 20$

1.5.4. Use the graph of $y = 1.5^x$ on the right of Figure 1.18 to estimate the following logarithms:

(a) $\log_{1.5} 2$

(b) $\log_{1.5} 3$

(c) $\log_{1.5} 4$

(d) $\log_{1.5} 5$

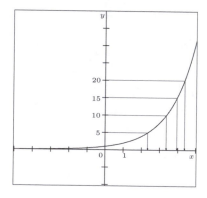
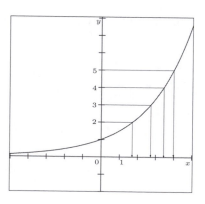

Figure 1.18. Left: Graph of $y = 2^x$. Right: Graph of $y = 1.5^x$.

1.5.5. Use formula (1.32) to find the number of half steps separating the following pairs of frequencies:

(a) $\nu_o = 262$ Hz, $\nu_o' = 300$ Hz.

(b) $\nu_o = 600$ Hz, $\nu_o' = 650$ Hz.

(c) $\nu_o = 440$ Hz, $\nu_o' = 500$ Hz.

(d) $\nu_o = 110$ Hz, $\nu_o' = 140$ Hz.

1.5.6. Use formula (1.32) to find the number of half steps separating the following pairs of frequencies:

(a) $\nu_o = 200$ Hz, $\nu'_o = 300$ Hz.

(b) $\nu_o = 440$ Hz, $\nu'_o = 550$ Hz.

(c) $\nu_o = 320$ Hz, $\nu'_o = 440$ Hz.

(d) $\nu_o = 110$ Hz, $\nu'_o = 150$ Hz.

1.5.7. Use formula (1.34) to find the number of cents separating the following pairs of frequencies:

(a) $\nu_o = 262$ Hz, $\nu'_o = 268$ Hz.

(b) $\nu_o = 110$ Hz, $\nu'_o = 112$ Hz.

(c) $\nu_o = 320$ Hz, $\nu'_o = 330$ Hz.

(d) $\nu_o = 115$ Hz, $\nu'_o = 110$ Hz.

(e) $\nu_o = 300$ Hz, $\nu'_o = 262$ Hz.

(f) $\nu_o = 444$ Hz, $\nu'_o = 440$ Hz.

1.5.8. Use formula (1.34) to find the number of cents separating the following pairs of frequencies:

(a) $\nu_o = 262$ Hz, $\nu'_o = 265$ Hz.

(b) $\nu_o = 262$ Hz, $\nu'_o = 260$ Hz.

(c) $\nu_o = 325$ Hz, $\nu'_o = 320$ Hz.

(d) $\nu_o = 110$ Hz, $\nu'_o = 105$ Hz.

(e) $\nu_o = 262$ Hz, $\nu'_o = 250$ Hz.

(f) $\nu_o = 880$ Hz, $\nu'_o = 885$ Hz.

1.5.9. Which of these two pairs of frequencies would you say is closer in pitch, $\nu_o = 262$ Hz and $\nu'_o = 265$ Hz, or $\nu_o = 240$ Hz and $\nu'_o = 242$ Hz? Use the cents measure to justify your answer.

1.5.10. Which of these two pairs of frequencies would you say is closer in pitch, $\nu_o = 440$ Hz and $\nu'_o = 445$ Hz, or $\nu_o = 262$ Hz and $\nu'_o = 265$ Hz? Use the cents measure to justify your answer.

1.5.11. In Exercise 1.3.5 on page 15, you were asked to use Table 1.2 to find the closest note corresponding to the tone from the female speaking the vowel ā (long a). How many cents difference is there between the actual fundamental and this closest fundamental from the table?

1.5.12. In Exercise 1.3.5 on page 15, you were asked to use Table 1.1 to find the closest note corresponding to the tone from the female speaking the vowel ā (long a). How many cents difference is there between the actual fundamental and this closest fundamental from the table?

1.5.13. (a) In just tuning systems, a value of $\frac{16}{9}\nu_o$ is typically used for the fundamental of $B^\flat$ (when the fundamental for the initial note C is ν_o). Find the cents difference between $\frac{16}{9}\nu_o$ and the fundamental for $B^\flat$ on the chromatic scale. (b) The value of $\frac{16}{9}\nu_o$ is close to $\frac{7}{4}\nu_o$. How close is it in cents? (c) Explain why $\frac{16}{9}\nu_o$ is generally preferred as a fundamental for $B^\flat$, rather than $\frac{7}{4}\nu_o$.

1.5.14. Given that the note C has fundamental ν_o, find the closest note on the chromatic scale to the fundamental $\frac{11}{8}\nu_o$ that is octave equivalent to the 11th harmonic $11\nu_o$ of C. How many cents difference is there between the fundamentals of these two notes?

Decibel measure of loudness

The **loudness** L of sound of intensity I is measured in **decibels** by the formula

$$L = 10 \log\left(\frac{I}{I_o}\right) \tag{1.35}$$

where I_o is the minimum audible sound intensity.[7] Formula (1.35) only roughly corresponds to human perception of loudness, as such perception is also dependent on sound frequency and, in a more complex manner, on I. Nevertheless, we shall use (1.35) as an objective measure, as commonly done by audio engineers. The following six exercises relate to formula (1.35).

1.5.15. If $I = 10^8 I_o$, then find L.

1.5.16. If $I = 10^4 I_o$, then find L.

1.5.17. If $I = 10^{12} I_o$, then find L.

1.5.18. If $I = 10^{16} I_o$, then find L.

1.5.19. Suppose that a comfortable listening level is a range of loudness between 20 and 100 decibels. What range of multiples of I_o correspond to this range of decibels?

1.5.20. Suppose that a dangerous loudness level, capable of causing hearing damage, is a loudness above 150 decibels. What multiple of I_o corresponds to 150 decibels?

---

[7]The sound intensities I and I_o are measurements of acoustical energy. They are proportional to the sum of the squares of all the amplitudes of harmonics in a sound wave. The minimal audible sound intensity I_o is an average computed for large numbers of test subjects reporting the smallest sound intensity I that they could hear.

Chapter 2

Basic Musical Notation

The ability to read music is sometimes regarded with awe, as though it offers privileged insights into the world of music from which non-initiates are excluded.... But it is not hard at all, at least to the extent of becoming able to name a note.

—Philip Ball

In this chapter we describe some basic musical notation. This material will be quite familiar to musicians, but we would like this book to be helpful to all who enjoy music. Our goal is to provide enough background so that non-musicians can follow our discussions of basic musical scores. We shall not try to explain the huge amount of notation used for all musical scores, that would require a separate book in itself.[1] In the first section we discuss how notes are scored in terms of pitch. In the next section we discuss how notes are scored in terms of their relative durations. If you are a musician, you may wish to just lightly skim these sections. The third section is on key signatures. Although key signatures will be familiar to musicians, we shall also describe the mathematics that underlies the Circle of Fifths. All readers should find that material of interest.

2.1 Staff Notation, Clefs, and Note Positions

In this section we discuss scoring of notes in terms of their pitch. We describe the three scoring graphs that are used most often: the treble clef staff, the bass clef staff, and the grand staff.

2.1.1 Treble Clef Staff

A musical score can be thought of as a mathematical graph in which pitch is indicated by vertical position and time is indicated by horizontal position. In Figure 2.1 we show the most frequently used scoring graph. It is called a *treble clef staff*. The treble clef (or G-clef) is the elaborately shaped letter G that sits on the left side of this staff. The long lines of the staff mark off a region of pitches indicated on the piano keyboard shown on the left side of Figure 2.1. Those pitches, along with a few pitches above or below that are scored using short line segments called *ledger lines,* cover a pitch range for many instruments. For example, violins play notes with pitches lying in this range, and so do flutes. Sopranos sing notes with pitches lying in this range as well.

Figure 2.1 shows that the staff lines and ledger lines can be thought of as markings on a vertical axis for a mathematical graph. This vertical axis indicates positions of higher or lower pitch for notes that are scored on the treble clef staff. You can see in the figure how the staff lines, and ledger lines (if they were extended), meet up with white keys on the piano keyboard. Ledger lines are used for marking pitches that are above or below the pitch range covered by the staff lines. The spaces between the lines also meet up with white keys on the piano keyboard. If only notes from a C-major scale are played, corresponding to the white keys on the piano keyboard, then we have a good analogy

---

[1]One standard reference is Gardner Read, *Music Notation, 2nd Edition,* Taplinger Publishing, New York, 1979.

between musical scoring of notes on the treble clef staff and pitches of these notes on a vertical axis. Unfortunately, the accidentals (the sharps and flats) do not correspond so well. Using a frequency axis instead of a pitch axis would resolve this problem of handling accidentals. But it can still be helpful to think of the treble clef staff as a kind of mathematical graph as we have done in the figure.

Figure 2.1. Treble clef staff and its relation to pitch and time. Notes in order from left to right are G_3, C_4, F_4, $A_4^\sharp$, B_4, $D_5^\flat$, G_5, C_6. The position of C_4, called middle C, is one ledger line below the full lines for the staff. On the right side of the staff, we show labeling of staff lines (EGBDF $\sim$ Every Good Boy Does Fine), and labeling of spaces between the staff lines (FACE). (The division of the time axis is described by a time signature, which we discuss in the next section.)

The notes that are marked on the staff lines of the treble clef staff are E, G, B, D, F. A mnemonic device for remembering that sequence is Every Good Boy Does Fine. Or, as is more commonly said in Canada and England, Every Good Boy Deserves Favour.[2] The notes in the spaces between the staff lines are F, A, C, E. They can be remembered with another mnemonic: FACE.

Here is an example of scoring notes with the treble clef staff:

The notes are $G, B, B^\flat, E, G^\sharp, F, B^\flat, C, A, C, E, F, A, F^\sharp, B, C^\sharp$.

2.1.2 Bass Clef Staff

Many instruments play notes in a pitch range that lies below the staff lines of the treble clef staff.

---

[2]There was an album with that title, *Every Good Boy Deserves Favour,* by The Moody Blues produced in the 1960s.

Such notes would require lots of ledger lines if scored on a treble clef staff. A widely used scoring graph for such lower pitch instruments is the *bass clef staff*. In Figure 2.2, we show the bass clef staff as a mathematical graph of pitch versus time. The pitch range shown in the figure contains the pitch ranges for cellos and trombones, for example. Baritones sing notes with pitches in this range as well.

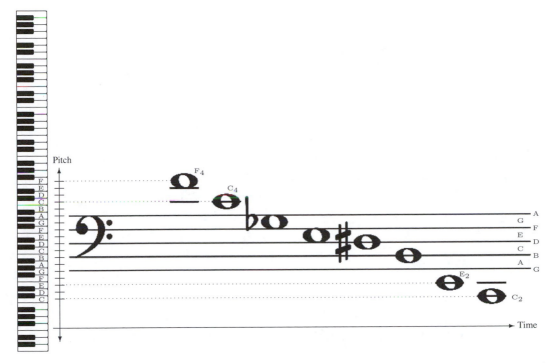

Figure 2.2. Bass clef staff and its relation to pitch and time. Notes in order from left to right are F_4, C_4, $G_3^\flat$, E_3, $D_3^\sharp$, B_2, E_2, C_2. The position of C_4 is one ledger line above the full lines for the staff. On right side of the staff, we show labeling of staff lines (GBDFA $\sim$ Good Boys Do Fine Always) and labeling of spaces between staff lines (ACEG $\sim$ All Cows Eat Grass).

A mnemonic for the notes marked on the staff lines of the bass clef staff is Good Boys Do Fine Always. A mnemonic for the notes marked between the lines is All Cows Eat Grass.

Here is an example of a bass clef staff with several notes marked:

The notes are $C, F, A, C^\sharp, E, G^\flat, B, C, D, E$.

2.1.3 Grand Staff

For many keyboard instruments like the piano, and other instruments like the harp with its larger pitch range, the treble clef staff and bass clef staff can be combined to create the *grand staff*. In Figure 2.3 we show the grand staff with several notes marked on it. The ledger lines that lie between the upper and lower staff are independent of each other. In fact, in the score shown in the figure, the pitches of the beginning notes on the upper staff actually overlap with the pitches of the final notes on the lower

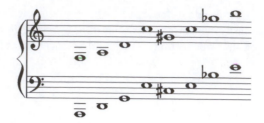

Figure 2.3. The grand staff with several notes scored. Notes on upper staff (treble clef staff) are F, A, D, C, G$^\sharp$, C, G$^\flat$, B. Notes on lower staff (bass clef staff) are A, D, G, E, C$^\sharp$, E, B$^\flat$, E.

staff.

This overlapping of pitches on the grand staff may seem odd to non-musicians, until one remembers that the grand staff is typically used for keyboard instruments like the piano. For keyboard instruments, the notes on the upper staff are usually played with the right hand, while the left hand is used for notes on the lower staff. So, generally speaking, the potential for overlapping of pitches is mediated by the fact that separate hands are used for playing the notes from each staff. Although with some pieces, the hands actually have to physically cross over each other at the keyboard.

Exercises

2.1.1. For the following staff

identify the notes: _____ .

2.1.2. For the following staff

identify the notes: _____ .

2.1.3. For the following Grand Staff, identify the notes:

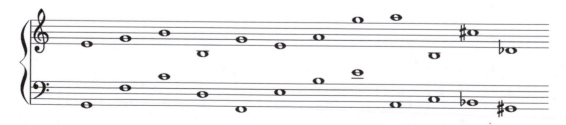

Upper staff: _____   Lower staff: _____ .

2.1.4. For the following Grand Staff, identify the notes:

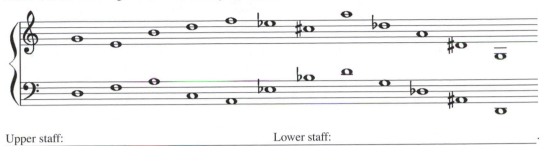

Upper staff: _____   Lower staff: _____ .

2.1.5. Alto clef. The *alto clef* is a clef that is used as the primary clef for the viola. Here is a staff with an alto clef, where we have indicated on the left side where middle C is located:

Identify the notes on this staff: _____ .

2.1.6. Tenor clef. The *tenor clef* is another clef that is sometimes used. (For example, it is used for the upper range of the cello.) Here is a staff with a tenor clef, where we have indicated on the left side where middle C is located:

Identify the notes on this staff: _____ .

2.1.7. The following score uses all of the clefs that we have discussed. Identify the notes that are scored.

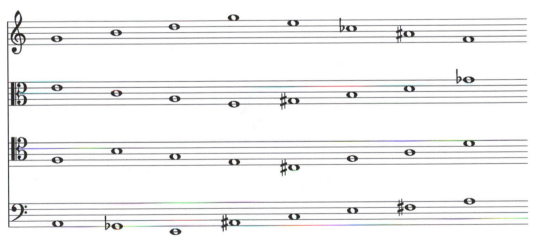

Treble clef staff: _____   Alto clef staff: _____

Tenor clef staff: _____   Bass clef staff: _____

2.2   Time Signatures and Tempo

The second aspect of musical scoring is the description of how notes are played over time. This is related to the timing of the playing of notes, and the relative durations of the notes, which are important aspects of the music's rhythm. The overall pace of the music, its *tempo*, is also important. We will discuss each of these aspects—timing, relative duration, and tempo—in this section.

In Western music, notes are played in discrete time durations and are organized in groups within a score. These groups are called *measures* (they are also called *bars*). The notes will have various relative durations, indicated by graphing the notes as whole notes, half notes, quarter notes, eighth notes, sixteenth notes, or $1/32^{nd}$ notes. These are the six most frequently used notes. See Figure 2.4.

Symbol	Relative Duration	Term	Symbol Parts	
			pitch marker	
○	1	**Whole note**	○	
♩ (half)	$\frac{1}{2}$	**Half note**	pitch marker (open)	stem
♩	$\frac{1}{4}$	**Quarter note**	pitch marker	stem
♪	$\frac{1}{8}$	**Eighth note**	pitch marker	stem and 1 flag
♪ (2 flags)	$\frac{1}{16}$	**Sixteenth note**	pitch marker	stem and 2 flags
♪ (3 flags)	$\frac{1}{32}$	**$1/32^{nd}$ note**	pitch marker	stem and 3 flags

Figure 2.4. The six basic note symbols are shown on the left, followed by their relative time durations, the terms identifying them, and their parts. The pitch markers (note heads) are used on the score to mark the pitches for notes. The number of flags on a stem indicate how many times a quarter note duration is divided by 2.

In addition to notes, there can also be *rests* between the notes. The rests on a score indicate pauses (silences) of relative durations equal to a corresponding note. For example, a whole rest is a pause in playing of duration equal to a whole note, while a quarter rest is a pause of duration equal to a quarter note. Figure 2.5 summarizes the notation used for the six most frequently used rests.

Whole rest	Half rest	Quarter rest	Eighth rest	Sixteenth rest	$1/32^{nd}$ rest
▬	▬	𝄾	𝄾	𝄾	𝄾
1	1/2	1/4	1/8	1/16	1/32

Figure 2.5. The six basic rest symbols and their relative time durations. The symbol for both whole rest and half rest is a solid rectangle. For the whole rest it is placed below a staff line, while for the half rest it is placed above a staff line. Rest symbols for eighth rests, sixteenth rests, and $1/32^{nd}$ rests, have flags which point to the left. The number of flags indicate how many times a quarter rest duration is divided by 2.

In what is called *common time,* whole notes and whole rests have time durations equal to a whole measure, half notes and half rests have time durations equal to half a measure, quarter notes and quarter rests have time durations equal to a quarter of a measure, and so on. In common time, four quarter notes (or quarter rests) will have the duration of one measure. The following score shows how the basic notes and rests appear in a score written in common time:

The measures are separated by vertical line segments, called *bar lines.* The end of the score is marked by two bar lines. In the first measure in this score, there is just one whole note. In the second measure there is just a single whole rest. The third measure contains a half note followed by a half rest. The fourth measure contains a quarter note, a quarter rest, another quarter note, and another quarter rest. Notice that the stems of the quarter notes are in opposite directions.[3] The fifth measure begins with an eighth note, followed by an eighth rest. After that, two eighth notes are connected by a *beam.* There follows an eighth rest, and then three beamed eighth notes. The sixth measure begins with a sixteenth rest, followed by a sixteenth note, a sixteenth rest, and another sixteenth note. After that, there are four beamed sixteenth notes, followed by a $1/32^{nd}$ rest, and then two beamed $1/32^{nd}$ notes. Then there is a sequence of a $1/32^{nd}$ rest, a $1/32^{nd}$ note, and a $1/32^{nd}$ rest. After that, there are three beamed $1/32^{nd}$ notes and a $1/32^{nd}$ rest. The final three notes are two $1/32^{nd}$ notes and one eighth note beamed together.

This last example can also be scored in this way:

The letter **c** appearing on the left of the score is a *time signature.* It indicates that common time is to be used for this score. Common time is also known as $\frac{4}{4}$ time. The notation $\frac{4}{4}$ is another instance of a time signature.

2.2.1 Time Signatures

Time signatures describe the rhythmic timing of the music, its division into relative durations and the timing of note onsets. We will only describe the two most common time signatures: the $\frac{4}{4}$ time signature and the $\frac{3}{4}$ time signature. Some additional time signatures will be explored in the exercises.

The $\frac{4}{4}$ time signature

The notation $\frac{4}{4}$ looks like a fraction, but without the fraction bar. We shall see that there is some relation between the time signature, $\frac{4}{4}$, and the fraction, $\frac{4}{4}$.

The top number in the time signature, in this case a 4, indicates that there will be four fundamental beats (notes or rests) in each measure. The bottom number, also a 4, indicates that each of those fundamental beats will be quarter notes, or rests of duration equal to a quarter note. In the following score, we show an example of a score in $\frac{4}{4}$ time, and how the relative time durations of notes and rests add up to the fraction $\frac{4}{4}$ in each measure:

---

[3]If the pitch marker lies on the middle staff line, or above it, then the stem is downward. If the pitch marker lies below the middle staff line, then the stem is upward.

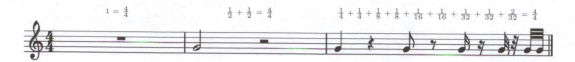

In the first measure in this score, one whole rest takes up the whole measure. This gives the equation $1 = \frac{4}{4}$. In the second measure, there is one half note followed by one half rest, giving the equation $\frac{1}{2} + \frac{1}{2} = \frac{4}{4}$. Finally, in the third measure, there is a more complex sequence of notes and rests obtained by further division of duration by 2. The sum of all the note and rest durations in the third measure produces this equation:

$$\frac{1}{4} + \frac{1}{4} + \frac{1}{8} + \frac{1}{8} + \frac{1}{16} + \frac{1}{16} + \frac{1}{32} + \frac{1}{32} + \frac{2}{32} = \frac{4}{4}.$$

The $\frac{3}{4}$ time signature

Another time signature that is frequently used is the $\frac{3}{4}$ time signature. The basic ideas are the same as with $\frac{4}{4}$ time, so we shall just briefly discuss this case. The fundamental beat-length is $1/4$, either a quarter note or a quarter rest, and there are 3 beats to a measure. In the following score, we show an example of how durations add up in $\frac{3}{4}$ time:

In the first measure in this score, there is a quarter note followed by a quarter rest and a quarter note. Since there are 3 beats in a row, we get the equation $\frac{1}{4} + \frac{1}{4} + \frac{1}{4} = \frac{3}{4}$. The second measure begins with a quarter note, followed by an eighth rest, an eighth note, a $1/32$ rest, an eighth note, another $1/32$ rest, and then two beamed $1/32$ notes. This gives the equation $\frac{1}{4} + \frac{1}{8} + \frac{1}{8} + \frac{1}{32} + \frac{1}{8} + \frac{1}{32} + \frac{2}{32} = \frac{3}{4}$ for the measure. Again we can see the connection of time signature notation with fractions: for this time signature of $\frac{3}{4}$, the sum of the note and rest durations for each measure produces the fraction $\frac{3}{4}$. The final measure in the example above provides further confirmation of this connection. It consists of a sixteenth rest, followed by an eighth note and a sixteenth rest, followed by a quarter rest, and finishing with a beamed eighth note and sixteenth note along with a sixteenth rest. Altogether these durations produce the equation: $\frac{1}{16} + \frac{1}{8} + \frac{1}{16} + \frac{1}{4} + \frac{1}{8} + \frac{1}{16} + \frac{1}{16} = \frac{3}{4}$.

Dotted notes, ties, and slurs

Some other aspects of scoring that we need to discuss are dotted notes, ties, and slurs. Dotted notes have a small dot on the right of the note symbol. A dotted note is 1.5 times the duration of the same note without the dot. For example, a dotted eighth note has duration $\frac{1}{8} + \frac{1}{16} = \frac{3}{16}$. A *tie* indicates that two notes with the same pitch should be played as one, with a duration equal to the sum of the duration of the individual notes. Slurs, which resemble ties in the score notation, are indications that notes of differing pitch should be played together, without audible pauses between the notes. In the following score, we show an example of these notations:

In the first measure there is a dotted half note. It has duration $\frac{1}{2} + \frac{1}{4}$. This gives $\left(\frac{1}{2} + \frac{1}{4}\right) + \frac{1}{4} = \frac{4}{4}$ as the equation for the note and rest durations in the first measure. The second measure begins with a double dotted quarter note, which has duration $\frac{1}{4} + \frac{1}{8} + \frac{1}{16} = \frac{7}{16}$. The second measure continues with a tie that connects a dotted quarter note and a $1/32$ note, both having pitch E_4. This indicates that a single note, of pitch E_4, should be played with a duration of $\left(\frac{1}{4} + \frac{1}{8}\right) + \frac{1}{32} = \frac{13}{32}$. Notice that this duration of $\frac{13}{32}$ for a single note *cannot* be specified using the basic note forms (either dotted or undotted), but it *can* be specified using the tie. Another use of a tie is shown at the end of the second measure. There is a tie of two eighth notes at pitch F_4. This indicates one note, of pitch F_4 and duration $\frac{1}{8} + \frac{1}{8} = \frac{1}{4}$, that extends across the measure line into the third measure. This note has exactly the same duration as a quarter note, but in order to stay within the constraints of $\frac{4}{4}$ time, it must be written as two tied eighth notes. The third measure then continues with an eighth note rest and then two eighth notes connected with a slur. This slur has no effect on the duration. The total duration of the two slurred eighth notes is $\frac{1}{8} + \frac{1}{8} = \frac{1}{4}$. As its name implies, the slur indicates that the two eighth notes should be slurred together as they are played so that no silence is audible between the individual notes. The measure then continues with a quarter note followed by four slurred sixteenth notes. The slurred sixteenth notes have total duration $\frac{4}{16} = \frac{1}{4}$ (the length of a beat for this time signature), and they are to be all slurred together as they are played.

Tuplets

Tuplets involve grouping notes whose duration is **not** a power of 2 division of a beat. The most common tuplet is a *triplet,* which has 3 notes in a group. An example of four triplets in $\frac{3}{4}$ time is given in the following score:

The first measure in this score begins with a triplet of eighth notes. Although the triplet is notated with eighth notes, their durations add up to the duration of a quarter note. So each of the notes in the triplet actually has a duration of $\frac{1}{3} \cdot \frac{1}{4} = \frac{1}{12}$. In the equation below the measure, we have listed the total duration of the triplet as $\left(\frac{1}{4}\right)$. Likewise, in the second measure, there is a triplet denoted using sixteenth notes. But the duration of the entire triplet is equal to the duration of an eighth note, which we have denoted by $\left(\frac{1}{8}\right)$ in the equation below the measure. The convention is that **the duration of the triplet is twice as long as the duration of the note that is repeated to form the triplet.** For example, in the third measure there is a triplet of $1/32^{\text{nd}}$ notes, so the length of the whole triplet is $1/16$ which we have written as $\left(\frac{1}{16}\right)$ in the equation for the measure.

In general, a **tuplet** is a set of several notes with total duration equal to some single note. An example of three different tuplets in $\frac{4}{4}$ time is given in the following score:

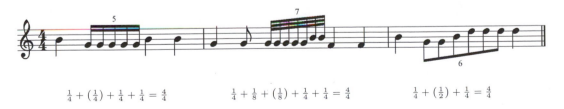

The first measure in this score has a quintuplet (or 5-tuplet) of total duration $1/4$, so each of the notes in the 5-tuplet has duration $1/20$. The second measure has a 7-tuplet of total duration $1/8$, so each of the notes in the 7-tuplet has duration $1/56$. Finally, the third measure has a 6-tuple of total duration

1/2, so each of the notes in the 6-tuplet has duration 1/12. The total duration of a tuplet can usually be inferred from context, using the durations of non-tuplet notes and the time signature.

2.2.2 Tempo

The time signature specifies only the *relative* duration of notes. Their *absolute* duration is specified by indicating the *tempo*. The tempo indicates the overall speed of playing. With the invention of the metronome, a tempo could be specified very accurately. Often in scores a metronome marking is given for the tempo. For example, when the following metronome marking

$$\quad = \mathit{120}$$

appears at the beginning of the score, then the tempo is set at 120 beats/minute (bpm). In this case, the beat is a quarter note and 120 quarter notes have a duration of 1 minute. Therefore, the duration of a quarter note is 0.5 seconds.

2.2.3 Rhythmic Emphasis

In this section we have emphasized the relative durations of notes within measures, specified by the time signature. This is an important aspect of musical rhythm, but not its only aspect. Another aspect of rhythm, best appreciated by listening to or performing music, is the relative emphasis of notes within each measure, the **meter** of the music. For example, the most common rhythmic emphasis in $\frac{4}{4}$ is "ONE-two-THREE-four." However, there can also be a *syncopated* emphasis of "one-TWO-three-FOUR." In rock music, these two emphases are often combined. The melody has the first type of emphasis. The drums or bass part has the second emphasis, called the "back beat." (See Exercise 2.2.15.) An excellent discussion of rhythmic emphasis can be found in "The World of Jazz" chapter in Leonard Bernstein's *The Joy of Music*, pp. 106–131.

Exercises

2.2.1. For the following score, for each measure write the equation with fractions that corresponds to the note durations, given the $\frac{4}{4}$ time signature:

2.2.2. For the following score, for each measure write the equation with fractions that corresponds to the note durations, given the $\frac{4}{4}$ time signature:

2.2.3. For the following score, for each measure write the equation with fractions that corresponds to the note durations, given the $\frac{3}{4}$ time signature:

2.2.4. For the following score, for each measure write the equation with fractions that corresponds to the note durations, given the $\frac{3}{4}$ time signature:

2.2.5. For the following score, for each measure write the equation with fractions that corresponds to the note durations, given the $\frac{3}{4}$ time signature:

For each triplet, what is the duration of each individual note in the triplet?

2.2.6. For the following score, for each measure write the equation with fractions that corresponds to the note durations, given the $\frac{3}{4}$ time signature:

For each tuplet, what is the duration of each individual note in the tuplet?

2.2.7. For the following score, for each measure write the equation with fractions that corresponds to the note durations, given the $\frac{4}{4}$ time signature:

For each triplet, what is the duration of each individual note in the triplet?

2.2.8. For the following score, for each measure write the equation with fractions that corresponds to the note durations, given the $\frac{4}{4}$ time signature:

For each tuplet, what is the duration of each individual note in the tuplet?

2.2.9. For each of the tempo marks below, find the length in seconds of a quarter note.

$$\quarternote = 60 \qquad\qquad \quarternote = 256 \qquad\qquad \halfnote = 60$$

2.2.10. Some musical scores use Italian terms to indicate tempo, such as *Adagio* or *Allegro* or *Andante*. Precise bpm values are not always specified, a metronomically precise tempo being up to the performer(s) or conductor. Suppose the performance is to be played at tempos that use these values for bpm (where a beat is a quarter note):

$$\text{Adagio} = 40 \qquad\qquad \text{Allegro} = 120 \qquad\qquad \text{Andante} = 80$$

In each case, find the length in seconds of a quarter note.

2.2.11. $\frac{2}{4}$ **time.** For the following score, for each measure write the equation with fractions that corresponds to the note durations, given the $\frac{2}{4}$ time signature:

2.2.12. $\frac{6}{8}$ **time.** For the following score, for each measure write the equation with fractions that corresponds to the note durations, given the $\frac{6}{8}$ time signature. Write the equations for the lower staff below it, and the equations for the upper staff above it.

Note: This exercise shows that $\frac{6}{8}$ time can be written using 6 eighth notes (1/8 duration for beats), as in the lower staff. It also shows that it can be written with 2 groups of duration 3/8 (durations of 3/8 for the beats), as in the upper staff. Traditionally the latter method is used. This example—from Brahm's F-minor Piano Sonata Op. 5, measures 258 to 260—shows that either method can be used. The time signature $\frac{6}{8}$ relates to the timing pattern and note durations for the music, it does not absolutely require a particular style of grouping note durations (a particular meter). Although, again, the typical meter is 3/8 durations for each of 2 beats in a measure.

Automatic Music Scoring

In computer music scoring software, such as MUSESCORE, when a note is entered the rests remaining in the measure are automatically computed. Here is an example:

The next two exercises deal with the algebra underlying how the software calculates the remaining rests in a measure after entering a note.

2.2.13. In the example above, let x stand for the duration left in the measure after the $1/32^{\text{nd}}$ note is entered. Show that the following equation holds

$$\frac{1}{32} + x = \frac{3}{4}$$

and solve for x. Show that x can be expressed in a way that reflects the rests shown in the example above.

2.2.14. Use the method outlined in the previous exercise to find the sequence of rests that remain in the measure after entering the indicated notes:

(a) Add a dotted sixteenth note at the start of the measure:

(b) Add a dotted sixteenth note at the start of the measure:

(c) Add a dotted quarter note (right after the sixteenth note):

2.2.15. Back beat. In his song, *Rock and Roll Music,* Chuck Berry says that rock music has a back beat, and that therefore you can dance any way you want to it. What is meant by this?

2.3 Key Signatures and the Circle of Fifths

A great deal of Western music is written in specific keys. In this section we will describe the notation for these keys and how they are connected mathematically by the famous Circle of Fifths. The logic underlying the circle of fifths is explained using addition on the chromatic clock.

For music written in a particular key, a specific scale is used for the vast majority of the notes. This scale is indicated at the beginning of the score by marking the sharps, or flats, that occur on the scale's notes. For example, if a score is written using predominantly notes from the scale of A-major:

$$A \quad B \quad C^\sharp \quad D \quad E \quad F^\sharp \quad G^\sharp \quad A$$

then the score will begin with 3 sharp symbols marked at positions for the notes C, F, and G. It is then unnecessary to mark these sharps on each appearance of one of the three sharped notes in the A-major scale. Here is an example of a score with an A-major key signature:

The score has a sequence of notes containing several instances of $C^\sharp$, $F^\sharp$, and $G^\sharp$, but these notes are not marked with sharps. The key signature of 3 sharps tells us that they are sharped notes.

For another example, consider the $B^\flat$-major scale:

$$B^\flat \quad C \quad D \quad E^\flat \quad F \quad G \quad A \quad B^\flat.$$

This scale has two flatted notes: $B^\flat$ and $E^\flat$. These flatted notes are indicated in the key signature by two flat symbols marked at positions for the notes B and E. Here is an example of a score with a $B^\flat$-major key signature:

It has a sequence of notes containing several instances of $B^\flat$ and $E^\flat$. In the score, however, these notes are not marked with flats. Notice also that $D^\flat$, which is not part of the $B^\flat$-major scale, is marked by a flat in the score. Whenever a note is used that is not part of the scale specified by the key signature, then such a note is called an *accidental*. So, in this case, there is one accidental of $D^\flat$.

The C-major scale has no flats or sharps. So when no flats or sharps are written at the start, as shown in the score in Figure 2.6, then that indicates a C-major scale (for the key of C-major or A-minor). All notes in the score are then naturals, no flats or sharps, unless the notes are specifically marked as accidentals.

Marking Accidentals within Measures

The key signature method greatly reduces the number of sharps and flats that must be explicitly

Figure 2.6. Illustration of marking accidentals. For the first two measures, the key signature (or lack thereof) indicates C-major scale. For the last two measures, the key signature of two sharps indicates D-major scale.

marked in a score. Another convention that reduces explicit marking of accidentals consists in ***not marking an accidental if that accidental has already been marked earlier in the same measure***. This convention only lasts as long as one measure. In Figure 2.6 we illustrate this convention. This figure also illustrates that key signatures can be altered during the course of a score.

2.3.1 Circle of Fifths

Key signatures are connected in an organized way. This organization is called the circle of fifths. The circle of fifths for the key signatures of the major scales is shown in Figure 2.7.

This figure should be interpreted as follows. The top of the circle is the position 0 for the key of C-major. The 0 indicates that the C-major key has 0 sharps and 0 flats. Each time we move one hourly position *clockwise* (numbered by sharps), we reach a new key that shares the same notes as the previous key except for one new sharped note. This new key can be indicated on a staff with one more sharp than the preceding key on the circle of fifths. This pattern continues until we reach "6♭ or 6♯." This label marks the key of F♯-major, because F♯-major has 6 sharps. The key of F♯-major is enharmonic with the key of G♭-major, which has 6 flats. When we say these two keys are enharmonic, we mean that each note of F♯-major is enharmonic with the corresponding note of G♭-major. On the

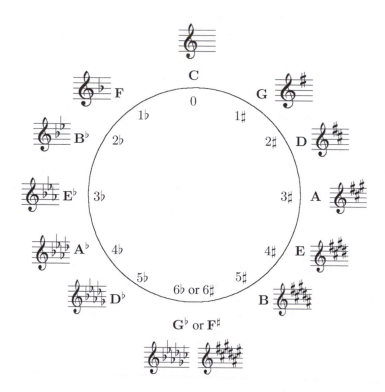

Figure 2.7. Circle of Fifths. Key signatures for major scales.

other hand, if we start again at position 0 for the key of C-major, and successively move one hourly position *counter-clockwise* (numbered by flats), each new key shares the same notes as the previous key except for one new flatted note. Each new key can be indicated on a staff with one more flat than the preceding key on the circle of fifths.

The mathematics underlying the circle of fifths can be explained using the chromatic clock. Suppose we begin with the key of C-major at the top of the circle of fifths. The sharpening of one note when we move clockwise one hour is based on the geometry of the chromatic clock, as shown on the left of Figure 2.8. The C-major scale, which begins at hour 0, is marked off by solid dots in the diagram. The G-major scale, which begins at hour 7, is marked off by open circles. All of the notes on the two scales match, except for F and F$^\sharp$. The arrow labeled by $\sharp$ in the diagram indicates the change of the single note F to F$^\sharp$.

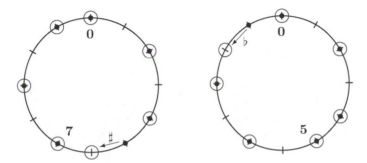

Figure 2.8. Left: C-major scale marked by solid dots, beginning and ending at hour 0; G-major scale marked by open circles, beginning and ending at hour 7. Arrow labeled $\sharp$ indicates how F on C-major scale moves to F$^\sharp$ on G-major scale. Right: C-major scale marked by solid dots, beginning and ending at hour 0; F-major scale marked by open circles, beginning and ending at hour 5. Arrow labeled $\flat$ indicates how B on C-major scale moves to B$^\flat$ on F-major scale. Flattening of single note, when going from C-major to F-major, is reversal of sharpening of single note, when going from F-major to C-major.

Similar reasoning applies each time we move another hourly position clockwise on the circle of fifths. We only need to observe that any major scale could be plotted in the same way as the C-major scale on the left of Figure 2.8, with its starting and ending note being placed at hour 0. The major scale created by going up a fifth is plotted with its starting and ending note being placed at hour 7. Consequently, we find that each time we go up a fifth and build a new major scale, this new major scale will have all the same notes as the preceding scale except for the note right before the end-note of the new scale. That note is a sharped version of the note right before this end-note *within the preceding scale*.

Remark 2.3.1. As reinforcement of our discussion, one can begin by writing down the C-major scale:

$$\text{C} \quad \text{D} \quad \text{E} \quad \text{F} \quad \text{G} \quad \text{A} \quad \text{B} \quad \text{C}.$$

Then construct the G-major scale by simply writing down the letters from the list above, starting with G, wrapping around to the start at C (which is only written once), and putting a sharp on the note right before the return to G. If you do that, you get the G-major scale:

$$\text{G} \quad \text{A} \quad \text{B} \quad \text{C} \quad \text{D} \quad \text{E} \quad \text{F}^\sharp \quad \text{G}.$$

We leave it as an exercise to check that this method generates all of the major scales correctly for positions 1$\sharp$ through 6$\sharp$ on the circle of fifths.

Going counter-clockwise on the circle of fifths will produce new key signatures by adding flats. Each additional hourly position counter-clockwise will add a flat to the previous key signature. When we go counter-clockwise on the circle of fifths we are going down a fifth from the preceding key. The flattening of one note when we move counter-clockwise one hour is based on the geometry of the chromatic clock, as shown on the right of Figure 2.8. The F-major scale, which begins at hour 5, is marked off by the open circles in the diagram. The C-major scale, which begins at hour 0, is marked off by the solid dots. All of the notes on the two scales match, except for B and $B^\flat$. The arrow labeled by $\flat$ in the diagram indicates the change of the single note B to $B^\flat$. Thus, the new scale has one flatted note ($B^\flat$), which is the flatted version of the note preceding the last note in the C-major scale.

Similar reasoning applies each time we move another hourly position counter-clockwise on the circle of fifths. We only need to observe that any major scale could be plotted in the same way as the C-major scale on the right of Figure 2.8, with its starting and ending note being placed at hour 0. The major scale created by going down a fifth is plotted with its starting and ending note being placed at hour 5. Consequently, each time we go down a fifth and build a new major scale, it will have all the same notes as the preceding scale, except for the note right before the end-note of the preceding scale. That note is flatted in the new scale.

Remark 2.3.2. To reinforce our discussion, one can begin by writing down the C-major scale:

$$C \quad D \quad E \quad F \quad G \quad A \quad B \quad C.$$

Then construct the F-major scale by simply writing down the letters from the list above, starting with F, putting a flat on the note just before the end-note C, wrapping around to the start at C (which is only written once), and continuing until F is reached. If you do that, you get the F-major scale:

$$F \quad G \quad A \quad B^\flat \quad C \quad D \quad E \quad F.$$

We leave it as an exercise to check that this method generates all of the major scales correctly for positions $1\flat$ through $6\flat$ on the circle of fifths.

Identifying scale names from key signatures

There is a simple method for identifying the names of the major key signatures, just using the key signatures on the score. For sharped positions on the circle of fifths, we saw that the new sharped note that is added as we move clockwise is the note right before the ending note for the new scale. That sharped note is always written as the rightmost sharp in the key signature. Therefore, *by moving up one position on the staff to the note above that last sharp in the key signature, we get the position for the note that names the key.* On the other hand, for flatted positions on the circle of fifths, the beginning note of a new key as we move counter-clockwise around the circle is the last flatted note from the previous key. Therefore, *the next to the last flat in the key signature marks the position for the note that names the key.* There is one exception here: For F-major there is only one flat in the key signature, so the rule does not apply. When there is just one flat in the key signature, we must simply remember that it indicates the F-major scale.[4]

2.3.2   Circle of Fifths for Natural Minor Scales

There is also a circle of fifths for the natural minor scales. A natural minor scale is obtained by using the following pattern of half steps from the initial note:

$$+2 \quad +1 \quad +2 \quad +2 \quad +1 \quad +2 \quad +2. \tag{2.1}$$

For example, if we start from the note A, which has hour 9 on the chromatic clock, then we perform these calculations:

$$9 \xrightarrow{+2} 11 \xrightarrow{+1} 0 \xrightarrow{+2} 2 \xrightarrow{+2} 4 \xrightarrow{+1} 5 \xrightarrow{+2} 7 \xrightarrow{+2} 9 \tag{2.2}$$
$$A \qquad B \qquad C \qquad D \qquad E \qquad F \qquad G \qquad A$$

---

[4]It may help to remember that F is the first letter of flat.

producing the natural A-minor scale:

$$A \quad B \quad C \quad D \quad E \quad F \quad G \quad A. \tag{2.3}$$

Notice that it has the same notes as the C-major scale. The natural A-minor scale is called the ***relative minor*** of the C-major scale.

This begs the question: Why is the natural A-minor scale referred to as minor, when it has the same notes as the C-major scale? The reason involves the different types of chords that are typically played with these two different scales (or keys). We will discuss this in the next chapter.

Using the pattern of half steps in (2.1) we can construct natural minor scales starting from any initial note. For example, if we were to go up a fifth from the natural A-minor scale, then we would begin on the note E and obtain these results:

$$4 \xrightarrow{+2} 6 \xrightarrow{+1} 7 \xrightarrow{+2} 9 \xrightarrow{+2} 11 \xrightarrow{+1} 0 \xrightarrow{+2} 2 \xrightarrow{+2} 4 \tag{2.4}$$
$$E \qquad F^\sharp \qquad G \qquad A \qquad B \qquad C \qquad D \qquad E.$$

Thus obtaining the natural E-minor scale:

$$E \quad F^\sharp \quad G \quad A \quad B \quad C \quad D \quad E. \tag{2.5}$$

This natural E-minor scale is the relative minor of the G-major scale.

This pattern looks familiar: If we go up a fifth on a natural minor scale, then we obtain a new natural minor scale with all the same note letters except for one that is sharped. We also are getting the same notes as in a related major scale, just in a different order. This produces a circle of fifths diagram that connects all of the natural minor scales. See Figure 2.9.

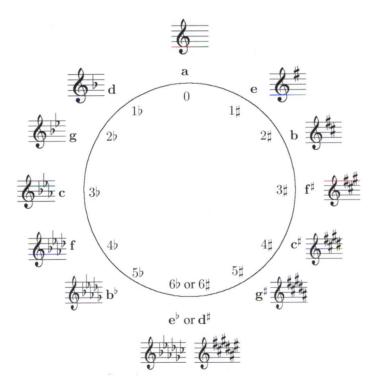

Figure 2.9. Circle of fifths for natural minor scales. Lowercase letters used to indicate starting notes for scales. For example, **a** indicates A-minor.

The mathematical explanation for the circle of fifths for natural minor keys is quite similar to what we described for the circle of fifths for the major keys. We leave it as an exercise for the reader to work out the details.

Exercises

2.3.1. 7 sharps or 7 flats. Occasionally, especially in classical music, one encounters key signatures with either 7 sharps or 7 flats. Place those two key signatures on the circle of fifths, and determine their names in both major and minor keys.

2.3.2. Begin by writing down the C-major scale. **(a)** Write down each major scale obtained by going up a fifth—using the description in Remark 2.3.1—until you reach the scale with all notes sharped. **(b)** Write down the C-major scale again. Then successively write down each major scale obtained by going down a fifth—using the description in Remark 2.3.2—until you reach the scale with all notes flatted.

2.3.3. Circle of fifths for minor keys. In Figure 2.9 we have shown the circle of fifths for the natural minor keys. Work out the details, similar to what we did for the circle of fifths for major keys, that explain why the natural minor keys can be organized in this way.

2.3.4. (a) Begin by writing down the natural A-minor scale (remember that it has the same notes as its relative major, the C-major scale). Then successively write down each natural minor scale obtained by going up a fifth, using an appropriate modification of the method described in Remark 2.3.1. **(b)** Write down the natural A-minor scale again. Then successively write down each natural minor scale obtained by going down a fifth, using an appropriate modification of the method described in Remark 2.3.2.

2.3.5. Circle of fifths for minor keys—another approach. Verify that adding -3 and $+7$ on the chromatic clock is the same as adding $+7$ and -3, and adding -3 and $+5$ is the same as adding $+5$ and -3. Use this fact to explain why there is a circle of fifths for natural minor keys.

2.3.6. Explain why the sharped notes that occur as one moves clockwise on the circle of fifths ($F^\sharp$, then $C^\sharp$, then $G^\sharp$, and so on) are also going up by fifths, starting from $F^\sharp$. Explain why the flatted notes that occur as one moves counter-clockwise on the circle of fifths ($B^\flat$, then $E^\flat$, then $A^\flat$, and so on) are also going down by fifths, starting from $B^\flat$.

Chapter 3

Some Music Theory

The musical language that made the classical style possible is that of tonality, which was not a massive, immobile system but a living, gradually changing language from its beginning.

—Charles Rosen

In this chapter we explore some music theory, concentrating on pitch and harmony. This will help us to appreciate some of the structural patterns in musical compositions. We begin by discussing musical intervals and chords, which describe how different pitches are related and combined. We then describe the basic guidelines that have been used for centuries to create music within the diatonic keys (the major and minor keys described in the previous chapter). Following those guidelines produces music that we shall refer to as diatonic music.[1] A commonly used technique for composing diatonic music is *tonal sequencing*. Tonal sequencing is a prime example of a mathematical procedure in musical composition. We conclude the chapter by examining sequencing when the chromatic scale is used, which is known as *real sequencing*.

3.1 Intervals and Chords

Most of Western music is composed from notes on the chromatic scale, often restricted to a specific diatonic scale (or key).[2] When comparing two notes, we can consider the change on a score that separates one note from another. This change is called an ***interval***. There are two types of intervals, melodic intervals and harmonic intervals. A melodic interval is the change on the score between two successive notes. A harmonic interval is the change on the score between two notes that are played simultaneously, as in a chord. Diatonic music uses intervals and chords for melody and harmony in a characteristic way, which we describe in the next section.

3.1.1 Melodic and Harmonic Intervals

When two notes occur successively, as in a melody, then the interval between them is called a ***melodic interval***. To compute a melodic interval, you count the number of spaces and lines on the score between the two notes, starting with 1 for the first note. For example, the note E_4 lies on the lowest staff line of the treble clef staff. If the next note is G_4, then it lies on the next staff line above. Counting ***all*** the lines and spaces between these two notes, ***inclusively***, we get a count of 3. So the interval between E_4 and G_4 is called a ***third***. Another example would be the notes C_4 and B_3, for these notes we count 1 ledger line and 1 space, inclusively. So the interval between C_4 and B_3 is a ***second***.

In Figure 3.1 we show some melodic intervals. Notice that the intervals are not dependent on whether the pitches are going up or down. To specify pitch direction we would say, in this example,

---

[1] It is also called *tonal music*, but we prefer to call it diatonic music.

[2] Although there has been a tendency within the last hundred years to avoid emphasis on diatonic scales, focusing exclusively on the chromatic scale (sometimes called *atonal* music or *post-tonal* music).

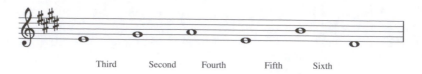

Figure 3.1. Some melodic intervals.

that the first two notes go *up a third*, while the last two notes go *down a sixth*.

There are two special cases of intervals. When two notes are exactly the same, then we say their interval is a **unison**. When two notes are separated by an octave, then we say their interval is an **octave** (rather than an eighth).

Major and Minor Intervals

If we consider the number of half steps between two pitches, then most intervals fall into two categories, called **major** and **minor**. For example, the interval between G_4 and B_4 is a third, and so is the interval between D_4 and F_4. However, the number of half steps in the former case is 4, while in the latter case it is 3. The interval between G_4 and B_4 is called a **major third**, while the interval between D_4 and F_4 is called a **minor third**. In Figure 3.2, we show the classification of intervals according to whether they are major or minor. Notice that the intervals of unison, fourth, fifth, and octave do not

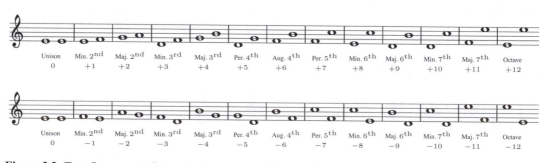

Figure 3.2. Top: Some ascending melodic intervals. Minor is abbreviated as Min., and Major is abbreviated as Maj. Perfect is abbreviated as Per., and Augmented is abbreviated as Aug. Bottom: Some descending melodic intervals. Beneath the label for each interval are the number of half steps from the first note to the second note.

have major or minor forms. The interval of a fifth is called a *perfect fifth*. An interval of a fourth can either be a *perfect fourth* (5 half steps), or an *augmented fourth* (6 half steps). The adjective *perfect* stems from ancient Greek theories of music. The augmented fourth is also referred to as a *tritone*.[3]

This categorizing of most intervals as major or minor is musically important. Because the number of half steps is related to the magnitude of frequency ratios between pitches, we will hear two notes that are a minor third apart as closer in pitch than two notes a major third apart.

We have described the basic forms for melodic intervals. There are some additional classifications for intervals which occur less frequently, but we shall not discuss them.[4]

Harmonic Intervals

When two pitches are played simultaneously then the interval between the notes is called a **harmonic interval**. The classification of harmonic intervals is essentially the same as for melodic intervals. The

---

[3]The precise definition of a tritone is an interval where the notes are 6 half steps apart on the chromatic clock. So, the notes D_4 and $G_4^\sharp$ are an augmented fourth, a tritone. However, the notes C_4 and $G_4^\flat$ are a diminished fifth, also a tritone. In both cases, the notes are separated by 6 half steps.

[4]Any book on music theory will describe the complete classification of intervals. See, for example, S. Kostka and D. Payne, *Tonal Harmony, with an introduction to twentieth century music, Sixth Edition*, McGraw-Hill, 2009.

counting of spaces and lines, and half steps, is always done from the lower pitch note to the higher pitch note. We summarize the main types of harmonic intervals in Figure 3.3.

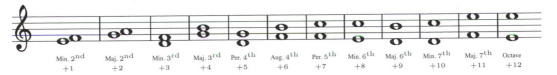

Figure 3.3. Some harmonic intervals.

In Figure 3.3, we did not include a unison interval because only one staff (for one instrument) was displayed. In scores where multiple instruments are employed, it is possible for two instruments to play the same notes simultaneously. The interval between these identical notes is a unison.

3.1.2 Chords

When multiple pitches are played simultaneously, we say that a *chord* is being played. While the two notes in a harmonic interval can be thought of as a chord (called a *dyadic* or 2-note chord), the most important chords in diatonic music are the *triadic* chords. Triadic chords are 3-note chords.

The most basic triadic chords in diatonic music are created by combining two harmonic intervals that are thirds. If the lowest pitch note and the middle pitch note form a major third, and the middle pitch note and the highest pitch note form a minor third, then the chord is called a *major chord*. If the lowest pitch note and the middle pitch note form a minor third, and the middle pitch note and the highest pitch note form a major third, then the chord is called a *minor chord*.

For example, here are two chords with their notes and the half steps separating them:

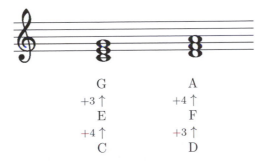

The first chord has a major third between its lowest pitch note and middle pitch note, and a minor third between its middle pitch note and its highest pitch note. Therefore it is a major chord. It is called a C-major chord. The second chord has a minor third between its lowest pitch note and middle pitch note, and a major third between its middle pitch note and its highest pitch note. So it is a minor chord. It is called a D-minor chord.

The seven ***fundamental triadic chords*** for the key of C-major are shown in Figure 3.4. In these fundamental triadic chords, the lowest pitch note is called the ***root***. With this terminology, the fundamental major chords are composed of a root note, a middle note a major third above the root, and a top note a minor third above the middle (a fifth above the root). The fundamental minor chords are composed of a root note, a middle note a minor third above the root, and a top note a major third above the middle (a fifth above the root). These chords are said to be in ***root position***. For example, chord **IV** in Figure 3.4, is an F-major chord in root position. Chord **vi** is an A-minor chord in root position.

The fundamental chords in any major key are arranged in the same way as in Figure 3.4. For example, in Figure 3.5 we show the fundamental triadic chords in the key of B-major. Notice that their form is identical with the form of the chords in Figure 3.4 for the key of C-major. The construction of

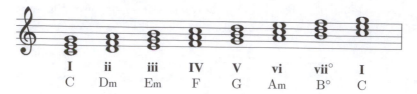

Figure 3.4. Fundamental triadic chords in key of C-major. Major chords are labeled by uppercase Roman numerals, **I**, **IV**, **V**, and also labeled by their root notes (C, F, and G). For instance, chord **V** is a G-major chord. Minor chords are labeled by lowercase Roman numerals, **ii**, **iii**, **vi**, and also by their root notes (D, E, and A). For instance, chord **iii** is an E-minor chord, and is notated as Em. The last chord, **vii°**, is called a *diminished* chord. The superscript ° indicates a diminished chord, which is constructed from two minor thirds.

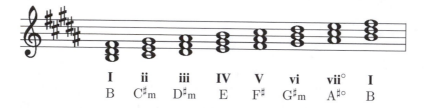

Figure 3.5. Fundamental triadic chords in key of B-major. Compare with Figure 3.4.

major and minor chords is the same. For instance, the **I** and **ii** chords shown in Figure 3.5 are B-major and $C^\sharp$-minor chords, respectively. They are both in root position, and the relationship between their notes and the half steps separating them is the following:

$$
\begin{array}{ccc}
F^\sharp & \qquad\qquad & G^\sharp \\
+3 \uparrow & & +4 \uparrow \\
D^\sharp & & E \\
+4 \uparrow & & +3 \uparrow \\
B & & C^\sharp \\
\text{B-major} & & C^\sharp\text{-minor}
\end{array}
$$

For both of these keys, and for all major and minor keys, the fundamental triadic chords are built using harmonic intervals of thirds. Since alternating staff lines and alternating staff spaces both span intervals of thirds, the staff system is perfectly adapted for displaying the fundamental triadic chords in any given key.

Inversions of Chords

Besides the fundamental triadic chords, which are in root position, there are also two other basic chord types obtained from these chords by a process called *inversion*. The *first inversion* of a triadic chord in root position, consists of raising its root note up by an octave. Its *second inversion* consists of raising the lowest pitch note of the first inversion up by an octave. See Figure 3.6. We shall see many examples of these inverted chords in the musical compositions discussed in this chapter. For instance, they occur frequently in the excerpt from Pachelbel's *Canon in D* shown in Figure 3.9 on p. 60.

In Figure 3.6, the first chord in the score is a C-major chord in root position. More precisely, it is the chord C_4-E_4-G_4. Its root note is C_4. The second chord is the 1st inversion of this C-major chord. Its root note is moved up an octave, so this chord is E_4-G_4-C_5. Its root note is still a C, but the lowest pitch note is now an E. The notation we use for indicating this chord is C/E, which specifies the root note C but also specifies the lowest pitch note E. The third chord is the 2nd inversion of the C-major chord. Moving the lowest pitch note from the 1st inversion up an octave has produced the

Figure 3.6. Inversions of triadic chords. First chord on left is C-major chord in root position, followed by its 1st and 2nd inversions. Fourth chord shown is a D-minor chord in root position, followed by its 1st and 2nd inversions. Notation for these chords is explained in the text.

chord G_4-C_5-E_5. Its root note is still a C, but the lowest pitch note is now a G. This chord is notated as C/G.

In Figure 3.6, we have also shown a D-minor chord and its 1st and 2nd inversions. The same operations are done on this chord as for the C-major chord, and the notations for the chords are similar as well. The D-minor chord shown in the figure is D_4-F_4-A_4. It is in root position, with a root note D. The 1st inversion of this D-minor chord is F_4-A_4-D_5. It has a root note D with lowest pitch note F, so it is notated as Dm/F. The 2nd inversion is A_4-D_5-F_5. It has a root note D with lowest pitch note A, so it is notated as Dm/A.

Typically in notating inverted chords in scores, the registers of the notes are **not** listed. Once inversion is performed, the chord can be shifted up or down by octaves as needed. For example, here is a score with a succession of triadic chords and their notations:

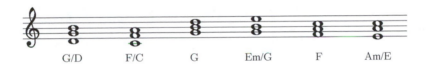

The reader should listen to these chords in order to hear the differences between them. An appreciation of the musical importance of inversions of triadic chords is gained by analyzing the flow between the separate notes in each chord (voice leading). We shall say more about this in the next section.

Seventh Chords

Seventh chords are created by adding another third, major or minor, on top of a triadic chord. The modifier, seventh, comes from the highest pitch note being a harmonic interval of a seventh above the root (when the chord is in root position). We will not discuss seventh chords in complete detail. Mostly, we will confine our discussion to the diatonic seventh chords in root position. See Figure 3.7 for a display of these fundamental diatonic seventh chords.

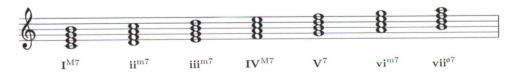

Figure 3.7. Diatonic seventh chords in key of C-major. Chord notations are explained in the text.

This figure displays the fundamental diatonic seventh chords for the key of C-major. The first chord, with C as its root, is called a ***major seventh chord***. It is denoted by $\mathbf{I}^{M7}$. It is also called a C-major seventh chord, and denoted C^{M7}. The second chord, with D as root, is called a ***minor seventh chord***. It is denoted by $\mathbf{ii}^{m7}$. It is also called a D-minor seventh chord, and denoted D^{m7}. The third chord $\mathbf{iii}^{m7}$, with E as root, is a minor seventh chord. It is also called an E-minor seventh chord, and denoted E^{m7}. The fourth chord $\mathbf{IV}^{M7}$, with F as root, is a major seventh chord. It is also called an

F-major seventh chord, and denoted F^{M7}. The fifth chord, with G as root, is called a ***major-minor seventh chord***. It is denoted by $\mathbf{V}^7$. It is also called a G-seventh chord, and denoted G^7. The sixth chord $\mathbf{vi}^{m7}$, with **A** as root, is a minor seventh chord. It is also called an A-minor seventh chord, and denoted A^{m7}. Finally, the last chord, with B as root, is called a ***half-diminished seventh chord***. It is denoted by $\mathbf{vii}^{\varnothing7}$. It is also called a B-half-diminished seventh chord, and denoted $B^{\varnothing7}$.

Here are some other examples that illustrate the half step changes that define these four fundamental types of seventh chords:

D♯	C	F	C
+4 ↑	+3 ↑	+3 ↑	+4 ↑
B	A	D	A♭
+3 ↑	+3 ↑	+4 ↑	+3 ↑
G♯	F♯	B♭	F
+4 ↑	+4 ↑	+3 ↑	+3 ↑
E	D	G	D
E-major seventh	D-seventh	G-minor seventh	D-half-dimin. seventh
E^{M7}	D^7	G^{m7}	$D^{\varnothing7}$

There are many ways of stacking harmonic thirds to built seventh chords. Consequently, there are several other seventh chords besides these four. One other important seventh chord that occurs is a diminished seventh chord. Here are two examples of diminished seventh chords:

C	D♭
+3 ↑	+3 ↑
A	B♭
+3 ↑	+3 ↑
F♯	G
+3 ↑	+3 ↑
D♯	E
D♯-dimin. seventh	E-dimin. seventh
$D^{\sharp o7}$	E^{o7}

Seventh chords can also be inverted in a variety of ways. In this book, we shall mainly be referring to triadic chords, only occasionally to seventh chords. We will generally examine seventh chords in the exercises.

Exercises

3.1.1. Identify the melodic intervals in this tune, and the number of half steps used. The first interval has been written for you as an example.

Per. 4th
+5

3.1.2. Identify the melodic intervals in this tune, and the number of half steps used. The first interval has been written for you as an example.

Per. 4th
+5

3.1.3. Name these chords, using Roman numerals for the key of C-major, and also using letter form (lead sheet notation). The first two chords have been done as examples.

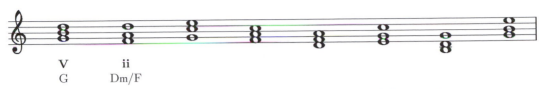

3.1.4. Name these chords, using Roman numerals for the key of D-major, and also using letter form (lead sheet notation).

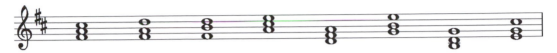

3.1.5. Name these chords, using Roman numerals for the key of A♭-major, and also using letter form (lead sheet notation).

3.1.6. Name these chords, using Roman numerals for the key of E-major, and also using letter form (lead sheet notation).

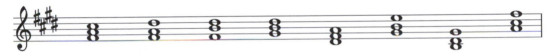

3.1.7. Consider the diminished chord **vii°** in root position, say in the key of C-major. Suppose the same number of half steps had been used, starting from the root note B, as was used for the minor chords, **ii, iii,** and **vi**. Show that the notes in the chord would have been B, D, F♯. Show that this chord is a minor chord, and find its Roman numeral in the key of G-major (the next adjacent key moving clockwise on the circle of fifths). What chord would you obtain if you used the same number of half steps for the root note B as was used for major chords?

3.1.8. For the staff shown below, fill out the chords with whole notes in order to create the indicated seventh chords (all chords in root position).

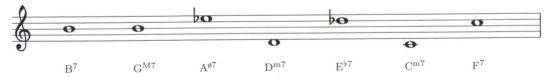

3.1.9. For the staff shown below, fill out the chords with whole notes in order to create the indicated seventh chords (all chords in root position).

3.2 Diatonic Music

Diatonic music is generally written in a specific key, although changes between keys do occur in more complex diatonic music. In this section, we briefly describe the main aspects of compositions written diatonically, a style developed during the *era of common practice* (from about 1650 to 1900). During the era of common practice, just about all compositions from the highest artistic endeavors of Haydn, Mozart, and Beethoven, to ordinary folk songs, were composed within the guidelines described here. In fact, the vast majority of popular music today is diatonic music. The most characteristic feature of diatonic music is the existence of *levels of importance* of the notes and chords in the music's key.

3.2.1 Levels of Importance of Notes and Chords

The most fundamental aspect of diatonic music is that not all notes in a key are equally important. There is a gradation of levels of importance of the notes on the scale for the key. Likewise, the triadic chords with their root notes taken from this scale inherit these levels of importance. We will discuss major keys first, then minor keys.

3.2.2 Major Keys

The most important note in a major key is the first note on its scale, which is called the ***tonic***. The note an octave higher, which completes the scale, is also called the tonic. For example, in C-major, the tonic is C. While in B$^\flat$-major, the tonic is B$^\flat$. ***A triadic chord with the tonic as root note is also referred to as tonic**.* The tonic chord is the most important chord in the major key. It is sometimes said that the tonic, either note or chord, is like home. It is the note, or chord, that provides the least tension within the music, the most relaxed feeling (like being home). Since home can be an emotionally loaded term, music theorists prefer the term ***tonal center.*** In any case, when music is composed diatonically, it usually begins with the tonic note or tonic chord and generally emphasizes these tonics frequently. This emphasis can be done in various ways, including frequent use of the tonics, or longer duration tonics.

As a rudimentary example of tonic notes in a melody, we show a score for the traditional song, *Frère Jacques,* in Figure 3.8. It is written in the key of C-major. Notice how the tonic note C is emphasized frequently at the start, and at the end of the melody.

Figure 3.8. Traditional song, *Frère Jacques,* in key of C-major. Tonic note C is emphasized by four quarter notes out of eight in first two measures. Move to dominant note G is emphasized by two half notes G at the ends of the next two measures. The melody concludes by resolving to tonic C at the ends of the last four measures, including two half notes at the ends of the final two measures. A perfect example of a diatonic melody.

The next most important note in a major key is just about equal in importance to the tonic. It is the fifth note on the scale, and is called the ***dominant***. For example, in the key of C-major, the dominant is G, while in the key of B$^\flat$ major, the dominant is F. On the chromatic clock, the dominant is always 7 hours ahead of the tonic (because the half-step changes on a major scale from tonic to dominant are $2 + 2 + 1 + 2 = 7$). For reasons that we will explain below, instrumental tones created from the dominant create a feeling of anticipation—anticipation of a return to the tonic.

There is an interplay occurring in diatonic music. The tonic note is reinforced at the start, the dominant is emphasized after it in order to create anticipation, and then the reappearance of the tonic confirms this anticipation. A perfect example of this is the charming melody from *Frère Jacques*, shown in Figure 3.8. As explained in the caption of that figure, the melody begins by establishing the tonic, then moves to the dominant, and finally returns to the tonic.[5]

A triadic chord, with root note equal to the dominant, is also referred to as **dominant**. The playing of a dominant chord in a major key seems to create anticipatory tension. A tension which can be resolved by returning to the tonic chord. We will explore this point in more detail below.

We have explained that the first, last, and fifth notes of a major scale have special names. Here we show how all the notes are named in a major scale, using a C-major scale as an example:

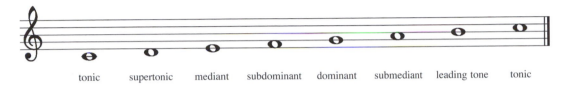

| tonic | supertonic | mediant | subdominant | dominant | submediant | leading tone | tonic |

Most of these terms have some reasoning behind their names. The term *mediant* is used for the third note in the scale because it is a middle note between the tonic and dominant notes. *Subdominant* is used as a term for the fourth note because it is a fifth *down* from the tonic at the end of the scale (also a whole step below the dominant). The sixth note is called the *submediant* because it is the middle note between the subdominant and the upper tonic (also a fifth down from the mediant lying in the next higher octave).

The seventh note on a major scale is called the *leading tone,* and this name requires some careful consideration. If you listen to a major scale played in ascending order, the leading tone occurs just before the ending tonic. There is only a half step change between those two tones, so they are significantly closer in pitch than the tones that immediately preceded them on the scale. This leads most people to hear the leading tone as *anticipating,* or *leading to,* the ending tonic. For example, on an ascending C-major scale, the notes F, G, A, and B are each separated by two half steps. But then the notes B and C are separated by only one half step, so B is close enough to C in pitch that many listeners will feel that B leads into C. This is true of all the major scales, since the numbers of half steps separating their notes are the same. For instance, F♯ leads into G on an ascending G-major scale, and A leads into B♭ on an ascending B♭-major scale.

Another aspect of the leading tone is that it helps explain why the dominant triadic chord creates a tension that is anticipating a resolution to the tonic. In the key of C-major, for example, the dominant chord is the **V** chord G-B-D. This chord contains the leading tone, B. So a listener will hear that tone and its consequent anticipation of the tonic note, C. The chord also contains the supertonic D, which is just above C on the C-major scale. Consequently, there is the anticipation of the dominant **V** chord being followed by a chord with C in it. The tonic chord **I** (C-E-G), contains C, so it is a prime candidate for resolving this anticipatory tension. This provides some rationale for the widely used *chord progression:* **V** → **I**.

Another chord that contains C is the **vi** chord. Following **V** with **vi** is said to be *deceptive* in the sense that **vi** contains the notes C and E from the **I** chord, but instead of containing the note G it has A instead. The progression **V** → **vi** is occasionally used, and is called the *deceptive progression* (or *deceptive cadence*).

---

[5]This interplay can even be done at a higher level in more elaborate compositions, such as the classical sonata. The composition may begin in one key (called the tonic), and then after some period of time, make a transition to the key up a fifth on the circle of fifths (the dominant key), and then eventually finish with a transition back to the tonic key.

Remark 3.2.1. In considering chord progressions it is important to keep in mind that the chords can appear in any one of their three forms: root, 1st inversion, or 2nd inversion. For example, consider the progression $\mathbf{V} \to \mathbf{I}$ in the key of D-major. In the key of D-major, the $\mathbf{V} \to \mathbf{I}$ chord progression describes an A-major chord followed by a D-major chord. The A-major chord could be in any one of its three forms. Likewise, the D-major chord could be any one of its three forms. For instance, a 1st-inversion $\mathbf{V}$ chord (an A/C$^\sharp$ chord) could be followed by a root position $\mathbf{I}$ chord (a D-major chord). Or, a root position $\mathbf{V}$ chord (an A-major chord) could be followed by a 2nd-inversion $\mathbf{I}$-chord (a D/A chord).[6] Often these inverted chords appear when multiple melodies are played simultaneously, with the different notes from the separate melodies forming chords. The excerpt from Pachelbel's *Canon in D* shown in Figure 3.9 can be interpreted as simultaneous melodies creating chords.

Example 3.2.1. As an illustration of chord progressions in a musical score, we show in Figure 3.9 a portion of the harpsichord part of *Pachelbel's Canon in D*. Although other instruments are playing as well, this portion of the score contains the notes for all the chords being played, and provides a good illustration of our points. In this score we have a variety of chord progressions:

$$\mathbf{I} \to \mathbf{V}, \mathbf{V} \to \mathbf{vi}, \mathbf{vi} \to \mathbf{iii}, \mathbf{iii} \to \mathbf{IV}, \mathbf{IV} \to \mathbf{I}, \mathbf{I} \to \mathbf{IV}, \mathbf{IV} \to \mathbf{V}, \text{and } \mathbf{V} \to \mathbf{I}.$$

It is interesting that this sequence of chord progressions covers a good many of the typical chord progressions in diatonic music.

Figure 3.9. Chord progressions for harpsichord part in measures 3 through 6 of Pachelbel's *Canon in D*. After first chord, the notes can be thought of as four independent melodies combining to create the chords shown. A melody on the lower staff, plus three melodies for the 3-note chords in the top staff: one from the top pitches, one from the middle pitches, and one from the lowest pitches. These separate melodies are called *voices*.

3.2.3   Chord Progressions

The typical chord progressions in diatonic music for major keys, according to a model of Kostka and Payne,[7] are listed in Figure 3.10. Any number of these progressions can be linked together in sequence. For example, this sequence of progressions is a typical one:

$$\mathbf{I} \to \mathbf{V} \to \mathbf{vi} \to \mathbf{iii} \to \mathbf{IV} \to \mathbf{I}.$$

It is interesting that this sequence, which is found in Pachelbel's canon, contains the deceptive progression $\mathbf{V} \to \mathbf{vi}$.

There is an underlying principle behind most of the chord progressions shown in the figure. Most of them can be described as descending fifths for the chord roots. For example, the chord progression we discussed above, $\mathbf{V} \to \mathbf{I}$, is an example of a descending fifth progression. Likewise $\mathbf{iii} \to \mathbf{vi}$

---

[6]Although all three forms are possible, in diatonic music the root and 1st-inversion chords are the most frequently used.

[7]S. Kostka and D. Payne, *Tonal Harmony, with an introduction to twentieth century music, Sixth Edition*, McGraw-Hill, 2009, p. 113.

can be viewed as descending by fifth. To see this most easily, consider the key of C-major. The progression $iii \to vi$ is then Em $\to$ Am. Going down by a fifth from E on the chromatic clock, would be going 7 hours counter-clockwise from hour 4 to hour 9, which yields A. Equivalently, counting down in pitch on the C-major scale from E as 1st note, yields D as 2nd note down, C as 3rd note down, B as 4th note down, and then A as the 5th note down.

Kostka and Payne point out that their listing of typical chord progressions also includes diatonic seventh chords, in addition to the fundamental triadic chords. For example, the chord progression $G^7 \to C$ is a very common one; because the seventh chord G^7 creates even more anticipation of the C-major chord than the triadic chord G (see Exercise 3.2.3). If the key is C-major, then we could write this progression in Roman numeral notation as $V^7 \to I$. This progression is included in Kostka and Payne's model when we allow diatonic seventh chords as well as the fundamental triadic chords.

It is fascinating to contemplate the astronomical variety of chord progression sequences obtained by linking together the chord progressions shown in Figure 3.10. It is no wonder that diatonic music contains such a wealth of compositions, with no sign of their abatement, at least in the popular realm. See Remark 3.2.4 and the three exercises that precede it, for more details.

$$I \to IV \text{ or } vii^\circ \text{ or } iii \text{ or } vi \text{ or } ii \text{ or } V$$
$$ii \to V \text{ or } vii^\circ$$
$$iii \to vi \text{ or } IV$$
$$IV \to vii^\circ \text{ or } ii \text{ or } V \text{ or } I$$
$$V \to I \text{ or } vi$$
$$vi \to ii \text{ or } IV$$
$$vii^\circ \to I$$
$$(V \to) \ vi \to ii \text{ or } IV \text{ or } vii^\circ \text{ or } iii \text{ or } V \text{ or } I$$

Figure 3.10. Typical chord progressions in major key, according to Kostka and Payne's model. The order of the chords in the progressions is mostly based on descending fifths for their roots. Last line describes progressions that follow deceptive progression, $V \to vi$, e.g., $V \to vi \to iii$ occurs in the Pachelbel example in Figure 3.9.

Remark 3.2.2. One thing not provided in Figure 3.10 is any information on how frequently different chord progressions are used. For instance, the chord progressions $I \to V$ and $V \to I$ are far more commonly employed than the rarely used progression $vii^\circ \to I$. However, labeling the progressions listed with probability values indicating how frequently each progression is likely to be used in a piece of diatonic music seems a very daunting task. Some related work, for Bach chorales and Mozart piano sonatas, has been reported by Tymoczko. See, for example, his tables of the frequency of chord progressions for major and minor keys in his book on the geometry of music.[8] His tables for major keys can be downloaded from his web site.[9] It is interesting that Tymoczko has found that, for major keys, Bach uses the $iii \to IV$ progression 52% of the time when moving from chord iii to another fundamental triadic chord. While Mozart, on the other hand, *never* uses $iii \to IV$.

3.2.4 Relation of Melody to Chords

Within major keys, and also minor keys, the fundamental triadic chords provide a guide for how notes are selected by a composer to form melodies. A basic guideline in diatonic music is to choose notes for melodies that belong to the fundamental triadic chords[10] for that key, often outlining those chords over short intervals of time. For example, in the *Frère Jacques* melody:

---

[8] Dmitri Tymoczko, *A Geometry of Music*, Oxford, 2011, p. 230.
[9] See the entry for D. Tymoczko (2010) in the bibliography on p. 308.
[10] Also seventh chords, in more complex compositions.

the notes are mostly chosen from the triadic chord **I** (the tonic chord C-E-G). The last eight notes, in particular, just travel a path down and up through the notes E_4, C_4, and G_3, a 2nd-inversion form C/G of the tonic chord **I**. Moreover, the last seven notes travel a melodic course from the tonic C to the dominant G and back to the tonic C (a common path for melodies in diatonic music that we referred to previously). The other notes, mostly D's and F's, and a couple of A's, are used to "fill in" the melody so that it is less "jumpy in pitch" (as it would be if notes were chosen exclusively from the tonic C-chord). The notes D and F in the melody are referred to as *passing tones*, as they are functioning like stepping stones between the notes from the C-chord. Passing tones are commonly used in diatonic music, typically to *reduce the frequency of large jumps in pitch between successive tones in the melody*. The note A in the melody is also a kind of stepping stone, and is usually referred to as a *neighbor tone*. Notice the difference between the contour of the melody when the passing tones are used versus when the neighbor tone is used. When a passing tone is used, the melody goes up through the passing tone, or goes down through the passing tone. When a neighbor tone is used, the melody reverses the course of its pitch at the neighbor tone.

Thus, the *Frère Jacques* melody illustrates another basic guideline of melody design for diatonic music. The intervals between its successive notes are kept small: major and minor seconds, major and minor thirds, perfect fourths. Generally, they are the intervals most frequently used in diatonic melodies. With the exception of major and minor seconds, these intervals are found in the fundamental triadic chords in their three forms (root, 1st inversion, and 2nd inversion). Some genres of music, however, use larger intervals. Rock music, in particular, makes frequent use of fifths.

It is also important to keep in mind the relation between tonic and dominant notes. As we mentioned above, a dominant note will create an anticipation of the tonic note. One explanation for this is that an instrumental tone contains many harmonics, in addition to its fundamental. In the case of the dominant note, say G in the key of C-major, its 5th harmonic corresponds to the frequency for a note B for a leading tone. To see this, take the case of the note G_4. We can write its fundamental as $r^7 \nu_C$, where ν_C is the fundamental for C_4. Multiplying by 5 to get the 5th harmonic, we obtain a frequency of $5r^7 \nu_C$. The tone with that fundamental is octave equivalent to the note with fundamental $\frac{5}{4} r^7 \nu_C$. Since $r^4 = 2^{4/12} = 1.259921$ is approximately $\frac{5}{4}$, this latter note has a fundamental of approximately $r^4 r^7 \nu_C = r^{11} \nu_C$. In other words, this latter note is very close in pitch to B_4. This shows that the dominant note has a 5th harmonic that closely approximates a leading tone. Likewise, its 3rd harmonic closely approximates the frequency for the supertonic D. Consequently, the tone of an individual dominant note does create some anticipation of the tonic (provided, of course, that the listener can perceive the harmonic content of the tones).

Perhaps a more widely applicable explanation for a dominant note anticipating the tonic note can be found in the guideline that, *in diatonic music,* notes within melodies are taken from the fundamental triadic chords for the key. Since the dominant chord anticipates the tonic chord, then the roots of these chords within melodies will inherit that same relationship. Thus, in diatonic melodies the dominant note in the key will anticipate the tonic note.

3.2.5 Minor Keys

In minor keys, the basic aspects of diatonic music are quite similar to what we have described for major keys. Therefore, we will concentrate on how these two types of keys differ.

One difference is that the fundamental major and minor chords are arranged in a different sequence. The seven *fundamental triadic chords* based on the natural A-minor scale are shown in Figure 3.11. In this case, the tonic chord **i** is a minor chord. The tonic plays the most important role

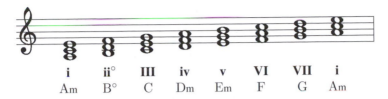

i	**ii°**	**III**	**iv**	**v**	**VI**	**VII**	**i**
Am	B°	C	Dm	Em	F	G	Am

Figure 3.11. Fundamental triadic chords based on natural A-minor scale. Root notes of chords are notes of natural A-minor scale. (Compare with C-major triadic chords in Figure 3.4.)

in minor keys, just as for major keys. So the fact that **i** is a minor chord is a crucial factor in diatonic music composed in a minor key. Another difference is the lack of a leading tone in a natural minor scale. Since there are two half steps between the seventh note and the eighth note on a natural minor scale, the seventh note is not a leading tone.

Here are the names for the notes on a natural minor scale (A-minor):

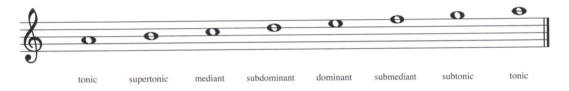

tonic	supertonic	mediant	subdominant	dominant	submediant	subtonic	tonic

The names are the same as for the major scales, except for the seventh note which is referred to as the *subtonic*.

The lack of a leading tone creates another difference in natural minor keys. The fifth tone, although it is called a dominant, does not play the same role of anticipating the tonic. This is especially true of the fundamental triadic chord **v**. *It does not contain a leading tone, so it does not create the anticipation of the tonic chord that the* **V** *chord does in a major key.* In music theory, the **v** chord in a natural minor key is said to *not* have a dominant function. It is sometimes referred to as the *minor dominant*.

These differences explain why, in a minor key, the natural minor scale is not used exclusively. One crucial aspect of a minor key is that the seventh tone, the subtonic, of the natural minor scale is allowed to move up in pitch by a half-step in certain instances. The main instance is when playing the seventh tone would correspond to a **v** chord. By raising the seventh tone a half step, a **V** chord occurs instead. For example, in the key of A-minor, the seventh tone is G. Raising G by a half step gives G♯ which is a leading tone for A. This allows for the dominant **V** chord, in this case the E-major chord, E-G♯-B. It also allows for the **V** → **i** progression. In Figure 3.12 we show the fundamental triadic chords in the key of A-minor when the raised seventh G♯ is used. In addition to the **V** chord, there is also a **vii°** chord, which is occasionally used. The root notes of the chords shown in this figure are the notes of the harmonic A-minor scale. (For the definition of a harmonic minor scale, see Exercise 1.4.5 on p. 22.) In Section 3.2.6, we will show a musical example that makes use of the raised seventh in a minor key.

In spite of these differences, however, the basic guidelines for diatonic music that we outlined above for major keys also apply to minor keys. In Figure 3.13, we list the typical chord progressions in minor keys. It is quite similar to the list for major keys. As with major keys, this list for minor keys mostly contains chord progressions based on descending fifths for the chord roots and allows for substitution of diatonic seventh chords in conjunction with fundamental triadic chords.

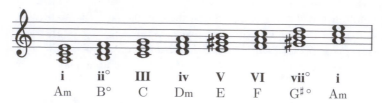

Figure 3.12. Fundamental triadic chords based on harmonic A-minor scale, with seventh note raised to create **V** chord and **vii°** chord. (Compare with chords for natural A-minor scale in Figure 3.11.)

i → iv or **VII** or **vii°** or **III** or **VI** or **ii°** or **V**

ii° → **V** or **vii°**

III → **VI** or **iv**

iv → **vii°** or **ii°** or **V** or i

V → i or **VI**

VI → **ii°** or **iv**

vii° → i

VII → **III**

Figure 3.13. Typical chord progressions in a minor key, according to Kostka and Payne's model.

We now turn to some examples of music written in minor keys. Consider the melody for the traditional ballad, *House of the Rising Sun,* shown in Figure 3.14. The tonic note, D, for the key

Figure 3.14. Melody of traditional ballad, *House of the Rising Sun.* Key signature could be for F-major, or for D-minor (relative minor of F-major). A large number of D notes are used, and notes in the melody mostly trace up and down through the notes of D-minor chord (D-F-A). So the song is in D-minor.

of D-minor is established very early with a half note tied with a dotted eighth note D. Moreover, the melody largely consists of tracing up and down through the notes of the tonic chord **i** (the chord D-F-A) for the key of D-minor. These tracings up and down in pitch, along with the rhythm of the changing note durations, combine to create a striking melody. Our next example is a famous piece of music in A-minor, Beethoven's *Für Elise*. This composition also uses chromatic notes.

3.2.6 Chromaticism

We have briefly outlined the main guidelines for diatonic music. We end by pointing out that within diatonic music a composer can occasionally use other notes from the chromatic scale that are not within the key for the music.

For example, in Figure 3.15 we show the beginning of the beloved piano piece by Beethoven, *Für Elise*. The piece is written in A-minor, so the key signature shows no sharps or flats. Beethoven, however, uses $D^\sharp$ and $G^\sharp$ several times. The $G^\sharp$ is considered part of the A-minor key, while the $D^\sharp$ is not. Both notes clearly enhance the melody. For example, the alternation of E_5 with its closest lower pitch neighbor on the chromatic scale $D^\sharp_5$, is an important motif that begins the piece and is played periodically throughout. These two notes, separated by only one half step, are relatively short duration sixteenth notes that are slurred together. This can create an extra tonality having a pitch lying midway between the pitches for E_5 and $D^\sharp_5$, an effect similar to playing blue notes in jazz. See Remark 3.2.3 for more discussion of blue notes. The use of $G^\sharp$ in the fourth measure creates two *arpeggiated* chords (chords whose notes are played "strung out" rather than completely at once) of type E-$G^\sharp$-B. Here Beethoven is employing the raised seventh note that is allowed in a minor key. The **v** chord in the key of A-minor is E-G-B. By raising the G by a half step to $G^\sharp$, Beethoven creates a major **V** chord. The $G^\sharp$ note in this chord is the leading tone for the tonic A, thereby anticipating the arpeggiated tonic chord A-C-E that follows in the next measure. This shows the melody following the lines of the typical chord progression **V** $\rightarrow$ **i**. Although Beethoven has repeatedly used the $D^\sharp$ note, which is called *chromatic* because it lies outside the key, *Für Elise* is still regarded as following the guidelines for diatonic music. An example that uses a great many more chromatic notes is shown in the passage by Mendelssohn in Figure 6.14, page 184. (See also Exercise 3.2.7.)

Figure 3.15. Introduction to Beethoven's *Für Elise* (Bagatelle No. 25 in A-minor, WoO 59). The tempo instruction, ***poco moto*** ("little motion"), is not precise, but does indicate a smoothly flowing speed. The dynamic marking, ***pp***, stands for *pianissimo* (very soft). The angle symbol, opening up from left to right, indicates *crescendo* (smooth increase in volume for notes lying above it). Reversed angle symbol, closing from left to right, indicates *diminuendo* (smooth decrease in volume for notes lying above it). Cursively written expression "Ped" and its connected line indicate the piano should be *pedaled* for notes lying above the extent of this notation.

Remark 3.2.3. Blue notes in jazz. In jazz, blue notes are often created by pairing two notes differing by a single half step. If a major scale is being used, then a jazz musician will commonly use flatted third, fifth, and seventh notes in addition to the notes of the scale. For example, if the scale is C-major, then the musician would be using this collection of notes:

scale:	C	D	E	F	G	A	B	C	
flatted notes:		$E^\flat$		$G^\flat$		$B^\flat$			(3.1)

This collection of notes allows for many pairings of notes separated by a single half step. When such notes are played together quickly, as in jazz piano playing, then extra mid-pitch tonalities will emerge.[11] Why this happens will be discussed in Chapter 4. It is interesting that the E_5, $D^\sharp_5$ notes in *Für Elise* are enharmonic with E_5, $E^\flat_5$ for the notes in (3.1). We will explore blue notes further in the next chapter.

---

[11] Instruments with continuous pitch ranges, like saxophones, can play these mid-pitch tonalities as single notes.

To better appreciate the guidelines for diatonic music, we close with an example of a modern piece that breaks totally from these guidelines. In Figure 3.16 we show a typical passage from Dmitri Tymoczko's *Piano Games*. A large percentage of the intervals in this passage are tritones, minor or

Figure 3.16. Movement 3, measures 1 through 3, of Dmitri Tymoczko's *Piano Games*. © Dmitri Tymoczko, used by permission. All notes are sixteenth notes. Some sixteenth notes are shown with open note heads. According to composer's instructions, these notes are played with more emphasis than the other sixteenth notes.

major sevenths, or large intervals of an octave or more. Such intervals are used very sparingly in diatonic music. Moreover, the main melody for the passage is expressed by the open note-headed sixteenth notes lying in the middle pitch range:

$$F^\sharp_3, \ G_3, \ A^\flat_3, \ G_3, \ F^\sharp_3, \ G^\sharp_3, \ A_3, \ A_3, \ A_3, \ A^\sharp_3, \ A^\sharp_3, \ B_3, \ldots, \ F_4, \ F^\sharp_4, \ E_4.$$

This melody consists mostly of increases, by minor second intervals, through the chromatic scale from $F^\sharp_3$ to $F^\sharp_4$. It is definitely not emphasizing any triadic chord from one of the diatonic keys, major or minor. Surrounding this main melody are the solid note-headed sixteenth notes, which are all $F^\sharp$ notes. These notes are not emphasizing a triadic chord with root $F^\sharp$. Their purpose seems to be to provide dissonant contrasts to the notes from the melody. For all of these reasons, Tymoczko's piece is not classified as diatonic music. It is classified as *atonal* or *post-tonal* music.[12] The full score of *Piano Games*, along with sound recordings of each of its movements, can be accessed from the book's web site (see Chapter 3/Scores).

Exercises

3.2.1. Within the first three measures of *Für Elise* (see Figure 3.15), find the percentage of notes in both upper and lower staves belonging to the A-minor chord (A-C-E).

3.2.2. The song *Frère Jacques* is usually sung as a round. Here are measures 3 and 4 when it is sung as a round:

What chord is being emphasized here?

3.2.3. Explain why the seventh chord G^7 anticipates the triadic chord C.

3.2.4. Kostka and Payne point out that in a major key, the major seventh chord I^{M7} is not a "harmonic goal." That is, it contains a significant amount of anticipatory tension, or need to resolve to a different chord. Explain what the source of that tension is, and which is the most likely chord for it to resolve to.

3.2.5. Using the passage from *Piano Games* shown in Figure 3.16, complete parts **(a)** and **(b)**:

(a) Make a tally of how often each of the open-head sixteenth notes occurs for the following chromatic scale,

---

[12]In these terms, the use of *tonal* alludes to diatonic music rather than musical tones per se.

that begins with $F_3^\sharp$ and ends with $F_4^\sharp$. The first two notes have been done for you.

$$F_3^\sharp: 2, \quad G_3: 2, \quad G_3^\sharp (A_3^\flat): \quad, \quad A_3: \quad,$$

$$A_3^\sharp: \quad, \quad B_3: \quad, \quad C_4: \quad, \quad C_4^\sharp: \quad,$$

$$D_4: \quad, \quad D_4^\sharp: \quad, \quad E_4: \quad, \quad F_4: \quad, \quad F_4^\sharp: \quad.$$

(b) For all of the notes in the passage, make a tally of the number of times each of the following intervals occurs:

Interval, half steps	Number of Times
Tritone, ± 6	
Minor 7th, ± 10	
Major 7th, ± 11	
$\pm$Octave or more	

What percentage of these intervals occurs out of the total of 47 intervals in the passage?

3.2.6. For the Bach passage shown in Figure 3.22, page 73, explain why the key is G-minor.

3.2.7. For the Mendelssohn passage shown in Figure 6.14, page 184:

(a) Explain how you know the music begins in the key of G-major. Your explanation should include more than just referencing the key signature.

(b) For the first line of the passage, how many notes marked with accidentals are part of the key, and how many are chromatic (not part of the key)?

(c) These accidentals occur in three out of the last four chords at the end of the first line. Name these last four chords, and explain how they are related to each other. [Hint: the first chord is $B^{\phi 7}$, and they are related by how their roots are changing.]

3.2.8. A large amount of popular music is based just on the chords $\mathbf{I}, \mathbf{IV}$ and $\mathbf{V}$. Suppose we consider only chord progressions in a major key, using only those three chords. Using the model in Figure 3.10, these progressions are

$$\mathbf{I} \to \mathbf{IV}, \quad \mathbf{I} \to \mathbf{V}, \quad \mathbf{IV} \to \mathbf{V}, \quad \mathbf{IV} \to \mathbf{I}, \quad \mathbf{V} \to \mathbf{I}. \tag{3.2}$$

If you link together these chord progressions, then how many chord progression sequences are possible from linking 3 chords? 4 chords? (Note: for 1 chord, there are just the 3 chords themselves, for 2 chords we have shown that there are 5 possible progressions.)

3.2.9. Define F_n to be the number of chord progressions involving n chords, linking together the chord progressions shown in (3.2). Prove that $F_{n+1} = F_n + F_{n-1}$ for $n \geq 3$. (This shows that $\{F_n\}$ belongs to the *Fibonacci sequence*.)

3.2.10. As in the previous two exercises, suppose that we consider only the chord progressions shown in (3.2). How many different chord progressions are possible using precisely n chords, if we distinguish between the three different forms for a triadic chord (root, 1st inversion, and 2nd inversion)? [Note: This question only concerns the possibility, not the likelihood, of progressions. For example, in practice, 2nd inversion chords are restricted to only certain progressions.]

Remark 3.2.4. What the previous three exercises show is that, even just considering only the three chords $\mathbf{I}, \mathbf{IV}$ and $\mathbf{V}$, there is a truly enormous number of typical chord progressions. For example, the total number of typical chord progressions, using from 1 to 20 chords and all three forms of chord, is $125, 846, 607, 502, 890$. Allowing for all of the fundamental triadic chords only increases that number even more. Consequently, there is an almost incomprehensibly large number of possible *melodies* that follow the guidelines for diatonic music.

3.2.11. Pentatonic scale. In Chapter 1 we discussed the pentatonic C-major scale:

$$\text{C} \quad \text{D} \quad \text{E} \quad \text{G} \quad \text{A} \quad \text{C.} \tag{3.3}$$

(a) Mark the notes of the pentatonic C-major scale on the chromatic clock, and show that the half steps between its notes follow this pattern:

$$2 \quad 2 \quad 3 \quad 2 \quad 3. \tag{3.4}$$

(b) The following score is for the beginning of a traditional Chinese folk melody (trans. *Jasmine Flower*):

Write out the notes used for this melody. (The double bar with colon means that the first two measures are repeated before the third measure is played.)

(c) For the notes in part (b), you should have found only letters from this 5-member set: $\{A, C, D, F, G\}$. The melody appears to be based on a pentatonic scale, but the notes are not exactly the same as for the pentatonic C-major scale (since they include F and do not include E). Show that the five notes that are used for the melody are another pentatonic scale obtained from the pentatonic C-major scale by adding a fixed number to each hour on the chromatic clock, and that the half steps between this pentatonic scale's notes follow the same pattern as in (3.4). What would be a good name for this pentatonic scale?

3.2.12. Chords in pentatonic scales. Generally, music that is written in pentatonic scales does not employ many chords. This exercise will provide some explanation for this. Suppose that we are using the pentatonic C-major scale in (3.3). Show that only one major triadic chord and only one minor triadic chord can be created from the notes of this scale. Explain why this result implies that little harmony is possible (using more than 2-note chords).

3.2.13. (a) Show that the middle pitch notes for the fundamental chords of the natural A-minor scale, shown in Figure 3.11, are the notes for the key of C-major. (b) Write out the fundamental triadic chords, in root position, for the natural E-minor key. Verify that the middle pitch notes for these chords are the notes for the key of G-major, the relative major key for E-minor. (c) Explain why the middle pitch notes of the fundamental triadic chords for each natural minor key consist of the notes for its relative major key.

3.2.14. On the staff shown below,

write out the fundamental triadic chords for E-Lydian mode. Explain why music written in E-Lydian mode will tend to sound like music in a major key.

3.2.15. On the staff shown below,

write out the fundamental triadic chords for C-Dorian mode. Explain why music written in C-Dorian mode will tend to sound like music in a minor key.

3.3   Diatonic Transformations — Scale Shifts

For the remainder of the chapter we pursue an in-depth exploration of one of the more mathematically oriented set of composing techniques. These techniques use transformations of small groups of notes, called *motives*.[13] Transformations of motives enjoys a long history in diatonic music, from the music

---

[13] The term, motive, comes from *motif* in art. As in the repetition and transformation of a motif in artistic design, there can be a repetition and transformation of musical motives.

of Bach and Beethoven, right up to today's popular music. The motive transformations that composers frequently use are (1) *diatonic scale shifts*, (2) *diatonic scale inversions*, and (3) *retrograde*. In this section, we describe diatonic scale shifts. In the next section, we describe the other two types.

Diatonic scale shifts extend the idea of melody from single notes to motives. A melody within the scale for a given key is a sequence of notes, which either go up or down on the scale (or repeat occasionally) as the melody progresses. If we apply this idea to a group of notes—moving the whole group up or down on the scale by the same amount (or repeating the whole group occasionally)—then we are using diatonic scale shifts. Looking at a score as a graph in time and pitch, as we did in the last chapter, a diatonic scale shift is a *rigid motion* or *translation* of a motive in the score. As used by great composers, this is far more than just mathematical trickery. We shall show in the examples that follow that it provides a basis for arpeggiating chord progressions in melodies, and for combining separate melodies (separate voices) that flow through a chordal framework.

Figure 3.17. Score illustrating diatonic scale shifts.

To see how a motive can be shifted about, consider the score shown in Figure 3.17. In the first measure, there is a motive consisting of three notes. Suppose we take each one of those notes and move them up by a melodic interval of a third, *without considering whether each third is major or minor*. In other words, each note is shifted up by three score positions inclusive. Doing this, we get the notes in the second measure. We call this group motion a **diatonic scale shift**, and denote it by S_2. We have used the subscript 2, even though we have shifted up by a third, for the following reason. When we successively combine these diatonic scale shifts in a sequence, *the subscripts will simply add up*. See Figure 3.19 for an example. This adding of subscripts occurs because we are **not** counting the initial score position (score line or score space) when we count changes in score position. When the interval amount is calculated, the initial score position **is** counted, and so a succession of diatonic scale shifts would count those initial positions twice in each succeeding shift (once as initial and once as final). For example, if we go up a third and up a third again, then that is going up a fifth. But $3 + 3 \neq 5$. With our notation, we have S_2 followed by S_2 is equal to S_4, and $2 + 2 = 4$. Successive combinations of diatonic scale shifts will be employed frequently in the examples we describe.

Another, more geometrical, way to view these diatonic scale shifts is in terms of a clock with 7 equally spaced hours representing the distinct notes of a scale. See Figure 3.18. In this figure, we show how the diatonic scale shift S_2 can be viewed as a change of 2 hours. This provides an explanation for the subscript 2 that does not depend on musical staff notation. We will explore the mathematics of this 7-hour clock more in Chapter 6. In this chapter, we will use musical staff notation, since using the 7-hour clock at this time might lead to confusion with the 12-hour chromatic clock.

An important special case of a diatonic scale shift consists of just repeating a set of notes. For instance, in measure 3 of the score in Figure 3.17, the notes from measure 2 are repeated. We denote this repetition by S_0. Applying S_0 has *zero* effect on the notes, so the subscript 0 makes sense for this *melodic unison* interval applied to each note in the set. Repetition in this way is a common practice in music, as it provides emphasis.

We can also shift a set of notes down by a common interval. For example, the three notes in the last measure of the score in Figure 3.17 have all been shifted down by a second from the notes in measure 3. This diatonic scale shift is denoted by S_{-1}. In Figure 3.19, we summarize these examples of diatonic scale shifts.

Example 3.3.1. Diatonic scale shifts in chords. Diatonic scale shifts occur in chord progressions.

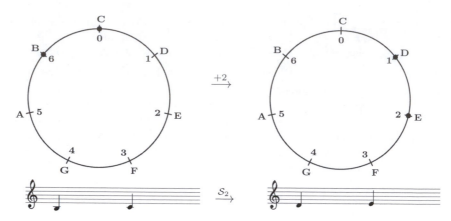

Figure 3.18. Geometric interpretation of diatonic scale shift. The 7 distinct notes of the C-major scale are represented by 7 equally spaced hours on a clock. The diatonic scale shift S_2 corresponds to adding 2 hours to each hour of the notes in a given motive.

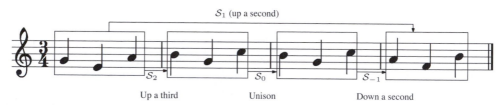

Figure 3.19. Sequence of diatonic scale shifts of three note motive. Since $2 + 0 - 1 = 1$, the combination of three diatonic scale shifts is equal to the diatonic scale shift S_1, which produces notes in last measure from those in first.

For instance, in the following score

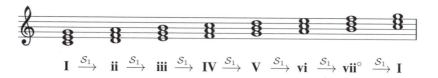

we show the fundamental triadic chords in the key of C-major, and how the diatonic scale shift S_1 is used to change from one chord to the next. Here we are viewing each chord as a (vertically stacked) motive. In Example 3.3.5 we discuss a case of applying diatonic scale shifts to chords.

The connection between diatonic scale shifts, melodic intervals, and chord progressions provides an important rationale for using these diatonic scale shifts when composing diatonic music. We now give some examples of diatonic scale shifts in a variety of musical compositions.

Example 3.3.2. In Figure 3.20 we show the first five measures of the piano introduction to a song by Schubert. It is written in the key of E-major.

In this figure, there is a triplet at the start of the first measure, consisting of a G$\sharp$, an E, and a G$\sharp$ again. We can see that a sequence of diatonic scale shifts is applied to this triplet. For instance, S_{-1} is applied first, followed by S_1, thereby reproducing the original triplet. It is clear that a large amount of the passage consists of applying diatonic scale shifts to this one triplet.

To see what these diatonic scale shifts are accomplishing, we need to take into account the notes on the lower staff as well. For example, in the first measure, the lower staff simply consists of two dotted half notes, an E note and a B note, played together throughout the whole measure. Each triplet

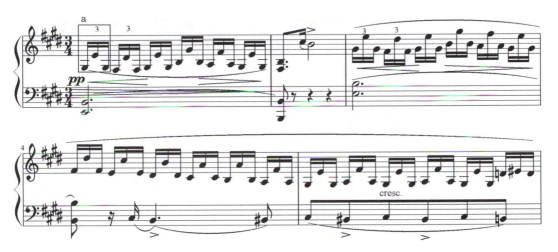

Figure 3.20. First 5 measures of piano introduction to Schubert song, *Der Lindenbaum* (The Lime Tree), fifth song in his song cycle *Winterreise* (Winter's Journey). Sequence of diatonic scale shifts is applied to triplet in rectangle labeled **a** at beginning of upper staff.

in the upper staff is played quickly, creating an *arpeggiated* chord. Combining the notes of the first triplet with the dotted half notes on the lower staff, we see that they comprise notes in the chord E-G$^\sharp$-B, which is chord **I** in the key of E-major. When the first diatonic scale shift S_{-1} is applied to the triplet, the notes on both upper and lower staves (except for an E) comprise notes of the chord B-D$^\sharp$-F$^\sharp$, which is chord **V** in the key of E-major. So the application of the diatonic scale shift S_{-1} corresponds to a melodic implementation of the chord progression **I** → **V**.[14] The next diatonic scale shift is S_1, which goes back to the initial triplet. Consequently, applying S_1 corresponds to a melodic implementation of the chord progression **V** → **I**.

Thus, the first half of measure 1 corresponds to the chord progression sequence:

$$\mathbf{I} \to \mathbf{V} \to \mathbf{I}$$

implemented melodically via a sequence of diatonic scale shifts. Many of the diatonic scale shifts in this passage are linked to standard chord progressions in this fashion.

There are exceptions, however, such as the fifth triplet in the first measure, which along with the lower staff notes, does not comprise a fundamental triadic chord. Of course, Schubert is composing music, not mechanically implementing mathematical transformations.[15] So, these exceptions are not surprising. Nevertheless, one of the guidelines for diatonic music is to mostly use melodic intervals that stem from the fundamental triadic chords (thirds, perfect fourths), as well as major seconds and minor seconds for transitional notes. With diatonic scale shifts, that guideline is extended to groups of notes. ***In diatonic music, groups of notes may be moved by diatonic scale shifts mostly in intervals of seconds, thirds, and perfect fourths.*** We will see a brilliant example of this in the excerpt from Bach's *Inventio 11* described in Example 3.3.4.

Combining a succession of diatonic scale shifts, as in the Schubert piece, is known as ***tonal sequencing*** in music theory. The whole succession is called a ***tonal sequence***.[16] We now describe another example of tonal sequencing, perhaps the most famous opening in all of Western music.

---

[14]The extraneous note, E in this case, is an example of a *pedal note*. It is held throughout as the lowest bass note. The terminology comes from the use of a pedal in organ music to hold a note, typically a low bass note, through long stretches of music. It is important that the pedal note in this example is the tonic.

[15]Here is an interesting comment by the composer, Aaron Copland, in this regard: "Music would be dull indeed if composers were not able to disguise, vary, and adorn the bare harmonic frame" (A. Copland, *What to Listen for in Music*, p. 56, Signet Classic, 2002).

[16]S. Kostka and D. Payne, *Tonal Harmony, with an introduction to twentieth-century music, Sixth Edition*, McGraw-Hill, 2009, p. 103.

Example 3.3.3. Beethoven's *Fifth Symphony*. An example of using diatonic scale shifts occurs at the start of Beethoven's Fifth Symphony. In Figure 3.21, we show the opening measures for the string section of the orchestra. In the score we have marked the many instances of diatonic scale shifts used

Figure 3.21. Opening measures of Beethoven's *Symphony Number 5 in C-minor* (Op. 67), for string section of orchestra. Almost the entire passage is created from tonal sequencing. Semicircles with center dot are *fermatas*. A *fermata* tells the conductor to hold the note for as long as desired. Staff for the violas uses an *alto clef*, described in Exercise 2.1.5.

to manipulate the four-note groups of three repeated eighth notes and a half note (beginning with three G's and an $E^\flat$, the famous "Da-Da-Da-Dum"). Besides the diatonic scale shifts, Beethoven also extends the length of the final half note by using ties and fermatas.

One interesting exception to an exact diatonic scale shift is marked as S_9' in the Violas section. We have used a prime symbol on S in this notation to indicate that there is a slight modification of the diatonic scale shift S_9. If S_9 had been employed at this point, then the ending half note for the violas in the next to the last measure would have been F_4. Beethoven instead uses the note G_4. This note fits in perfectly with all the half notes played at this same moment: C_4 (Violoncellos), $E_4^\flat$ (Violins 2), G_4 (Violas), *forming precisely the tonic chord* **i** *in the key of* C-*minor*. Here we have an elementary case of what Leonard Bernstein has said: "Beethoven, more than any other composer, before or after him, had the ability to find exactly the right notes that had to follow his themes."[17]

Example 3.3.4. J.S. Bach's *Inventio 11*. J.S. Bach used diatonic scale shifts frequently as well. In Figure 3.22 we show the opening measures of Bach's *Inventio 11 (BWV 782)* with many different diatonic scale shifts marked. There are a few diatonic scale shifts that we left unmarked, because the diagram would have been too cluttered. The figure clearly shows that almost all of the notes in the score are involved in some way with diatonic scale shifts. Notice that in several of the diatonic scale shifts, such as the one labelled S_1', there are various *accidentals* for the notes. That is, some of the notes are marked with a $\flat$, or $\natural$, or $\sharp$. For these diatonic scale shifts, the sequences are referred to as *modified sequences* in music theory. They are only approximate diatonic scale shifts, but it still helps to consider them as part of the method of diatonic scale shifts, even if they are only approximating exact versions. It is important to keep in mind, however, that not all the sharps, flats, or naturals in the passage are accidentals. In some cases they indicate notes that are on the scale specified by the key signature. For example, the second transformation S_{-1} on the bass clef staff is a diatonic scale shift,

---

[17]L. Bernstein, *The Joy of Music*, chapter on *Beethoven's Fifth Symphony*, pp. 85–105, Amadeus Press, 2004.

Figure 3.22. Opening measures of Bach's *Inventio 11, BWV 782*. Numerous diatonic scale shifts, and modified diatonic scale shifts, are used.

since the natural and flat there are indicating notes on the scale determined by the key signature. To indicate that diatonic scale shifts are modified, we have marked them with a prime symbol (e.g., $\mathcal{S}_4'$ rather than $\mathcal{S}_4$), as we did in the previous example from Beethoven.

Example 3.3.5. Alicia Keys' *If I Ain't Got You.* Besides classical music, diatonic scale shifts occur in popular music as well. A notable example occurs in the first 10 measures of the song, *If I Ain't Got You,* by Alicia Keys. These measures can be viewed, and played, by going to the book's web site at Chapter 3/Scores. The sequencing in these measures is reminiscent of the Schubert passage that we considered previously. At the beginning of the upper staff there is triplet of three notes, an E, B, and G. The notes in the upper staff for this entire passage consist of repeated applications of diatonic scale shifts to this triplet. This creates in the upper staff a sequence of arpeggiated chord progressions, which one can hear as the bass notes fade out. When these bass notes are first struck, the effect is to have 4-note chords heard in arpeggiated form at the start of each measure. For example, the notes of the very first triplet arpeggiate a chord E_5-B_4-G_4 which is a first inversion of an E-minor chord (Em/G). Including the bass note, C_4, produces notes from a chord of type C-E-G-B, which is a C-major seventh chord. There is an interesting alternation between the arpeggiated major seventh chord that starts measure 1 and the arpeggiated E-minor chords that are more clearly audible as the bass notes fade away. Like the Schubert piece described above, the tonal sequencing in this passage is related to chord progressions that provide the harmonic frame for the notes in the flowing melody. In fact, the chords are listed in the score given at the book's web site. They form these chord progressions:

$$C^{M7} \to B^{m7} \to A^{m7} \to G^{M7} \to A^{M7} \to B^{m7} \to C^{M7}.$$

At the end of measure 10, the sequencing has returned the melody to its starting point at measure 1.

It is interesting to observe the patterning of the chords in this sequence. On a G major scale, they are diatonic seventh chords whose roots are descending down one scale position at a time, starting from C and finishing at G, and then ascending one scale position at a time until C is reached again. In Roman numeral notation, for the key of G-major, we have

$$\mathrm{IV}^{\mathrm{M7}} \xrightarrow{S_{-1}} \mathrm{iii}^{\mathrm{m7}} \xrightarrow{S_{-1}} \mathrm{ii}^{\mathrm{m7}} \xrightarrow{S_{-1}} \mathrm{I}^{\mathrm{M7}} \xrightarrow{S_{1}} \mathrm{ii}^{\mathrm{m7}} \xrightarrow{S_{1}} \mathrm{iii}^{\mathrm{m7}} \xrightarrow{S_{1}} \mathrm{IV}^{\mathrm{M7}}. \tag{3.5}$$

Remark 3.3.1. The tonal sequencing of chords in (3.5) is related to a composing technique known as *fauxbourdon*. A good discussion of *fauxbourdon*, with several examples from classical music, can be found in the beginning of Tymoczko's lecture notes on music theory.[18]

Exercises

3.3.1. For the set of three notes below, perform the diatonic scale shift S_{-1}, and write the resulting notes on the remainder of the staff:

3.3.2. For the set of three notes below, perform the diatonic scale shift S_2, and write the resulting notes on the remainder of the staff:

3.3.3. Find at least five diatonic scale shifts in the Mozart passage in Figure 3.23.

Figure 3.23. Introductory passage from 4th movement of Mozart's *Divertimento No. 14, K.V. 270.*

---

[18]See the entry for D. Tymoczko (2011a) in the bibliography on p. 308.

3.3.4. For the following passage by Beethoven, mark a sequence of diatonic scale shifts on the upper staff and lower staff.

3.3.5. The following passage is from J.S. Bach's *Toccata and Fugue in* D-*minor, BWV 565*, measures 8 and 9. Identify the diatonic scale shifts in this passage, and their related chord progressions.

3.3.6. For the third measure in the Schubert passage in Figure 3.20, find what fundamental triadic chords the notes belong to and what chord progression is related to the tonal sequencing.

3.3.7. The following passage is a single measure from the piano part of Mozart's *Piano Concerto No. 26 in* D-*major, Mvmt. 1, K.V. 537*. Identify the diatonic scale shifts in this passage, and what are the related chord progressions (two of the chords will be seventh chords).

3.3.8. The following passage is the first three measures from Beethoven's *Moonlight Sonata* (Piano Sonata in C♯-minor, *Sonata quasi una Fantasia, Op. 27 No. 2*):

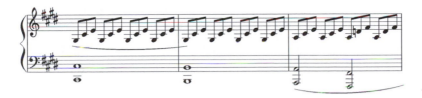

Identify the diatonic scale shifts in this passage, and their related chord progressions. Indicate the chords using note letters (lead sheet notation). Note: one of the chords is a seventh chord.

3.3.9. The following passage is the first measure of J.S. Bach's *B-flat Major Prelude (BWV 866)*. Identify the diatonic scale shifts in this passage, and their related chord progressions.

3.3.10. The following passage is from Handel's *Suite No. 7 in G-minor* (HWV 432, Andante movement, measures 4 and 5). Find what fundamental triadic chords the notes belong to, and what chord progressions are being implemented.

Hint: The first two beats consist of G_5, C_5, C_5, D_5, $E^\flat_5$, C_5 on the top of the upper staff, and F_4, $E^\flat_4$, $E^\flat_4$ on the bottom of the upper staff. These notes largely belong to the chord C-E$^\flat$-G, the C-minor chord that is chord **iv** for the piece. (The D note is a passing tone, while the F note is a neighboring tone accenting the E$^\flat$ note that follows it.)

3.4   Diatonic Transformations — Inversions, Retrograde

The other two types of diatonic scale transformations are inversions and retrogrades. We will now describe examples of their use in music.

3.4.1   Diatonic Scale Inversions

A *diatonic scale inversion*[19] consists in reversing (or inverting) the direction of the melodic intervals separating a group of notes. The basic form of diatonic scale inversion is to keep the starting note of the group unchanged and just reverse the direction of the intervals. We will denote this diatonic scale inversion by $\mathcal{I}$. See Figure 3.24.

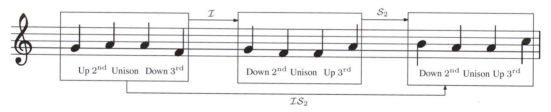

Figure 3.24. Diatonic scale inversion. Basic diatonic scale inversion $\mathcal{I}$ is performed first. By following it with the diatonic scale shift $\mathcal{S}_2$, the general diatonic scale inversion $\mathcal{I}\mathcal{S}_2$ is performed.

The general form of a diatonic scale inversion is to perform $\mathcal{I}$ on a collection of notes, and then follow it by applying a diatonic scale shift, so that the starting note is changed. See Figure 3.24. This general form of diatonic scale inversion is denoted by $\mathcal{I}\mathcal{S}_k$ for some integer k. If a modified diatonic scale shift $\mathcal{S}'_k$ is applied after $\mathcal{I}$, then we have a *modified diatonic scale inversion, $\mathcal{I}\mathcal{S}'_k$*. Like diatonic scale shifts, these diatonic scale inversions can be applied in sequences, as the following example shows.

Example 3.4.1. Tchaikovsky's *Pathetique* symphony. An extensive use of sequencing of diatonic scale inversions occurs in Tchaikovsky's Symphony No. 6 (Pathetique). In Figure 3.25, we show an excerpt from the score for the string sections.

We begin by looking at the notes for the first violins in the uppermost staff. In measure 1, a sequence of diatonic scale inversions, $\mathcal{I}\mathcal{S}_{-1}$ alternated with $\mathcal{I}\mathcal{S}_1$, is applied to the triplet enclosed in the box labeled a. For the remaining measures, a sequence of (modified) diatonic scale inversions

---

[19]The term *inversion* is used in music for multiple purposes. It is also used in mathematics. To avoid confusion, we will always refer to the inversion described here by the somewhat cumbersome phrase, *diatonic scale inversion*.

Figure 3.25. Sequencing of diatonic scale inversions in Tchaikovsky's *Symphony No. 6 (Pathetique)*, Movement III, measures 53 to 56. Triplets within rectangles labeled a and b are the key to this sequencing. Vns 1 stands for first violins, Vns 2 stands for second violins, Vlas stands for violas, Vcs stands for violincellos, and Cbs stands for contrabasses. The dynamic instruction ***crescendo poco a poco*** indicates that all of the strings, except for the contrabasses, are to increase gradually in volume. In contrast, the contrabasses are to change relatively abruptly in volume from *pianissimo* (very soft) to *mezzopiano* (medium soft) to *mezzoforte* (medium loud). Dots above or below the pitch markers of notes indicate that those notes should be played *staccato*.

could be thought of as applied. However, we could also view the remaining measures as obtained by a sequence of modified diatonic scale shifts applied to the whole of the first measure. This latter method might correspond better to what we actually hear, since our brains detect the similarity of each measure to the preceding one.

A different sequence of diatonic scale inversions is applied for the second violins. In measure 1, a sequence of diatonic scale inversions, $\mathcal{IS}_{-2}$ alternated with $\mathcal{IS}_2$, is applied to the triplet enclosed in the box labeled b. The second measure is obtained by applying the diatonic scale shift $\mathcal{S}_1$ to the first measure, and the third measure is obtained by applying the diatonic scale shift $\mathcal{S}_1$ to the second measure. The fourth measure begins with a modified version of the triplet b, followed by a sequence of modified diatonic scale inversions of this new triplet.

For the violas, the sequencing that was used for the first violins is repeated. For the violincellos, the sequencing for the second violins is repeated an octave lower.

This extensive sequencing of diatonic scale inversions and diatonic scale shifts, with its gradual increase in volume, is set off by a contrasting series of rhythmic changes for the contrabasses, along with their more abrupt volume changes. We have left out the other parts of the orchestral score, where the remaining instruments (such as trumpets, trombones, tubas, and tympani) are repeating the notes played by the contrabasses *after transposition to higher registers*. The contrast between the sonority of the violins, violas, and violincellos, with the sonorities of all these other instruments, creates a complex texture of sound.

Here is another example, which shows that diatonic scale inversions appear in popular music as well. In this case, a diatonic scale inversion plays an important part in a well-known traditional melody.

Example 3.4.2. *Down in the Valley.* In Figure 3.26 we show a diatonic scale inversion, $\mathcal{IS}_{-3}$, acting on four notes in the melody of the traditional song, *Down in the Valley*. It is interesting to observe how this diatonic scale inversion emphasizes the corresponding lyrics of the song. The four notes that $\mathcal{IS}_{-3}$ is applied to correspond to the lyrics: "Valley so low." While the resulting four notes after applying $\mathcal{IS}_{-3}$ correspond to the lyrics: "Hear the wind blow." The musical contrast produced by the diatonic scale inversion emphasizes a contrast in the thoughts expressed by the lyrics. See Exercise 3.4.5 for some further analysis of the melody of *Down in the Valley*.

Figure 3.26. Diatonic scale inversion in melody of traditional song, *Down in the Valley*.

Remark 3.4.1. There is a relation between diatonic scale inversions and mirror reflections in geometry. If we imagine that a horizontal mirror is drawn through the first note in a group of notes, then the diatonic scale inversion $\mathcal{I}$ can be thought of as reflecting the group of notes through this mirror. Since the first note is on the mirror, it stays put. The intervals between the notes in the group will reverse their direction due to reflection through the mirror. The more general diatonic scale inversions $\mathcal{IS}_k$ can also be thought of as reflection through a horizontal mirror, but not through the first note (when $k \neq 0$). It is probably easier to view them, however, as a combination of mirror reflection using $\mathcal{I}$ followed by rigid motion using $\mathcal{S}_k$.

3.4.2 Retrograde

Our final type of scale transformation is simply called retrograde. The **retrograde** R applied to a sequence of notes will produce them in reverse order. For example, here is a retrograde transformation:

Although a retrograde can be used in the way shown here, it is much more commonly employed in diatonic music as an immediate application. This can be done in two ways. Either a set of notes is immediately followed by their retrograde version, or one intervening note is placed between the set of notes and its retrograde. Both cases create a **palindrome symmetry.** See Figure 3.27. A palindrome

Figure 3.27. Two types of palindrome symmetry. The palindrome symmetry, labeled P1, is obtained by following three notes by their retrograde. Second type of palindrome symmetry, labeled P2, is obtained by following first two notes by their retrograde *but with one intervening note.* Both palindromes have left-right mirror symmetry. Vertical mirror locations are marked by m.

symmetry can be thought of as using a vertical mirror to reflect notes, creating a left-right mirror symmetry of notes within a score.

Example 3.4.3. Chopin's *Valse, Op. 64 No. 3.* In Figure 3.28, we show a palindrome symmetry on the bottom right of the score for a passage from Chopin's *Valse, Op. 64 No. 3.* We have also indicated a sequence of diatonic scale shifts (some of which are modified diatonic scale shifts). It is interesting that these two mathematical aspects of musical composition are played simultaneously in the passage, as indicated by their alignment within the score.

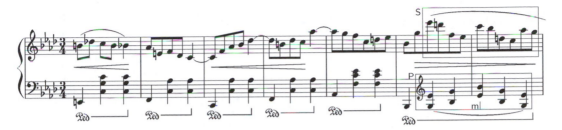

Figure 3.28. Measures 8 to 14 of Chopin's *Valse, Op. 64 No. 3*, illustrating palindrome symmetry, labeled by P with its vertical mirror position labeled m. There is also a sequence of diatonic scale shifts labeled by S, which aligns with palindrome in the score. Treble clef appearing right before the palindrome's notes changes the pitches for those notes (i.e., they are not the same notes as they would be if scored with the bass clef).

Example 3.4.4. Mendelssohn's *Violin Concerto in* E-*minor, Op. 64*. In Figure 3.29 we show a passage from Mendelssohn's *Violin Concerto in* E-*minor*. The notes in each measure are palindromes, each palindrome formed from a succession of two palindromes. These palindromes are formed from notes belonging to chords. For example, the notes in the first measure of the passage successively trace out E, G, and B notes. These are notes from the **i** chord, the Em chord. We leave the identification of the other chords to the reader as an exercise.

Figure 3.29. Measures 347 to 349 of Mendelssohn's *Violin Concerto in* E-*minor, Op. 64*, for principal violinist. There is a sequence of palindromes for each measure (and each measure's palindrome is formed from two successive palindromes).

Example 3.4.5. J.C. Bach's *Prelude and Fugue in* E$^\flat$-*major*. In this piece by Johann Christian Bach—one of Johann Sebastian Bach's sons—there are a number of palindrome symmetries. In Figure 3.30 we show a passage of his that contains several of them. It is also notable in that it contains a concatenation of palindrome symmetries, a chain of palindrome symmetries. There are other palindrome symmetries that we did not mark on the score in Figure 3.30, as well as several diatonic scale shifts. See Exercise 3.4.12.

Exercises

3.4.1. For the set of three notes below, perform the diatonic scale inversion $\mathcal{IS}_2$, and write the resulting notes on the remainder of the staff:

3.4.2. For the set of three notes below, perform the diatonic scale inversion $\mathcal{IS}_{-1}$, and write the resulting notes

Figure 3.30. Measures 78 to 83 of J.C. Bach's *Prelude and Fugue in* E$^\flat$-*major, BWV Anh. 177,* illustrating two palindrome symmetries. Palindromes are labeled P1 and P2. Mirror location for each palindrome is labeled by m. A concatenation of palindromes, is labeled CP, and marked with a succession of mirror positions. There is also a mirror symmetry of the two sixteenth notes that precede and follow this concatenation.

on the remainder of the staff:

3.4.3. For the set of three notes below, perform the retrograde transformation R, and write the resulting notes on the remainder of the staff:

3.4.4. For the set of three notes below, create a palindrome symmetry, positioning the mirror at the third note, and write the resulting notes on the remainder of the staff:

3.4.5. For the melody of *Down in the Valley* shown in Figure 3.26, show how it can be built up from applications of $\mathcal{S}'_0$, $\mathcal{S}_1$, and $\mathcal{IS}_{-3}$ to three different collections of notes.

3.4.6. For the melody of *Down in the Valley* shown in Figure 3.26, find what fundamental triadic chords the notes belong to and what chordal progression is related to the melody.

3.4.7. For the passage from Mendelssohn's *Violin Concerto,* shown in Figure 3.29, find what chords the notes belong to for the second and third measures of the passage.

3.4.8. In Figure 3.31, we show the beginning of the first variation from Mozart's *Twelve Variations on "Ah vous dirais-je, Maman"* (also known as "Twinkle Twinkle, Little Star"), K.V. 265. Mark all of the diatonic scale shifts and diatonic scale inversions that you find there.

3.4.9. In Figure 3.32, we show the first two measures of Variation 23 of Beethoven's *Diabelli Variations,* Opus 120. Mark all of the diatonic scale shifts and diatonic scale inversions that you find there.

3.4.10. In Figure 3.33, we show the beginning of the eleventh variation from Mozart's *Twelve Variations on "Ah vous dirais-je, Maman,"* K.V. 265. Mark three palindromes on the upper staff. Also mark a sequence of diatonic scale shifts on the lower staff in measures 2 through 4.

Figure 3.31. Beginning of a Mozart variation on "Twinkle Twinkle, Little Star." Variation 1 from *Twelve Variations on "Ah vous dirais-je, Maman," K.V. 265.*

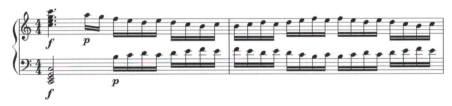

Figure 3.32. First two measures of Variation 23 of Beethoven's *Diabelli Variations,* Op. 120.

3.4.11. For the Mozart passage in Figure 3.33, find an example of multi-timescale sequencing of diatonic scale shifts (diatonic scale shifts within diatonic scale shifts). Hint: It occurs on the upper staff at the end of the score.

3.4.12. In the J.C. Bach passage shown in Figure 3.30, find a diatonic scale shift and a modified diatonic scale shift. Also find another palindrome. Discuss how these diatonic scale shifts, combined with the palindromes, produce a sense of motion in the passage.

3.5 Chromatic Transformations

The diatonic transformations discussed in the previous section have equivalent forms when working on the chromatic scale. A scale shift on the chromatic scale is called a *transposition*. An inversion on the chromatic scale is called a *chromatic inversion*. A retrograde is the same on all scales, including the chromatic scale, since it simply reverses the order of a set of notes (no matter what scale they are written with). These chromatic transformations have been used by classical composers, even within diatonic music. But it is within the realm of atonal music that they have played a preeminent role. While atonal music mostly abandons the diatonic approach to harmony, it retains the transformational approach of using scale shifts, inversions, and retrograde in creating musical compositions.

3.5.1 Transpositions

A *transposition* raises or lowers the pitch of all the notes in a group by the same number of half steps. If each note is changed in pitch by k half steps, then the transposition T_k has been performed on those

Figure 3.33. Beginning of another Mozart variation on "Twinkle Twinkle, Little Star." Variation 11 from *Twelve Variations on "Ah vous dirais-je, Maman," K.V. 265.*

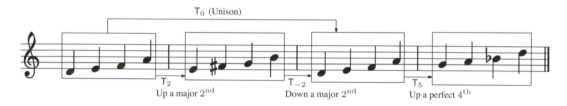

Figure 3.34. Sequence of transpositions of four notes. Compare with Figure 3.19.

notes. To see how this works, consider the score shown in Figure 3.34. When the transposition T_2 is applied to the four notes in the first measure, they are all shifted up in pitch by 2 half steps. On the chromatic clock, this can be done by adding $+2$ to all of the hours for the notes:

$$2 \quad 4 \quad 5 \quad 9 \quad \xrightarrow{+2} \quad 4 \quad 6 \quad 7 \quad 11$$

$$D_4 \quad E_4 \quad F_4 \quad A_4 \quad \xrightarrow{T_2} \quad E_4 \quad F_4^\sharp \quad G_4 \quad B_4$$

The note $F_4^\sharp$ is used here, rather than the enharmonic note $G_4^\flat$. If a $G_4^\flat$ had been used in the score, the contour of the note positions in the second measure would look different than in the first measure. By using $F_4^\sharp$ these contours have identical shapes.

This discussion shows one way of implementing a transposition. There is, however, a second way. If we specify the beginning note for the transposed group of notes, say E_4 in the second measure in Figure 3.34, then all subsequent notes are separated by the same number of half steps as in the original set of notes. In Figure 3.34, the half steps that separate the notes in the first measure are $+2, +1$, and $+4$. After applying T_2, the notes in the second measure are separated by that same number of half steps. Here is how this implementation of T_2 works out in this case:

$$D_4 \xrightarrow{+2} E_4 \xrightarrow{+1} F_4 \xrightarrow{+4} A_4$$

$$\downarrow +2$$

$$E_4 \xrightarrow{+2} F_4^\sharp \xrightarrow{+1} G_4 \xrightarrow{+4} B_4.$$

When reading from a score, the register subscripts can be dispensed with as well. A musician, of course, would describe the calculations above in terms of intervals (going up a major second, a minor second, and a major third for both the original melody and its transposition).

To get the third measure in Figure 3.34, the transposition T_{-2} is applied. Just like for diatonic scale shifts, this undoes the effect of T_2 and the original four notes are returned to in the third measure. With transpositions, we can just add half step changes. In this case, $+2 - 2 = 0$ so $\mathsf{T}_2\mathsf{T}_{-2} = \mathsf{T}_0$. The transposition T_0 adds 0 half steps in pitch to any group of notes, thereby leaving them unchanged.

Finally, to get the last measure in Figure 3.34, we applied the transposition T_5. Using the method of preserving the number of half step changes between notes, we apply T_5 in this way:

$$D_4 \xrightarrow{+2} E_4 \xrightarrow{+1} F_4 \xrightarrow{+4} A_4$$

$$\downarrow +5$$

$$G_4 \xrightarrow{+2} A_4 \xrightarrow{+1} B_4^\flat \xrightarrow{+4} D_5$$

Notice that $B_4^\flat$ is used, in preference to $A_4^\sharp$, in order to preserve the contour of the notes on the score. Also, for the last note, we have gone past the top hour of the clock. So we have entered a higher register. In terms of frequency multipliers, the frequency for $B_4^\flat$ is $\nu_c \cdot \mathbf{r}^{10}$, where ν_c is the frequency for C_4 and $\mathbf{r} = 2^{1/12}$. Adding 4 half steps to the pitch of this note corresponds to multiplying $\nu_c \cdot \mathbf{r}^{10}$ by the factor $\mathbf{r}^4$, producing the frequency $\nu_c \cdot \mathbf{r}^{14}$. Since $\mathbf{r}^{12} = 2$, we have a frequency of $2\nu_c \cdot \mathbf{r}^2$, which is the frequency for D_5.

Transpositions are used throughout music to express melodies in different keys, and to write scores for instruments that play in different pitch ranges. Here are some examples illustrating these important applications of transpositions.

Example 3.5.1. Consider the following melody for a portion of the traditional song, *Down in the Valley:*

It is written in the key of C-major. We can transpose this melody to notes in the A-major scale in the following way. We observe that the first note of the melody, a G, is the fifth note in the C major scale. Therefore, the transposed melody should begin on the fifth note in the A-major scale, which is an E (just add 7 and 9 on the chromatic clock). Therefore, we can begin the transposed melody on an E, say E_4, and then use the half-step changes from the original melody to generate the transposed melody:

$$G_3 \xrightarrow{+5} C_4 \xrightarrow{+2} D_4 \xrightarrow{+2} E_4 \xrightarrow{-4} C_4 \xrightarrow{+4} E_4 \xrightarrow{-2} D_4 \xrightarrow{-2} C_4 \xrightarrow{+2} D_4$$

$$\downarrow \mathsf{T}_9$$

$$E_4 \xrightarrow{+5} A_4 \xrightarrow{+2} B_4 \xrightarrow{+2} C_5^\sharp \xrightarrow{-4} A_4 \xrightarrow{+4} C_5^\sharp \xrightarrow{-2} B_4 \xrightarrow{-2} A_4 \xrightarrow{+2} B_4$$

If we retain the key signature for C-major, then the transposed melody is scored this way:

However, if we use the A-major key signature instead, then we have this score:

Comparing this to the original score, we see that transposing to A-major looks just like a diatonic scale shift **provided** we change the key signature.

Example 3.5.2. Transposing pitch for instruments. Not all instruments play on the same scales. For example, in orchestras, the $B^\flat$-clarinet will play notes that are scored on the C-major scale with notes having pitches on the $B^\flat$-major scale. This instrument "reads in C but plays in $B^\flat$." What this means is that any notes that are scored for a $B^\flat$-clarinet in a given key, must undergo a transposition T_{-2} in order to obtain the notes that are actually sounded (just as going back 2 hours on the chromatic clock changes hour 0 for C to hour 10 for $B^\flat$). For example, in Figure 3.35 we show the beginning of Beethoven's Fifth symphony, which includes notes scored for $B^\flat$-clarinets. When the transposition T_{-2} is applied to the notes that are scored for these clarinets, they match the notes for the first violins. So these instruments are, in fact, playing in unison. The key signature shown for the $B^\flat$-clarinets has just one flat, which indicates the key of D-minor. Remember, this symphony is in a minor key.

Applying T_{-2} to the D-minor scale produces the C-minor scale, since $D \xrightarrow{T_{-2}} C$ and T_{-2} preserves the number of half steps between notes on the two scales. Therefore, the $B^\flat$-clarinets are actually playing in the key of C-minor, the key for the symphony.

Figure 3.35. Score for $B^\flat$-clarinets and first violins in beginning of Beethoven's *Symphony Number 5 in C-minor* (Op. 67). Transposition T_{-2} is applied to notes scored for $B^\flat$-clarinets in order to obtain the pitches actually played by them. In fact, they play in unison with the violins.

Composers also use transpositions for producing repetitions in varying pitch ranges of groups of notes. As we said above, they apply transpositions in a similar way as diatonic scale shifts. An example is shown in Figure 3.36.

Example 3.5.3. J.S. Bach's *Inventio 11*. In the passage from Bach given in Example 3.3.4, we saw that there were several modified diatonic scale shifts. In Figure 3.36 we show that one of these is, in fact, a transposition. One other modified diatonic scale shift is a transposition as well. See Figure 3.37. In this figure, we have also left some of the modified diatonic scale shifts labeled in the form of S'_k, in order to indicate that they are **not** transpositions. .

This last example shows that there is no exact relationship between diatonic scale shifts, modified diatonic scale shifts, and transpositions. This stems from the fact that diatonic scale shifts and

Figure 3.36. Lower staff for first two measures of Bach's *Inventio 11, BWV 782.* Transposition T_{-2} is used. It moves $F_3^\sharp$ down 2 half steps to $E_3^\natural$, and preserves all the half steps between notes (written below the two rectangles). Compare with Figure 3.22 on p. 73, where this transposition T_{-2} is labeled as a modified diatonic scale shift S'_{-1}.

Figure 3.37. Opening measures of Bach's *Inventio 11, BWV 782,* with both diatonic scale shifts, modified diatonic scale shifts, and transpositions marked.

modified diatonic scale shifts are related to the diatonic scales, while transpositions are related to the chromatic scale. The non-uniform changes in half-steps on the diatonic scales, versus the perfectly uniform half-step changes on the chromatic scale, is the cause of this inexact relationship. In atonal music, this issue does not arise, since the chromatic scale is preeminent with no specific diatonic scale emphasized. Consequently, transpositions are the only types of scale shifts in atonal music.

3.5.2 Chromatic Inversions and Retrograde

The two other types of chromatic transformations are chromatic inversions and retrograde. Chromatic inversions are similar to diatonic scale inversions, but the scale used is the chromatic scale. The most basic chromatic inversion, denoted as I, keeps the beginning note of a group unchanged and reverses

Figure 3.38. Chromatic scale inversion. The basic chromatic scale inversion I is performed first. It reverses directions for half steps between each note. By following it with the diatonic scale shift T_4, the general chromatic scale inversion IT_4 is performed.

the sign of the number of half steps between each subsequent note. See Figure 3.38.

The general form of a chromatic scale inversion consists in performing I on a collection of pitches, and then following it with a transposition, so that the starting pitch is changed. See Figure 3.38. This general form of a chromatic inversion is denoted by IT_k for some integer k. Like transpositions, these chromatic inversions can be applied in sequences, as the following examples show.

Example 3.5.4. Mozart's *Piano Sonata No. 5 in G-major*. In Figure 3.39, we show a passage from a Mozart piano sonata. In this passage we see that Mozart has applied a sequence of inversions: a chromatic inversion, followed by a diatonic scale inversion, and then another chromatic inversion.

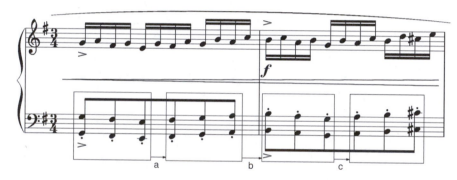

Figure 3.39. Measures 20 and 21 of Mozart's *Piano Sonata No. 5 in G-major, K.V. 283.* Sequencing of three inversions, *applied to the lowest pitch notes and octave-higher pitch notes simultaneously.* Inversion labeled a is chromatic inversion IT_{-1}. Next inversion, labeled b, is diatonic scale inversion $\mathcal{IS}_3$. Last inversion, labeled c, is chromatic inversion IT_{-2}.

Example 3.5.5. Tchaikovsky's *Pathetique* symphony. The Tchaikovsky passage described in Example 3.4.1 contained many diatonic scale inversions. See Figure 3.25. Several of the diatonic scale inversions that are applied to the triplets in that passage are also chromatic scale inversions. There is also a fascinating sequence of chromatic inversions in the score for the contrabasses. See Figure 3.40. In this figure, the two labels a indicate the chromatic inversion I, with a slight rhythm change. Each

Figure 3.40. Sequencing of chromatic inversions in passage by Tchaikovsky (*Symphony No. 6 (Pathetique)*, Movement III, measures 53 to 56). Labels a to d are explained in the text.

label b indicates the chromatic inversion IT_{-5}, but with a modification of rhythm (two eighth notes are changed to a sixteenth note and eighth note). Label c indicates a "staggered" chromatic inversion IT_5 with rhythm change, applied to the result of b, *but with the notes offset back by one position so as to begin on the end of the previous pair of notes.* This creates a brief palindrome symmetry for the pitches of the notes (not their durations). Finally, label d indicates a chromatic inversion I which returns to the first two pair of notes, and then the sequencing of a, b, and c is repeated. Alternatively, one could view the last two measures as S_0 applied to the first two measures. The combination of sequencing and rhythm alterations in this passage creates a remarkable sonority—especially when the contrabasses are played together with other instruments, such as trumpets, which are playing transposed notes in higher registers.

Retrogrades

The retrograde transformation, R, is no different than the one we previously described. Since R simply reverses the order of notes, it can be applied with any scale, including the chromatic scale.

3.5.3 Chromatic Transformations and Chords

We end our discussion by examining how transpositions and chromatic inversions can be applied to chords. Because this is an introduction to an idea that we will consider in more detail later, we will just focus now on fundamental triadic chords in root position. Since a chord is a group of notes, a transposition can be applied by simply shifting each note in the chord by the same number of half steps. For example, if the transposition is T_4, then each note in the chord is shifted up by 4 half steps. Applying a transposition to a major chord will produce another major chord. For instance, if we have a C-major chord, then applying T_4 to that chord will produce an E-major chord. Here is a diagram of the calculation (since the chord is in root position, we omit the subscripts indicating register):

$$
\begin{array}{ccc}
\text{G} & & \text{B} \\
+3 \uparrow & & +3 \uparrow \\
\text{E} & & \text{G}^\sharp \\
+4 \uparrow & \xrightarrow{+4} & +4 \uparrow \\
\text{C} & & \text{E} \\
\\
\text{(C-major)} & \xrightarrow{T_4} & \text{(E-major)}
\end{array}
$$

In this diagram we have shifted the root note C of the C-major chord by $+4$ half steps. That produces the note E. The transposition then preserves the number of half steps between the other notes, thereby producing a major chord with E as root note.

Here is another example. Suppose the transposition T_5 is applied to the A-major chord. We obtain the D-major chord as follows:

$$
\begin{array}{ccc}
\text{E} & & \text{A} \\
+3 \uparrow & & +3 \uparrow \\
\text{C}^\sharp & & \text{F}^\sharp \\
+4 \uparrow & \xrightarrow{+5} & +4 \uparrow \\
\text{A} & & \text{D} \\
\\
\text{(A-major)} & \xrightarrow{T_5} & \text{(D-major)}
\end{array}
$$

These examples illustrate that a transposition will always map a major chord to a major chord. In fact, given any two major chords, there is always a transposition that maps one of them to the other (just pick the transposition that maps the root of one to the root of the other). Likewise, a transposition will always map a minor chord to a minor chord, and every minor chord can be obtained from any other minor chord by a transposition.

In order to apply chromatic inversions to chords, we need to decide which note we will use as the starting note in applying the basic inversion I. The root note of the chord seems like a reasonable choice, so *we will always apply I to the root note of a chord.* After applying I to the root note, all the half-step changes in the chord are reversed. For example, here is the application of the basic chromatic inversion I to the C-major chord:

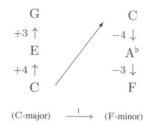

$$\text{(C-major)} \xrightarrow{\quad\text{I}\quad} \text{(F-minor)}$$

The root note C is mapped by I to the highest pitch note C, and the reversing of the directions of the half steps creates a minor chord with root note F.

Here is the application of a general chromatic inversion IT_4 to a B-major chord:

$$\text{(B-major)} \xrightarrow{\quad\text{IT}_4\quad} \text{(G}^\sharp\text{-minor)}$$

The root note B is mapped by I to the highest pitch note B, and T_4 then maps B to the highest pitch note $\text{D}^\sharp$. Reversing the directions of the half steps creates a minor chord with root note $\text{G}^\sharp$.

These examples show that a chromatic inversion will always map a major chord to a minor chord. In fact, given a major chord and minor chord, there is always a chromatic inversion that maps the major chord to the minor chord (just pick the chromatic inversion that maps the root of the major chord to the highest pitch note of the minor chord). Likewise, a chromatic inversion will always map a minor chord to a major chord, and every major chord can be obtained from any other minor chord by a chromatic inversion.

Our discussion in this section is summarized by the following theorem.

Theorem 3.5.1. *A transposition always maps major chords to major chords and minor chords to minor chords. A chromatic inversion always maps major chords to minor chords and minor chords to major chords. The set of transpositions and chromatic inversions* **acts transitively** *on the set of all major and minor chords in the following sense: For every pair of chords* $(\text{chord}_1, \text{chord}_2)$, *there is a transposition or chromatic inversion that maps* chord_1 *to* chord_2.

Remark 3.5.1. For some readers, these transformations of chords may be easier to follow if they are displayed graphically on the chromatic clock. We do this in the first two sections of Chapter 6. We also describe how important chromatic inversions of chords are for musical theory in Chapter 7.

Exercises

3.5.1. Transpose the first five measures of the melody for *Frere Jacques*, given in Figure 3.8 on page 58, to the key of G-major.

3.5.2. Transpose the first five measures of the melody for *Frere Jacques*, given in Figure 3.8 on page 58, to the key of $\text{B}^\flat$-major.

3.5.3. A D-trumpet is said to *read in* C *and sound in* D. That means that notes written with a C-major key signature will sound within the D-major key signature. Using that idea, if the following notes:

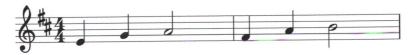

are to sound from a D-trumpet, then write how these notes should be scored.

3.5.4. The principle for scoring a D-trumpet was explained in the previous exercise. Suppose the following notes:

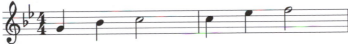

are to *sound from* a D-trumpet. Write how these notes should be scored.

3.5.5. For the Mozart passage in Figure 3.23 on p. 74, the notes for the two horns in B-flat are written with a key signature of C-major. However, these notes will sound with a $B^\flat$ key signature. For example, the first note for Horn I will sound as a $B^\flat$ (not a C). Likewise, the first note for Horn II will also sound as a $B^\flat$. Identify all the subsequent notes that sound for Horns I and II in this passage.

3.5.6. For the first four measures of the Mozart passage in Figure 3.23, find what fundamental triadic chords the notes belong to and what chord progressions are being implemented.

3.5.7. The following score fragment is taken from a composition by Mozart:

Find the chromatic inversion that occurs on the upper staff.

3.5.8. Find the chord that results from applying the transposition T_7 to the G-major chord G-B-D. Also find the chord that results from applying T_7 to the G-minor chord $G\text{-}B^\flat\text{-}D$.

3.5.9. Find the chord that results from applying the chromatic inversion IT_4 to the G-major chord G-B-D. Also find the chord that results from applying IT_4 to the G-minor chord $G\text{-}B^\flat\text{-}D$.

3.5.10. Find the chord that results from applying the transposition T_4 to the F-major chord F-A-C. Also find the chord that results from applying T_4 to the F-minor chord $F\text{-}A^\flat\text{-}C$.

3.5.11. Find the chord that results from applying the chromatic inversion IT_2 to the F-major chord F-A-C. Also find the chord that results from applying IT_2 to the F-minor chord $F\text{-}A^\flat\text{-}C$.

3.5.12. For the following chords:

apply the transformations given below, and write the resulting chord in the remaining portion of each measure.

Measure 1: Apply the transposition T_2.

Measure 2: Apply the chromatic inversion IT_7.

Measure 3: Apply the transposition T_{-4}.

Measure 4: Apply the chromatic inversion IT_5.

3.5.13. For the following chords:

apply the transformations given below, and write the resulting chord in the remaining portion of each measure.

Measure 1: Apply the transposition T_3.

Measure 2: Apply the chromatic inversion IT_5.

Measure 3: Apply the transposition T_{-5}.

Measure 4: Apply the chromatic inversion IT_7.

3.5.14. (a) For the C-major seventh chord shown in Figure 3.41, find the chords resulting from the transposition T_4 and the chromatic inversion IT_5. **(b)** Explain why every transposition and every chromatic inversion maps a major seventh chord to a major seventh chord.

3.5.15. (a) For the G-seventh chord shown in Figure 3.41, find the chords resulting from the transposition T_4 and the chromatic inversion IT_5. **(b)** Explain why every transposition maps a seventh chord to a seventh chord. **(c)** Explain why every chromatic inversion maps a seventh chord to a half-diminished seventh chord.

B	F	C	A
+4 ↑	+3 ↑	+3 ↑	+4 ↑
G	D	A	F
+3 ↑	+3 ↑	+4 ↑	+3 ↑
E	B	F	D
+4 ↑	+4 ↑	+3 ↑	+3 ↑
C	G	D	B
C-major seventh	G-seventh	D-minor seventh	B-half-dimin. seventh
C^{M7}	G^7	D^{m7}	$B^{ø7}$

Figure 3.41. Seventh chords for Exercises 3.5.14 through 3.5.17.

3.5.16. (a) For the D-minor seventh chord shown in Figure 3.41, find the chords resulting from the transposition T_4 and the chromatic inversion IT_5. **(b)** Explain why every transposition and every chromatic inversion maps a minor seventh chord to a minor seventh chord.

3.5.17. (a) For the B-half-diminished seventh chord shown in Figure 3.41, find the chords resulting from the transposition T_5 and the chromatic inversion IT_5. **(b)** Explain why every transposition maps a half-diminished seventh chord to a half-diminished seventh chord. **(c)** Explain why every chromatic inversion maps a half-diminished seventh chord to a seventh chord.

3.5.18. "Circle of Fourths" for pentatonic major scales. Consider the pentatonic C-major scale. **(a)** Apply the transposition T_5 to this scale, and show that it is also a pentatonic major scale, which shares all its notes with the pentatonic C-major scale, except for one note that has moved one hour clockwise on the chromatic clock. [Note: Use only the note letters A, C, D, E, and G, with sharps as needed. For example, the first note of the new scale will be $E^{\sharp}$ rather than F.] **(b)** If we call the application of T_5 to be "moving up a fourth" and T_{-5} to be "moving down a fourth," then explain how all the pentatonic major scales can be organized into a "Circle of Fourths." Illustrate your explanation with diagrams similar to the ones in Figure 2.8 on p. 47. [Note: Use only the note letters A, C, D, E, and G, with sharps or flats as needed.]

3.6   Web Resources

For additional study of basic music theory, there are a number of free sources available on the web. At the book's web site, we provide links for accessing some of this material, including the following:

1. Sites that provide more discussion of music theory, with animation and sound examples, historical and cultural background, and other features.

2. Sites that provide access to a huge number of free, public domain, music scores. You can download scores from classical music, traditional folk music, and jazz.

3. Lecture notes on mathematically oriented music theory, made freely available by Prof. Dmitri Tymoczko, Princeton University Music Department.

To access these resources, please visit the book's web site and click on the topic *Links*.

Chapter 4

Spectrograms and Musical Tones

One of the oldest and most successful parametric models for sound is the sinusoidal model....Perhaps the main reason for this is that the ear focuses most acutely on the peaks (amplitude maxima) in the spectrum of a sound.

—Scott N. Levine and Julius O. Smith III

In this chapter we discuss how mathematics can be used to model the tones we hear in music. The principal mathematical tool we will use is called a *spectrogram*. Spectrograms allow us to create visual representations of the structure of music relative to both time and frequency. We begin the chapter by showing how various musical tones appear in spectrograms. This will lay the groundwork for our use of spectrograms to analyze music in the next chapter. The remainder of the chapter describes the mathematics used to create spectrograms. This mathematics involves the trigonometric functions, sine and cosine. The sine and cosine functions provide a basic model for pure tones. By taking finite sums of sine and cosine functions, we will have a basic model for tones from musical instruments. After laying the foundations of analyzing music in a time-frequency manner in this chapter, we will apply time-frequency analysis to a wide variety of music in the next chapter.

4.1 Musical Gestures in Spectrograms

In this section, we describe how musical gestures are displayed in spectrograms. Instrumental tones, chords, and percussion, each have a particular style of appearance in spectrograms. Recognizing those styles will make it easier to identify these musical gestures when they appear in the many spectrograms we discuss in the rest of the book.

Example 4.1.1. Piano notes in an arpegiatted chord. In Color Figure 1 we show the score of the first three measures of Beethoven's *Moonlight Sonata*, along with a spectrogram of a recording of these same three measures. Color Figure 1 provides good confirmation of the fundamental musical theory that we have described so far. The score shows a $C_2^\sharp$ note and a $C_3^\sharp$ note in the lower staff for the first measure, held throughout the measure. In the spectrogram, the fundamental for the $C_3^\sharp$ note is indicated by a horizontal band centered on the position of about 139 Hz on the vertical frequency axis. This is consistent with the frequency value for $C_3^\sharp$ in Table 1.2 on p. 13. The fundamental for $C_3^\sharp$ is a match for the second harmonic for $C_2^\sharp$. The fundamental for $C_2^\sharp$ is indicated by a horizontal band centered at about 69 Hz on the vertical frequency axis. One thing the spectrogram shows, which is not apparent from the score, is that the sound from the fundamental for $C_2^\sharp$ fades away more quickly than the sound from the fundamental for $C_3^\sharp$.

One can also see in the spectrogram the striking of the arpeggiated notes. They occur at the start of the horizontal bands for the fundamentals of each note. For example, the triple arrow in Color Figure 1 points to the strikes within the fundamentals for the first triplet of notes $G_3^\sharp$, $C_4^\sharp$, and E_4, on the upper staff. The fundamental for $G_3^\sharp$ matches with the 3rd harmonic for $C_2^\sharp$, since its frequency is approximately 3 times the frequency of the fundamental for $C_2^\sharp$. Therefore, all the harmonics for $C_2^\sharp$ and $G_3^\sharp$ are good matches. We know also that all of the harmonics of $C_4^\sharp$ are exact matches of harmonics of $C_2^\sharp$. Therefore, we see one ascending ladder of bands at multiples of 69 Hz, before the striking of the E_4 note.

The striking of the E_4 note at 330 Hz is pointed to by the rightmost arrow in the triple of arrows in the figure. It creates a fundamental that is one half-step below the 5th harmonic for $C_2^\sharp$ at $5 \cdot 69 = 345$ Hz. This one half-step difference is small enough in frequency that there is interference in the sound waves from the two pitches. This interference appears in the spectrogram as an alternation of light and dark rectangles immediately following the key strike. Due to the fading out of the $C_2^\sharp$ tone, it is difficult to see this interference. There is also interference between the 2nd harmonic of E_4 and the 5th harmonic of $C_3^\sharp$. The downward pointing arrow in Color Figure 1 points to this interference, and here the alternation of bright and dark regions is evident. A magnified view is shown at top right of Color Figure 1. This interference, which corresponds to an alternation of higher and lower volume, is called *beating*. Beating interference between closely spaced harmonics can be seen in several other parts of the figure. We will explore beating and its relation to music in some detail later in the chapter.

It is interesting that, due to the long lasting sound (slow decay) of many of the piano notes, their harmonics frequently merge into long horizontal bands along the spectrogram. The spectrogram visually shows us how these arpeggiated chords create a cascading, flowing effect in the music. The long decay of the sound from these piano notes contributes to the piano's ability to create these beautiful arpeggiated chords. Here is an affirmation of the value of this aspect of piano sound from the musicologist and concert pianist, Charles Rosen, referring to a piano whose sound he admired:

> [the piano] produced a lovely tone with a remarkably long decay of sound that every piano ideally should have.[1]

Example 4.1.2. Trumpet notes. The horizontal bands that appear in the spectrogram of piano notes also appear in the spectrograms for other instruments playing notes of constant pitch. For example, we show in Figure 4.1 a solo trumpet passage from the Anthony Plog composition *Postcards*. The spectrogram confirms the accuracy and consistency of the timing of the notes, as well as their accurate pitches. Even without the score, which we have chosen not to display, we can identify the notes being played from the position of their fundamentals relative to the frequency axis. For example, the first six notes in succession are all at about 330 Hz so they are all E_4 notes. Moreover, the relative lengths of the fundamentals show that the first six notes are of equal duration, and the next three notes are of half that duration. The total duration of the passage then implies that the first six notes are most likely quarter notes, while the next three are eighth notes.

Comparing this spectrogram of trumpet notes with the piano spectrogram in Color Figure 1, we can see the differing visual brightness of the harmonics for the trumpet notes *vis a vis* piano notes. The visual brightness of the overtones of the trumpet notes is uniform over the first five or six harmonics, while the piano harmonics vary considerably in their brightness.[2] Another contrast between the two instruments is the difference in decay length of the sounds of the notes. The trumpet notes, in particular, die off extremely rapidly once the player's breath stops. Also illustrated in the spectrogram is the ability of a trumpet to *bend* pitches, which is not possible with a piano. In the spectrogram shown here, the notes are being played in a classical style so there is very little pitch bending. Much more elaborate bending of pitches occurs in jazz trumpeting. For example, see the spectrogram of a Louis Armstrong trumpet passage in Figure 4.3 on p. 97.

Example 4.1.3. Vocals. Our vocal chords—and the amplification of their sound through the chamber in our mouth and the position of our tongue—may be our most ancient musical instrument. In Color Figure 2 we show a spectrogram containing prominent vocals of the concluding portion of a live recording of the song, *And All That Jazz*, composed by Fred Ebb and John Kander.

---

[1]Charles Rosen, *Beethoven's Piano Sonatas*, Yale University Press, 2002, p. xi.
[2]It is interesting that the metaphor, *brightness*, is often used to describe the sound quality of musical tones, and we can see the visual brightness (non-metaphorical) in spectrograms as well.

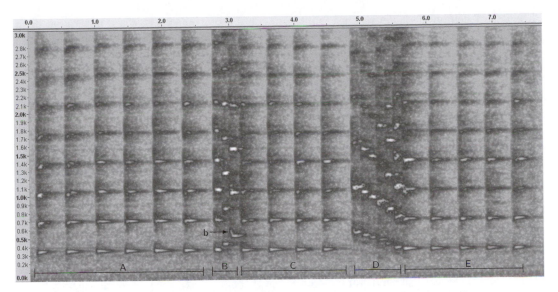

Figure 4.1. Spectrogram of passage from *Postcards*. Above the line segment A, there are six E_4 notes of equal duration. They appear to be quarter notes, followed by an ascending series of three, shorter duration notes above the line segment B. These would be eighth notes, followed by four more quarter notes, all E_4 notes, above the line segment C. Then there is a descending series of six eighth notes above the line segment D, finishing with five quarter notes (all E_4 again) above the line segment E. The trumpet notes also can *bend* in pitch, as shown by the pitch structure indicated by arrow b.

For each of the three sung notes—lying above the lyrics *All*, *That*, and *Jazz* in Color Figure 2— there is a different arrangement of relative volumes of the harmonics, indicated by different degrees of brightness. For *All*, the second and third harmonics are the most bright, while for *That* the amplitude is more nearly equal for the second, third, and fourth harmonics. For the concluding lyric, *Jazz*, the amplitude is highly concentrated in the second harmonic. If you listen closely to this concluding lyric, you can hear both the faint underlying pitch and the bright second harmonic. All of these differences in variation in volumes of different harmonics for the separate lyrics produce clearly distinct sonorities. These volume variations in the overtones are produced through changes in amplification created by altering the vocal chamber of the mouth.

In this passage, the singer is able to produce fairly straight tones, tones with constant pitch. For example, the tone she is singing for *All* is held at a constant pitch corresponding to $B_4^\flat$. However, she can also produce wavering pitch, known as ***vibrato,*** which is indicated in the figure at the end of the tone for *That*. When vibrato is produced, the tone we hear *softens* in its quality, compared to the straight tone portion that precedes it. This softening effect is perhaps why some singers employ vibrato. Another reason is that vibrato allows for a high volume of sound in singing. Vibrato occurs when the vocal tract is wide open, allowing more air to pass through, and this generates more sound volume. This explains the extensive use of vibrato in operatic style. It is used because a large volume is needed when singing without microphones in opera halls.

From the spectrogram we can accurately estimate two *quantitative* aspects of the singer's vibrato: **(1)** its ***frequency spread***, the amount of frequency change from peak to trough in the oscillating curve for the vibrato; **(2)** its ***rate***, how many oscillations per second are occurring in the curve for the vibrato. For this example, we leave the determination of the frequency spread and rate of this singer's vibrato to the reader as an exercise. We will discuss vibrato in more detail in the next chapter. For now, it is important to emphasize that determining frequency spread and rate of vibrato is one of the ways that spectrograms aid our analysis of music, beyond what listening alone can do.

Example 4.1.4. Piano Chords. In Color Figure 3 we show how chords appear in the opening notes from the piano composition, *Gymnopedie I*, by Erik Satie. The score of the passage is shown at the top, and the spectrogram of a recording of those notes played on a piano is shown at the bottom. In the score the first note is a G_2 on the lower staff, a dotted half note held throughout the first measure. The triadic chord that follows—played while G_2 is still sounding—consists of the three notes B_3-D_4-$F_4^\sharp$. The fundamentals and overtones of these notes, and their duration in time, are accurately shown in the spectrogram. We can see that the piano notes overlap each other, especially the bass notes which are intended to overlap with the sounding of the triadic chords. Using Table 1.2 in conjunction with the spectrogram, we can confirm *quantitatively* that the proper notes are being played, that the piano is in tune, and that many of the overtones from separate notes are near matches. Some near matches are indicated in the figure.

One final observation on the spectrogram in Color Figure 3 is that at the beginning of each of the notes and chords we can see vertical bands over the whole frequency range; one is marked by P. Previously we pointed out that these are due to a percussive effect when a piano hammer first strikes a piano string. Our final example illustrates how percussion appears in spectrograms.

Example 4.1.5. Percussion. Percussive effects appear in spectrograms as ***vertical bands***. For example, in Color Figure 4 we compare drum strikes in a spectrogram with clarinet tones. In this figure there are a large number of drum strikes. Two of these drum strikes are indicated in the spectrogram by the label P. The vertical bands for the drum strikes contrast markedly with the clarinet tones. We will explain why vertical bands occur for percussive strikes later in the chapter.[3] The fundamentals for the first five of the clarinet notes are marked by the label T_1. For the first two notes especially, these fundamentals show some pitch bending. Labels T_3 and T_4 mark the fundamentals for notes that exhibit more pitch bending, including vibrato for T_3 and a gentle pitch descent for T_4. The clarinet fundamental marked by T_2, however, shows almost no pitch bending. It appears to be nearly a straight tone. However, its higher harmonics show some bending, so this tone is indeed bending in pitch. Our ears do detect this subtle variation in pitch, which produces a slight crying or wailing effect in the sound. Notice that the drum strikes also exhibit some tonal quality. The label P' marks a bass tone for one of these drum strikes. All the drum strikes exhibit this bass tonality in addition to their percussive sound.

Exercises

4.1.1. On the left of Figure 4.2 we show the vibrato for the end of the singer's vocalization of the lyric, *That*, discussed in Example 4.1.3. Estimate the frequency spread and rate for this vibrato.

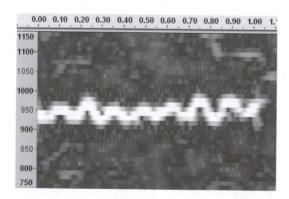 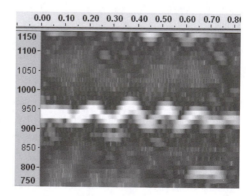

Figure 4.2. Left: Vibrato in singing. Right: Another example of vibrato by the same singer.

4.1.2. On the right of Figure 4.2 we show the vibrato for the end of a singer's vocalization of the lyric, *Jazz*, discussed in Example 4.1.3. Estimate the frequency spread and rate for this vibrato. Are your results the same as for Exercise 4.1.1?

---

[3]See Remark 4.6.1 on p. 121.

4.1.3. **(a)** The chord in both upper and lower staves of the first measure of Satie's *Gymnopedie I* (see top of Color Figure 3) is G_2-B_3-D_4-$F_4^{\sharp}$. Is this a minor or major chord? **(b)** Consider the chord in both upper and lower staves of the next measure. What are its notes, and is it a major or minor chord?

4.1.4. Identify in the spectrogram in Color Figure 3 the line segments that correspond to fundamentals for the notes in the second measure of the score. Also, identify the nearly matched overtones pointed to by arrow H_4.

4.1.5. In Figure 4.3 we show a spectrogram of a short passage from a recording of *La Vie en Rose* by Louis Armstrong. The spectrogram shows the trumpet notes played by Armstrong, some with vibrato and some without. Describe the difference in the qualities of the sounds you hear when the note has vibrato and when it does not. (Your answer should take into account that many notes begin with no vibrato at first, but then Armstrong increases the vibrato as the note progresses.)

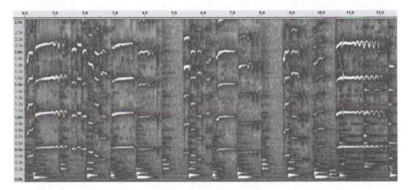

Figure 4.3. Spectrogram from a recording of the song *La Vie en Rose* by Louis Armstrong.

4.1.6. In Figure 4.4 we show a spectrogram from a recording of *Sylvie* by the choral group, *Sweet Honey in the Rock*. Describe all of the features (vibrato, constant tone, chords, etc.) that you see in the spectrogram.

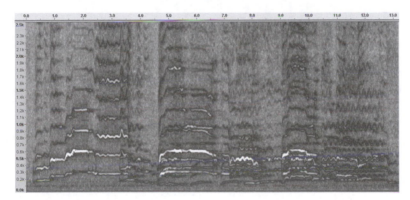

Figure 4.4. Spectrogram from a recording of the song *Sylvie* by the choral group *Sweet Honey in the Rock*.

4.1.7. A spectrogram of a clip from Freddie Mercury's song, *Bohemian Rhapsody,* is shown in Figure 4.5. Identify in this spectrogram at least one occurrence of each of these musical gestures: (1) sung tones, (2) vibrato, (3) drum strikes, and (4) piano tones.

4.1.8. Why do bass notes often sound richer than treble notes? On the other hand, why do treble notes often sound purer than bass notes?

4.2 Mathematical Model for Musical Tones

For the rest of the chapter, we shall discuss a mathematical model for the waveforms of musical

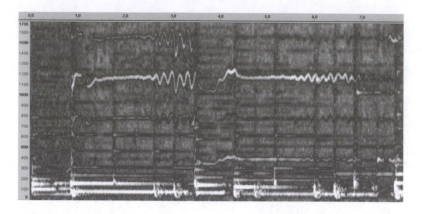

Figure 4.5. Spectrogram of a clip from a recording of the song *Bohemian Rhapsody* by the rock group *Queen*.

tones. We will be using ideas from trigonometry and pre-calculus, material that is covered in high school mathematics courses. We shall provide some important musical examples in the course of our discussion in this chapter, especially in Section 4.4, so we would like to encourage all readers to attempt to master this material. The most difficult mathematics occurs after Section 4.4. Readers who are primarily interested in music, and are not proficient in high school mathematics, may wish to proceed to the next chapter after Section 4.4.

4.2.1   Basic Trigonometry

We will mathematically model pure tones in this section, and then model more complex musical tones in the next section. To model a pure tone waveform we use the trigonometric functions of sine and cosine. Recall that for an acute angle θ within a right triangle, its sine and cosine are defined by the formulas:

$$\sin\theta = \frac{\text{opposite}}{\text{hypotenuse}}, \quad \cos\theta = \frac{\text{adjacent}}{\text{hypotenuse}}.$$

See the left of Figure 4.6.

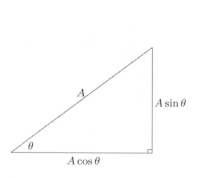
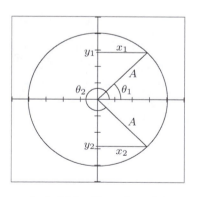

Figure 4.6. Left: Definition of sine and cosine for an acute angle θ. Right: Extension of sine and cosine to all possible angles, using a circle of radius A centered at the origin. For example, $y_1 = A\sin\theta_1$ and $y_2 = A\sin\theta_2$. Moreover, $x_1 = A\cos\theta_1$ and $x_2 = A\cos\theta_2$.

We will also need an extension of the definitions of $A\cos\theta$ and $A\sin\theta$ to all possible angles θ. The extension we need is shown in the circle diagram on the right of Figure 4.6. In this diagram, for all possible angles θ, where θ is measured from the positive x-axis, the ordinate value y is defined by $y = A\sin\theta$. We also have the abscissa value x defined by $x = A\cos\theta$, as shown in the diagram. The angle θ is said to be *positive* if it opens outward from the x-axis in a counter-clockwise direction, and *negative* if it opens clockwise. For example, if the angle θ marks off one-quarter of the circle's circumference in the counter-clockwise direction, then $\theta = 2\pi/4$ radians. Or, if θ winds around the entire circumference in a clockwise direction, so that it is co-extensive with an angle of 0 radians, then $\theta = -2\pi$ radians.

These sine and cosine functions satisfy several trigonometric identities. The most important one is the **Pythagorean Identity**:

$$(\cos \theta)^2 + (\sin \theta)^2 = 1 \tag{4.1}$$

which holds true for all angles θ. We will also need the ***Sine and Cosine Addition and Subtraction Identities***:

$$\sin(\theta + \phi) = \sin \theta \cos \phi + \cos \theta \sin \phi \tag{4.2a}$$

$$\sin(\theta - \phi) = \sin \theta \cos \phi - \cos \theta \sin \phi \tag{4.2b}$$

$$\cos(\theta + \phi) = \cos \theta \cos \phi - \sin \theta \sin \phi \tag{4.2c}$$

$$\cos(\theta - \phi) = \cos \theta \cos \phi + \sin \theta \sin \phi \tag{4.2d}$$

which hold true for all angles θ and ϕ.

4.2.2 Modeling Pure Tones

We are now ready to describe our model for a pure tone waveform. Our approach will be to mathematically model the oscillation of the tuning fork, shown in Figure 1.1 on page 2. To model this up and down oscillation, we create an up and down oscillation along a vertical axis, as shown in the graph on the left of Figure 4.7.

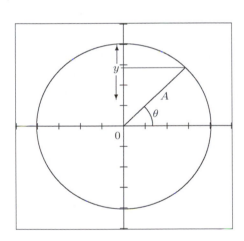 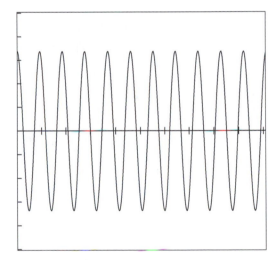

Figure 4.7. Mathematical model for tuning fork vibration. The tuning fork shown in Figure 1.1 oscillates up and down. The left graph indicates how a similar oscillation along a vertical axis is obtained by the motion of $y = A \sin \theta$, as θ changes at a uniform rate. A plot of the value of y with respect to time t is shown in the right graph. It can be formulated as a graph of $y = A \cos(2\pi\nu t - \phi)$ for a particular choice of constants $A, \nu,$ and ϕ.

Using radian measure, the angle θ will satisfy $\theta = 2\pi\nu t + \theta_0$ if θ changes at the rate of ν circumferences per second, and θ_0 is the initial value of θ when $t = 0$. We can then use the formula $y = A \sin(2\pi\nu t + \frac{\pi}{2})$ as a model for the tuning fork oscillation, ***provided*** we know the tuning fork is displaced by its maximum amount A at $t = 0$. (Because when $t = 0$, we have $y = A \sin(2\pi\nu \cdot 0 + \frac{\pi}{2})$, and so $y = A \sin(\frac{\pi}{2}) = A$.)

Although we could continue to use the sine function, it is more common in acoustical theory to use the cosine function. For the case of $y = A\sin(2\pi\nu t + \frac{\pi}{2})$, we make use of the Sine Addition Identity (4.2a), and obtain:

$$
\begin{aligned}
y &= A\sin(2\pi\nu t + \tfrac{\pi}{2}) \\
&= A\sin(2\pi\nu t)\cos(\tfrac{\pi}{2}) + A\cos(2\pi\nu t)\sin(\tfrac{\pi}{2}) \\
&= A\sin(2\pi\nu t)\cdot 0 + A\cos(2\pi\nu t)\cdot 1 \\
&= A\cos(2\pi\nu t).
\end{aligned}
$$

We have now shown that our model for a tuning fork vibration is

$$
y = A\cos(2\pi\nu t)
$$

provided we know that the tuning fork is displaced by its maximum amount A at time $t = 0$. Since we do ***not*** know that in general, we instead write $y = A\cos(2\pi\nu t - \phi)$ for some constant ϕ, where $y = A\cos(-\phi) = A\cos\phi$ represents the displacement of the tuning fork when $t = 0$. This waveform $y = A\cos(2\pi\nu t - \phi)$ is our model for the tuning fork oscillation. Here is a formal definition for this model.

Definition 4.2.1. *An* **harmonic oscillator** *is described by*

$$
y = A\cos(2\pi\nu t - \phi). \tag{4.3}
$$

where A, ν, and ϕ are constants. The positive constant A is called the **amplitude**, *the positive constant ν is called the* **frequency**, *and the real constant ϕ is called the* **phase**.

The phase constant ϕ in this definition can always be chosen so that it lies between $-\pi$ and π, inclusive. That is because the cycle length of $\cos\theta$ is 2π.

In Figure 4.8 we show a graph of a typical harmonic oscillator. We will now explain why $1/\nu$ is the cycle duration, and ν is the frequency, for this harmonic oscillator. Two successive peaks of $y = \cos\theta$ occur for $\theta = 0$ and $\theta = 2\pi$. Therefore, two successive peaks for the harmonic oscillator $y = A\cos(2\pi\nu t - \phi)$ occur for

$$
2\pi\nu t - \phi = 0 \qquad \text{and} \qquad 2\pi\nu t - \phi = 2\pi.
$$

Hence these two successive peaks occur at these time values:

$$
t = \frac{\phi}{2\pi\nu} \qquad \text{and} \qquad t = \frac{2\pi + \phi}{2\pi\nu}.
$$

Therefore, one cycle for the harmonic oscillator $y = \cos(2\pi\nu t - \phi)$ has duration

$$
\frac{2\pi}{2\pi\nu} = \frac{1}{\nu}.
$$

This shows that 1 cycle has a duration of $1/\nu$ sec. In other words, the harmonic oscillator $y = A\cos(2\pi\nu t - \phi)$ has a frequency of ν cycles/sec.

Testing the model

We now check our proposed model by comparison with an actual recording of a tone from a tuning fork. On the top left of Figure 4.9, we show a clip from a recording of a tuning fork sounding the pitch C_4. The vibration of the tuning fork creates an air vibration with matching frequency, and a sound

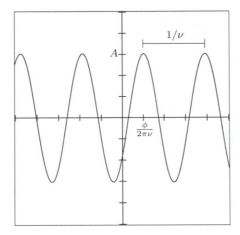

Figure 4.8. Graph of $y = A\cos(2\pi\nu t - \phi)$. It has amplitude A. It has a cycle length of $1/\nu$, so its frequency is ν. Its first peak to the right of $t = 0$ is at $t = \phi/(2\pi\nu)$, so its graph is the same as the graph of $y = A\cos(2\pi\nu t)$ shifted by $\phi/(2\pi\nu)$ along the t-axis.

recording registers a waveform from this air vibration. Therefore, we shall use a harmonic oscillator to model the waveform in this tuning fork recording. We use the following one:

$$y = 1803\cos(2\pi \cdot 262t). \tag{4.4}$$

The constants in this formula were obtained by a method we describe later in the chapter. A graph of this harmonic oscillator is shown on the top right of Figure 4.9.

The two graphs at the top of Figure 4.9 are very similar. We have graphed the two waveforms together at the bottom of Figure 4.9. This bottom graph confirms that the two waveforms are closely matched for the first 0.01 seconds. Later in the chapter we shall describe how we adjust the amplitude A and phase ϕ, by computing over short time intervals that slide forward in time, so that our model will closely match the recorded waveform in each short time interval. This localized modeling will be sufficient for accurate musical analysis.

This example has shown that a harmonic oscillator can provide an accurate model, at least over short time intervals, for a pure tone waveform. In the next section, we model the more complex waveforms from musical instruments.

Exercises

4.2.1. For the following harmonic oscillators, find their frequency, amplitude, and phase:

(a) $y = 32\cos(426\pi t - 0.3)$

(b) $y = 240\cos(328\pi t + 0.2)$

(c) $y = 120\cos(256\pi t)$

(d) $y = -340\sin(244\pi t)$

(e) $y = -240\cos(120\pi t)$

4.2.2. For the following harmonic oscillators, find their frequency, amplitude, and phase:

(a) $y = 84\cos(124\pi t - 0.1)$

(b) $y = 240\cos(440\pi t + 0.5)$

(c) $y = -320\cos(1200\pi t)$

(d) $y = 600\sin(512\pi t)$

(e) $y = -150\cos(240\pi t)$

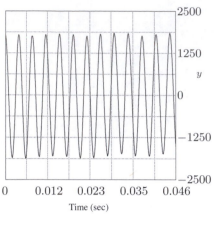
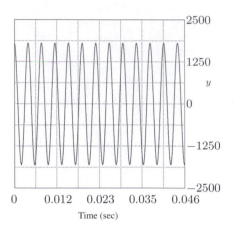

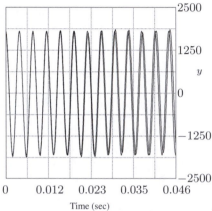

Figure 4.9. Comparison of sinusoidal model for a tuning fork waveform with actual waveform. Top Left: Portion of recorded waveform from tuning fork. Top Right: Mathematical model for tuning fork waveform. Bottom: Waveforms are a close match for the first 0.01 seconds, then slowly drift out of alignment. Over a succession of short duration time intervals, each around 0.01 seconds in length, it is possible to closely match the two waveforms by changing the phase ϕ and amplitude A in Equation (4.3) for each time interval.

4.2.3. For the following harmonic oscillators, use Table 1.2 to find the pitch for the tones they create:

(a) $y = 32\cos(440\pi t - 0.3)$

(b) $y = 240\cos(622\pi t + 0.2)$

(c) $y = 120\cos(1046\pi t)$

(d) $y = 340\sin(248\pi t)$

(e) $y = -240\cos(830\pi t)$

4.2.4. For the following harmonic oscillators, use Table 1.2 to find the pitch for the tones they create:

(a) $y = 100\cos(248\pi t - 0.3)$

(b) $y = 200\cos(124\pi t + 0.2)$

(c) $y = 150\cos(1480\pi t)$

(d) $y = 360\sin(208\pi t)$

(e) $y = 420\cos(350\pi t)$

4.3 Modeling Instrumental Tones

As described in Chapter 1, an instrumental tone is a combination of harmonics. Each harmonic is a multiple by a positive integer of a fundamental frequency that determines the tone's pitch. To model instrumental tones we will use, as a first approximation, sums of harmonic oscillators. The general form of the model is:

$$y = A_1 \cos(2\pi\nu t - \phi_1) + A_2 \cos(2\pi \cdot 2\nu t - \phi_2) + \cdots + A_M \cos(2\pi \cdot M\nu t - \phi_M). \qquad (4.5)$$

where M is a positive integer specifying the number of harmonics used for the model. The frequencies are all positive integer multiples of the fundamental, ν, and the amplitudes and phases can vary for different harmonic oscillators.

The sum in (4.5) is cumbersome to write. Therefore, we shall use the following more compact notation for sums.

Definition 4.3.1. *A sum of M numbers*

$$y_1 + y_2 + y_3 + \cdots + y_M$$

is expressed as $\displaystyle\sum_{k=1}^{M} y_k$. *That is,*

$$\sum_{k=1}^{M} y_k = y_1 + y_2 + y_3 + \cdots + y_M.$$

For example, this definition says that

$$\sum_{k=1}^{5} k^2 = 1^2 + 2^2 + 3^2 + 4^2 + 5^2$$

$$\sum_{k=1}^{3} 8 \cos(2\pi k t) = 8 \cos(2\pi t) + 8 \cos(4\pi t) + 8 \cos(6\pi t)$$

$$\sum_{k=1}^{4} \frac{3}{k} \cos(2\pi k\nu t) = 3 \cos(2\pi\nu t) + \frac{3}{2} \cos(4\pi\nu t) + \cos(6\pi\nu t) + \frac{3}{4} \cos(8\pi\nu t).$$

Using our notation for sums, we rewrite (4.5) as

$$y = \sum_{k=1}^{M} A_k \cos(2\pi k\nu t - \phi_k). \qquad (4.6)$$

As an example of (4.6), we shall model the portion of the flute note shown on the left of Figure 4.10. For our model of this note we use a sum of just two harmonic oscillators:

$$y = 2556.48 \cos(2\pi \cdot 329t - 0.22962) + 2401.44 \cos(2\pi \cdot 658t - 1.96379). \qquad (4.7)$$

The fundamental here is 329 Hz, a close approximation of the fundamental for E_4. The second harmonic is 658 Hz, twice the fundamental. The graph of this sum is shown on the right of Figure 4.10. Even though we used only two harmonics for our model, it is still a decent approximation of the flute's waveform. Over short time intervals of around 0.01 seconds it is possible to match the two

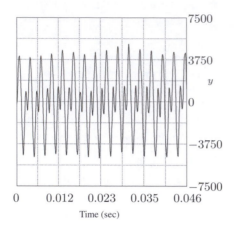
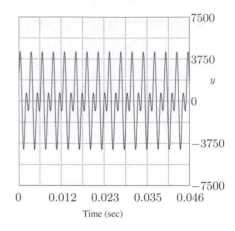

Figure 4.10. Left: Portion of recorded waveform from flute note. Right: Sinusoidal sum model for flute note waveform using only two terms (two harmonics).

waveforms closely by using more harmonic oscillators, with varying amplitudes and phase constants, within each time interval. By sliding these short time intervals forward in time, we can produce a localized modeling that is sufficient for musical analysis. We will discuss this localized modeling in Sections 4.5 and 4.6.

4.3.1 Beating

We close this section by describing an important phenomenon that occurs when two harmonic oscillators are added. Suppose we have the waveform described by

$$y = A\cos(2\pi\nu t) + B\cos(2\pi\omega t)$$

which is the addition of two harmonic oscillators of amplitudes A and B and frequencies ν and ω. To simplify matters, we have assumed that the phase constants are both 0. When ν and ω are fairly close to each other, an interference phenomenon occurs that is called **beating**. When two tones of nearly equal frequency are sounding, they interfere with each other, alternately reinforcing each other and canceling each other.

To explain how beating arises, let's first consider the sum of two cosines: $\cos\theta + \cos\phi$. If we let $\theta_a = (\theta + \phi)/2$ and $\phi_d = (\theta - \phi)/2$, then we have

$$\cos\theta + \cos\phi = \cos(\theta_a + \phi_d) + \cos(\theta_a - \phi_d).$$

Using the cosine addition and subtraction formulas, given in Equations (4.2c) and (4.2d), we obtain

$$\begin{aligned}
\cos\theta + \cos\phi &= \cos(\theta_a + \phi_d) + \cos(\theta_a - \phi_d) \\
&= \cos\theta_a \cos\phi_d - \sin\theta_a \sin\phi_d + \cos\theta_a \cos\phi_d + \sin\theta_a \sin\phi_d \\
&= 2\cos\phi_d \cos\theta_a.
\end{aligned}$$

Thus,

$$\cos\theta + \cos\phi = 2\cos\left(\frac{\theta - \phi}{2}\right)\cos\left(\frac{\theta + \phi}{2}\right).$$

Applying this result to a sum of two harmonic oscillators with equal amplitudes, we have

$$\begin{aligned}
y &= A\cos(2\pi\nu t) + A\cos(2\pi\omega t) \\
&= 2A\cos\left(2\pi\,\tfrac{\nu-\omega}{2}\,t\right)\cos\left(2\pi\,\tfrac{\nu+\omega}{2}\,t\right).
\end{aligned} \tag{4.8}$$

The factor $2A\cos\left(2\pi\,\frac{\nu-\omega}{2}\,t\right)$ is thought of as *varying amplitude* (or *amplitude modulation*) for the cosine factor $\cos\left(2\pi\,\frac{\nu+\omega}{2}\,t\right)$. For example, in Figure 4.11 we show graphs for

$$y = 3\cos(2\pi \cdot 24t) + 3\cos(2\pi \cdot 20t)$$
$$= 6\cos(2\pi \cdot 2t)\cos(2\pi \cdot 22t).$$

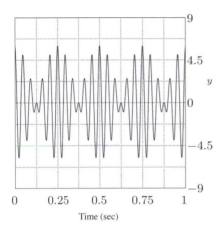
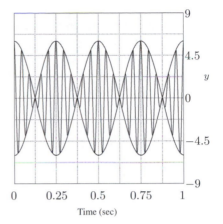

Figure 4.11. Graphs of beating for sum of two harmonic oscillators. Left: Sum of two harmonic oscillators with frequencies of 24 and 20 Hz: $y = 3\cos(2\pi \cdot 24t) + 3\cos(2\pi \cdot 20t)$. Right: Graph of sum of these harmonic oscillators together with graphs that indicate the amplitude modulation, $y = \pm 6\cos(2\pi \cdot 2t)$.

In the left graph of the figure, we can see the periodic damping down of the vibration. In the right graph of the figure, we have plotted the two graphs $y = \pm 6\cos(2\pi \cdot 2t)$, which provide an *envelope* for the graph of $y = 3\cos(2\pi \cdot 24t) + 3\cos(2\pi \cdot 20t)$. The successive zero values of the amplitude modulation factor $6\cos(2\pi \cdot 2t)$ mark off *lobes* in the graph of the sum of the two harmonic oscillators. These lobes are the **beats**. The number of beats per second is the **beat frequency**. Since the zero values of $6\cos(2\pi \cdot 2t)$ occur every half-cycle, they occur with *twice the frequency* of this cosine factor. Hence, the beat frequency is 4 beats/sec. In general, the beat frequency is $\nu - \omega$ beats/sec when $\nu > \omega$, and $\omega - \nu$ beats/sec when $\omega > \nu$.

The amplitude modulation factor $2A\cos\left(2\pi\,\frac{\nu-\omega}{2}\,t\right)$ explains the regular diminishment of loudness that creates the beats. However, the factor $\cos\left(2\pi\,\frac{\nu+\omega}{2}\,t\right)$ is also important. It has a frequency of $\frac{\nu+\omega}{2}$, so it corresponds to a pitch with frequency midway between the pitches of frequency ν and ω. It is interesting that often we hear only this middle pitch of frequency $\frac{\nu+\omega}{2}$ and **not** the two pitches of frequencies ν and ω. This occurs when the two frequencies fall within a single *critical band* in the ear's basilar membrane. Such critical bands are keyed to hearing specific frequencies. Hence our brains cannot distinguish the two frequencies that make up the tone. We hear only the middle frequency.[4]

If the amplitudes A and B are different, then the calculation above needs to be modified. If we assume that $B > A$, then we have

$$y = A\cos(2\pi\nu t) + B\cos(2\pi\omega t)$$
$$= A\cos(2\pi\nu t) + A\cos(2\pi\omega t) + (B - A)\cos(2\pi\omega t).$$

We then apply our previous result to obtain

$$y = A\cos(2\pi\nu t) + B\cos(2\pi\omega t)$$
$$= 2A\cos\left(2\pi\,\frac{\nu-\omega}{2}\,t\right)\cos\left(2\pi\,\frac{\nu+\omega}{2}\,t\right) + (B - A)\cos(2\pi\omega t). \tag{4.9}$$

---

[4] J.G. Roederer, *The Physics and Psychophysics of Music, 3rd Edition,* Springer, 2008. Sec. 2.4, pp. 34–42.

Equation (4.9) shows that there will still be beating, coming from the term

$$2A \cos\left(2\pi \, \tfrac{\nu-\omega}{2} \, t\right) \cos\left(2\pi \, \tfrac{\nu+\omega}{2} \, t\right).$$

But there is also a residual pure tone, of amplitude $B - A$ and frequency ω, coming from the term $(B - A)\cos(2\pi\omega t)$. Since we have not specified whether ν is smaller or larger than ω, the case of $B > A$ is completely general.

To ease the burden of notation here, we have assumed that the phases are all 0. Allowing for different phases is relatively straightforward, so we have left it to the reader to perform in the exercises.

Beating is used for tuning musical instruments. Suppose the frequency of a pure tone, from a tuning fork say, is a close match but not exactly equal to the fundamental frequency of a note played by a musical instrument. This will occur when the instrument is out of tune. A beating effect will then occur when the tuning fork and the instrument's note are played together. By bringing the instrument into proper tune, the time between successive beats will lengthen beyond the point of audibility. In the next section we will discuss cases where beating interference occurs in musical performance.

Exercises

4.3.1. For the following sums of harmonic oscillators, identify the amplitudes, frequencies, and phases, and also identify any pitches that the frequencies correspond to from Table 1.2.

 (a) $y = 300\cos(2\pi \cdot 262t - 0.3) + 600\cos(2\pi \cdot 330t - 0.3) + 100\cos(2\pi \cdot 392t - 0.29)$

 (b) $y = 400\cos(2\pi \cdot 82t - 0.5) + 200\cos(2\pi \cdot 98t - 0.5) + 100\cos(2\pi \cdot 110t - 0.5)$

 (c) $y = 1200\cos(2\pi \cdot 220t + 0.2) + 800\cos(2\pi \cdot 294t + 0.2) + 100\cos(2\pi \cdot 370t) + 90\cos(2\pi \cdot 740t + 0.3)$

4.3.2. For the following sums of harmonic oscillators, identify the amplitudes, frequencies, and phases, and also identify any pitches that the frequencies correspond to (sometimes only approximately) from Table 1.2.

 (a) $y = 500\cos(2\pi \cdot 110t + 0.4) + 800\cos(2\pi \cdot 220t + 0.4) + 300\cos(2\pi \cdot 330t + 0.9)$

 (b) $y = 1000\cos(2\pi \cdot 124t - 0.5) + 200\cos(2\pi \cdot 97t - 0.5) + 100\cos(2\pi \cdot 196t - 0.5)$

 (c) $y = 2000\cos(2\pi \cdot 164t) + 1200\cos(2\pi \cdot 785t) + 1000\cos(2\pi \cdot 882t) + 900\cos(2\pi \cdot 1979t)$

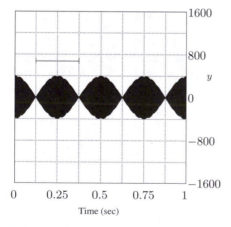 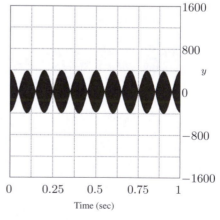

Figure 4.12. Graphs of beating interference for two closely spaced frequencies. Left: Frequencies 450 and 454 Hz. Length of one beat indicated by line segment above first full lobe. Right: Frequencies 450 and 460 Hz.

4.3.3. The graph on the left of Figure 4.12 shows beating when two pure tones of frequencies 450 and 454 are added. Given that the *length of a beat* is the amount of time spanned by a complete lobe (indicated in the figure), find the number of beats per second. Does your result match with the result described in the text?

4.3.4. The graph on the right of Figure 4.12 shows beating when two pure tones of frequency 450 and 460 are added. Find the number of beats per second. Does your result match with the result described in the text?

4.3.5. Modify the discussion in the text to show that

$$y = A\cos(2\pi\nu t - \phi) + A\cos(2\pi\omega t - \theta)$$

satisfies

$$y = 2A\cos(2\pi \tfrac{\nu - \omega}{2} t - \tfrac{\phi - \theta}{2})\cos(2\pi \tfrac{\nu + \omega}{2} t - \tfrac{\phi + \theta}{2}).$$

Explain how there will be the same beating phenomenon in this case, as for the case discussed in the text.

4.3.6. In this problem, you should modify the discussion in the text to analyze the case of

$$y = A\cos(2\pi\nu t - \phi) + B\cos(2\pi\omega t - \theta)$$

when $B > A$. In particular, answer these questions: **(a)** What is the phase shift, beat-frequency for the beating tone, and frequency for the pitch of the beating tone? **(b)** What is the amplitude, phase shift, and frequency for the residual pure tone?

4.4 Beating and Dissonance

In this section we show how the phenomenon of beating occurs in musical performance, and how it relates to the musical concept of dissonance. The presence of beating between closely spaced harmonics in different tones has been used as a scientific explanation for why some combinations of tones sound unpleasant to many listeners. Such unpleasant combinations sometimes correspond to the dissonant melodic or harmonic intervals referred to in music theory. In diatonic music, in addition to unison and octaves, the intervals that are generally considered consonant are intervals obtained from the fundamental triadic chords (major and minor thirds, and perfect fifths). In addition, major and minor sixths are also generally considered consonant. Perfect fourths are sometimes considered *dissonant* intervals, and sometimes considered *consonant* intervals. The classification used, consonant or dissonant, depends on the musical context. When a perfect fourth occurs alone, without lower pitch notes, then it is considered dissonant (and needs to resolve to a different interval). When there are lower pitch notes present, then the perfect fourth interval is considered consonant. However, we shall see that, according to Helmholtz's theory, perfect fourths are always *acoustically* consonant because of a relative lack of beating interference between harmonics. The question of when intervals are dissonant is a subtle one. It will help to look at a couple of musical examples. The most widely accepted scientific theory for when intervals are acoustically dissonant is due to Helmholtz. In Helmholtz's theory, an interval between two tones will be acoustically dissonant when there is significant beating—with a high enough beat frequency to create a pitch-like tension—between one or more of the harmonics for the tones. The term *significant* is a subjective one, but some objective evidence for the beating can be obtained from spectrograms.

At the top of Color Figure 5 we show a score for two harmonic intervals that are classed as dissonant in music theory: a minor second and a tritone (diminished fifth). For the minor second interval, the notes are separated by just one half step. The beating between the 1st, 2nd, and 3rd harmonics of the violin tones is clearly visible in the figure. It appears as alternating dark and light spots on the bands for these harmonics. As the frequencies go up from lower to higher harmonics, the beating frequencies of the interference also increase (see Exercise 4.4.3). Consequently, the alternation of dark and light spots is occurring at a faster rate between the 2nd harmonics than between the 1st harmonics, and occurring faster still between the 3rd harmonics.

On the other hand, for the tritone harmonic interval there is a different appearance to the beating interference. This tritone interval is between an E_4 note and a $B_4^\flat$ note. These notes are separated by 6 half steps. This is a large enough separation in frequency that no beating is evident between the 1st harmonics for these tones. However, the 2nd harmonic for the $B_4^\flat$ note and the 3rd harmonic for the E_4 note are close enough to each other that there is noticeable beating in the spectrogram. The separation of the harmonics is about one half step. To show this, we use the frequency factor $\mathbf{r} = 2^{1/12}$ discussed in Chapter 1. Let ν_o stand for the fundamental frequency of the lower pitch E_4 note. Then the $B_4^\flat$ note has fundamental frequency $\nu_o \mathbf{r}^6$, and the ratio between the frequencies of the harmonics satisfies

$$\frac{(3^{\text{rd}} \text{ harmonic of } E_4)}{(2^{\text{nd}} \text{ harmonic of } B_4^\flat)} = \frac{3\nu_o}{2\nu_o \mathbf{r}^6}$$

$$= \frac{3}{2} \frac{1}{\mathbf{r}^6}$$

$$\approx \mathbf{r}^7 \frac{1}{\mathbf{r}^6}$$

$$= \mathbf{r}^1.$$

We used the fact that $\mathbf{r}^7 = 1.4983 \approx \frac{3}{2}$ in this calculation. The calculation shows that these harmonics are separated by about 1 half step. Therefore, we expect interference to occur, just as in the case of the minor second harmonic interval. We do see this interference in the spectrogram for the violin tones, but it is not quite the same as between the fundamentals in the minor second interval. Because the amounts of frequency separation are greater in magnitude for these higher harmonics, the rates of beating oscillation are much faster. Consequently, the interference appears more as a darker shading between harmonics, corresponding to an additional tonality. Listening carefully to the sound from this tritone interval, compared to the minor second interval played just before it, one can perceive a difference between the two dissonances. The sound of the minor second interval seems to fluctuate in pitch, while the sound of the tritone harmonic interval seems to have a fairly steady pitch. The interference occurring for the tritone is in the timbre of the sound, rather than in the pitch. This interference within the timbre is sometimes called a *roughness* of the sound.

4.4.1 Some Uses of Dissonance in Music

We now consider two examples of the use of consonance and dissonance in musical compositions. Our first example is from the Beethoven piano composition, *Für Elise*. Our second example is from a marimba composition, *Hegira,* by Carrie Magin.

Example 4.4.1. In Color Figure 6 we show a three-measure passage from Beethoven's *Für Elise* (Bagatelle No. 25 in A-minor, WoO 59). In Helmholtz's theory, notes separated by acoustically consonant intervals have their harmonics sufficiently separated that beating interference is not an issue. For melodic intervals, this implies that dissonance will be much less noticeable acoustically due to the relative lack of time overlap of harmonics from separate notes. In *Für Elise,* however, the slowly decaying sound of the piano notes allows us to apply ideas of consonance and dissonance that are generally more noticeable for harmonic intervals. The first two melodic intervals in the passage, E_5 to B_4 and then B_4 to E_5, are both perfect fourths. Acoustically, these are consonant intervals. Following these perfect fourth intervals are a succession of minor second intervals, consisting of alternations of E_5 and $D_5^\sharp$. The beating interference of the fundamentals from these notes, which overlap due to the long lasting tones from the piano strings, is clearly shown in the spectrogram at around 650 Hz. A middle tone, with brief fundamental bands in between the fundamentals of these interfering notes, is also faintly audible during the time interval of 1.8 to 2.8 seconds. In jazz, that middle tone is called a blue note. The rapid on and off sounding of this middle tone imparts a slight ringing effect, an instance of acoustic dissonance.[5]

---

[5]In the case of *Für Elise*, this ringing/beating is dependent on an individual's performance technique. If the keys are not struck as hard, for instance, the beating will not be audible. In fact, many performers do not strike the keys hard for this passage. Our point here is that spectrograms allow us to visually examine this aspect of performance technique.

Beethoven follows these dissonant intervals with an interval of a perfect fourth, E_5 followed by B_4, and then an interval of a minor third, the previous B_4 followed by $D_5^\natural$. In this context, these intervals are musically consonant, since they follow the repeated dissonance of the minor seconds and are free of the beating effects of those repeated minor seconds. Moreover, they lead into the next two and a half measures, which consist of the arpeggiated triadic chords A-C-E and E-G$^\sharp$-B, composed of consonant intervals of major and minor thirds. (See Figure 4.13.) In the score shown there, Beethoven indicates that these arpeggios are to be played with pedal, thereby increasing the overlapping of sound in the chords and amplifying their consonant nature.

Figure 4.13. Alternation of dissonance with consonance in Beethoven's *Für Elise*.

After this period of consonant intervals, there is a return to the dissonant sequence of repeated E and D$^\sharp$ notes. The passage shown in Figure 4.13 occurs frequently throughout the piece, it provides an alternation between dissonance and consonance. The consonant passages providing a kind of release from the dissonant passages. Here we see that Beethoven's *Für Elise* is a good illustration of the following observation from a classic text on harmony:[6]

> It cannot be too strongly emphasized that the essential quality of dissonance is its sense of movement and not, as is sometimes erroneously assumed, its degree of unpleasantness to the ear.

Example 4.4.2. Another instance of beating and dissonance occurs in the composition for solo marimba, *Hegira,* by Carrie Magin. In Figure 4.14 we show a passage from the score and a spectrogram from the composer's performance of this passage. In the score, the higher pitch notes running along the top are referred to as one **voice**. The lower pitch notes below them, are referred to as another voice. So this passage has two voices in it. The intervals between successive notes in each voice are predominantly minor seconds, so they are musically dissonant. In the spectrogram the beating interference between the tones of these notes is quite evident. The result of this beating is a rich sounding, oscillating bass sonority, which is not at all unpleasant to the ear. Because these intervals are in the bass range, the beat frequency is low enough that it does not produce a pitch perception—instead, it produces a slow alternation of volume of this bass pitch. In Helmholtz's theory, these are acoustically consonant intervals.

In this passage from *Hegira,* as with the *Für Elise* example, we see alternation between musical consonance and dissonance. For instance, the intervals in the fourth measure are musically consonant intervals. This measure provides a brief musically consonant interlude between the surrounding musically dissonant measures. Similarly, there are brief interludes of musical consonance at the end of the second measure and the beginning of the last measure.

Hegira also provides some nice illustrations of the musical transformations that we discussed in Chapter 3. See Exercise 4.4.6. In addition, *Hegira* has an interesting rhythmic structure, such as the rapid changes in timing and meter in the passages shown in Figures 4.14 and 4.16.

---

[6]W. Piston and M. Devoto, *Harmony, 4th edition*, Norton, 1978, p. 7.

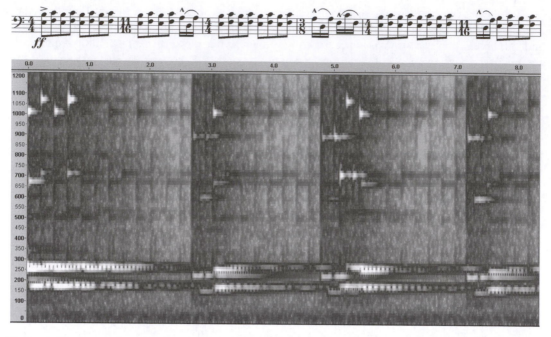

Figure 4.14. Passage from Carrie Magin's *Hegira*. © Carrie Magin, used by permission. Top: Score shows repeated E_3 and F_3 notes for the lower voice and repeated B_3 and C_4 notes for the upper voice. Bottom: Spectrogram shows extensive beating between the tones in these two voices.

Exercises

4.4.1. Beating in a musical recording. In Figure 4.15, we show a spectrogram of a very brief clip from Eric Clapton and Jim Gordon's song, *Layla*. The spectrogram shows beating in several places, one example is indicated by an arrow. For the beating indicated by this arrow, estimate the beating frequency by counting the beats *and* by estimating the difference between the two close frequency bands that seem to be creating the beating. Do your results match?

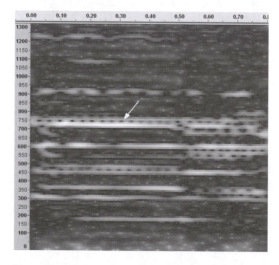

Figure 4.15. Spectrogram of very short clip from *Layla*, showing beating in guitar harmonics. One case of beating is pointed to by the arrow.

4.4.2. Verify that the 3rd harmonic of $B_4^\flat$ and the 4th harmonic of E_4 are separated by about one half step.

4.4.3. Show that if there is beating interference of β beats/sec between the 1st harmonics of two tones, then the beating interference will be 2β beats/sec between their 2nd harmonics, and 3β beats/sec between their 3rd harmonics.

4.4.4. For the *Für Elise* passage shown in Color Figure 6, assuming that the tones for the $D_5^\sharp$ and E_5 notes are of equal amplitudes, find the frequency for the middle tone created by their beating interference.

4.4.5. In the *Hegira* passage shown in Figure 4.14, there are two voices with dissonant intervals. Explain how this creates two subsidiary tones from beating interference. If it is assumed that the tones all have equal amplitude, then find the frequencies for these subsidiary tones. If the amplitudes are unequal for various tones, then how many different tones are occurring within this passage?

4.4.6. The score for the ending of the marimba composition, *Hegira,* is shown in Figure 4.16. In this score, find four examples of palindromes that involve more than six notes (and are not just constantly repeated notes). Also find a sequencing of diatonic scale shifts in the measures with $\frac{9}{8}$ time signature.

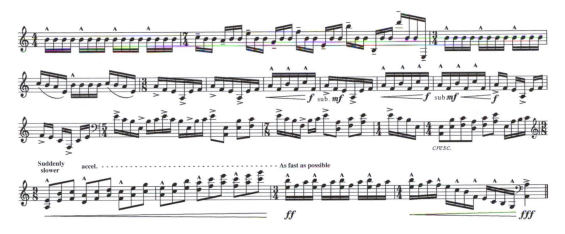

Figure 4.16. Score for ending of Carrie Magin's *Hegira.* © Carrie Magin, used by permission.

4.5   Estimating Amplitude and Frequency

The model of a musical waveform in (4.6) is a sum of harmonic oscillators of the form

$$y = \sum_{k=1}^{M} A_k \cos(2\pi \cdot k\nu t - \phi_k). \tag{4.10}$$

In this section we shall describe how to estimate the values of those amplitudes A_k, at frequency $k\nu$, which are the significant terms in this sum. Estimating the phase ϕ_k will not be discussed in any detail. In Remark 4.5.1, on page 113, we explain why we do not need to know the phase ϕ_k in order to do significant musical analysis. We will concentrate on how basic formulas are applied in this section. The justification of these formulas is somewhat complex, so we will take that up in the last section of the chapter. These formulas have been applied by a host of audio engineers for over 40 years. Their validity is well established. So we shall first show how effective they are, and then provide more mathematical justification for readers who wish to learn those details.

Our first step is to write the expressions for the harmonic oscillators in (4.10) in an alternative form that is easier to handle mathematically. By the Cosine Subtraction Identity (4.2d), we find that

$$A_k \cos(2\pi \cdot k\nu t - \phi_k) = A_k \cos(\phi_k) \cos(2\pi \cdot k\nu t) + A_k \sin(\phi_k) \sin(2\pi \cdot k\nu t).$$

Defining the constants α_k and β_k by

$$\alpha_k = A_k \cos(\phi_k), \quad \beta_k = A_k \sin(\phi_k) \tag{4.11}$$

we have the following equivalent form of (4.10):

$$y = \sum_{k=1}^{M} A_k \cos(2\pi \cdot k\nu t - \phi_k)$$

$$= \sum_{k=1}^{M} \left[\alpha_k \cos(2\pi \cdot k\nu t) + \beta_k \sin(2\pi \cdot k\nu t) \right]. \tag{4.12}$$

An important fact about the expression

$$\alpha_k \cos(2\pi \cdot k\nu t) + \beta_k \sin(2\pi \cdot k\nu t)$$

is that we can determine the amplitude A_k from it. Using (4.11) and the Pythagorean Identity (4.1), we have

$$\alpha_k^2 + \beta_k^2 = A_k^2 \cos^2(\phi_k) + A_k^2 \sin^2(\phi_k)$$

$$= A_k^2 \left(\cos^2(\phi_k) + \sin^2(\phi_k) \right)$$

$$= A_k^2.$$

Since we know that A_k is positive, it follows that

$$A_k = \sqrt{\alpha_k^2 + \beta_k^2}. \tag{4.13}$$

Our second step is to introduce functional notation $f(t)$ for the value y of the sum of harmonic oscillators. In other words, we write

$$f(t) = \sum_{k=1}^{M} A_k \cos(2\pi \cdot k\nu t - \phi_k)$$

$$= \sum_{k=1}^{M} \left[\alpha_k \cos(2\pi \cdot k\nu t) + \beta_k \sin(2\pi \cdot k\nu t) \right]. \tag{4.14}$$

This equation says that the value y for the sum of the oscillators depends on the time value t. At a given value of t, this value of y is written as $f(t)$. This functional expression for the sum of harmonic oscillators will play a crucial role in all of our subsequent work in this chapter.

4.5.1   How the Estimating Is Done

We now show how to estimate an amplitude A_k and frequency $k\nu$ for our harmonic oscillator model. Since we do not know in advance what frequencies $k\nu$ are present in a given musical waveform, we try to determine if a frequency is present in a set of test frequencies of the form $\{k\nu_o\}$, and what the amplitude for that frequency is. To do this, we use discrete values of the waveform, and multiply them by discrete values of harmonic oscillators created from our set of testing frequencies $\{k\nu_o\}$.

Let Ω stand for the time duration of the portion of the musical waveform that we want to analyze, so that we are working on a time-interval $[0, \Omega]$. To obtain discrete values from the waveform, we will use these N time values (where N is a large positive integer):

$$t_1 = 0, \; t_2 = \frac{\Omega}{N}, \; t_3 = \frac{2\Omega}{N}, \; t_4 = \frac{3\Omega}{N}, \ldots, \; t_N = \frac{(N-1)\Omega}{N}. \tag{4.15}$$

For the harmonic oscillators corresponding to our testing frequencies, we shall assume that Ω is the longest cycle duration. Consequently, $\nu_o = 1/\Omega$ is the smallest frequency. Therefore, our testing frequencies are

$$k\nu_o = \frac{k}{\Omega}, \qquad (k = 1, 2, 3, \ldots). \tag{4.16}$$

To test our range of frequencies we compute two quantities, $\mathcal{C}(k\nu_o)$ and $\mathcal{S}(k\nu_o)$, defined as follows:

$$\mathcal{C}(k\nu_o) = \frac{2}{N} \sum_{m=1}^{N} f(t_m) \cos(2\pi k\nu_o t_m)$$

$$\tag{4.17}$$

$$\mathcal{S}(k\nu_o) = \frac{2}{N} \sum_{m=1}^{N} f(t_m) \sin(2\pi k\nu_o t_m).$$

For each frequency $k\nu_o$, the value of $\mathcal{C}(k\nu_o)$ is an estimate of α_k, and $\mathcal{S}(k\nu_o)$ is an estimate of β_k. Because of Equations (4.13), we also compute the quantity $\mathcal{A}(k\nu_o)$, defined by

$$\mathcal{A}(k\nu_o) = \sqrt{[\mathcal{C}(k\nu_o)]^2 + [\mathcal{S}(k\nu_o)]^2}. \tag{4.18}$$

The quantity $\mathcal{A}(k\nu_o)$ provides an estimate of the amplitude A_k.

These three estimators are standard formulas in the mathematical field of Fourier analysis. They are rather complex formulas to explain in general terms. In this section, we will show how effective they are for some specific cases. More general discussion will be given in the last section of the chapter.

Example 4.5.1. Estimates for a sum of three harmonic oscillators. We examine how well our estimator in (4.18) performs for this sum of three harmonic oscillators:

$$f(t) = 40 \cos(2\pi \cdot 280t + 0.4) + 200 \cos(2\pi \cdot 360t)$$
$$+ 120 \cos(2\pi \cdot 440t - 0.2) \tag{4.19}$$

The three oscillators in this sum have frequencies of 280 Hz, 360 Hz, and 440 Hz. To compute the estimators in (4.17), we need to specify values for N and Ω. We choose these values in order to obtain a relatively simple set of test frequencies $\{k\nu_o\}$, close to the frequencies measured by AUDACITY when we use it to compute spectrograms. The spectrograms shown in this book were generally computed using AUDACITY with $N = 4096$ for recorded audio containing $44,100$ discrete values per second. In this case, 4096 values span a time-interval of length $\frac{4096}{44100} \approx 0.093$ seconds. To simplify our explanation, we chose a value of Ω that is near to 0.093 and for which $\nu_o = \frac{1}{\Omega}$ is a whole number. The value we chose was $\Omega = 0.125$, for which $\nu_o = 8$. Another reasonable choice would be $\Omega = 0.1$, which gives essentially the same results as we describe here.

We are computing $\mathcal{A}(k\nu_o)$ by using values of $f(t)$ over the time-interval $[0, 0.125]$, with $\nu_o = 1/0.125 = 8$. Our testing frequencies are multiples of 8 Hz, having the form $k\nu_o = 8k$. On the left of Figure 4.17, we have graphed the amplitude estimator $\mathcal{A}(8k)$. The graph shows three peak values at the frequencies 280 Hz, 360 Hz, and 440 Hz. It is also shown in the graph that these peaks have the values: $\mathcal{A}(280) = 40$, $\mathcal{A}(360) = 200$, and $\mathcal{A}(440) = 120$. For this example, we have determined the exact values of the frequencies and amplitudes for our sum of three harmonic oscillators.

The graph on the left of Figure 4.17 only shows amplitudes. In fact, the estimators $\mathcal{C}(k\nu_o)$ and $\mathcal{S}(k\nu_o)$ also provide the exact values of α_k and β_k. We will discuss that further in the last section of the chapter. For now, we shall just find amplitudes and frequencies for the waveform. These will be the essential quantities needed when we analyze music. The following remark explains why.

Remark 4.5.1. Why we ignore phase in musical analysis. To see why we do not need to find phases in musical analysis, we look at the problem of hearing the phase ϕ in a harmonic oscillator

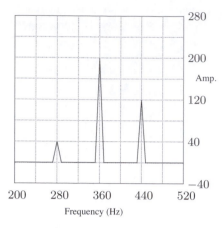 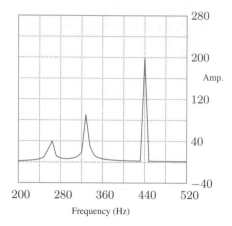

Figure 4.17. Left: Amplitude estimator for a sum of three harmonic oscillators, given in Equation (4.19). The peaks are located precisely at the three frequencies. Right: Amplitude estimator for a sum of harmonic oscillators containing the fundamentals for a first inversion of an A-minor chord, C_4-E_4-A_4, given in Equation (4.20). The peaks are located at 264 Hz, 328 Hz, and 440 Hz, which are good approximations of the frequencies used in the model (close enough to correctly identify the notes).

$A\cos(2\pi\nu t - \phi)$. Because humans cannot generally hear frequencies that are less than 20 Hz, we assume that $\nu \geq 20$ Hz. We have also required that $|\phi| \leq \pi$. We then have

$$A\cos(2\pi\nu t - \phi) = A\cos\left[2\pi\nu\left(t - \frac{\phi}{2\pi\nu}\right)\right]$$

with

$$\left|\frac{\phi}{2\pi\nu}\right| \leq \frac{1}{40}\,\text{sec.}$$

A shift of no more than $\pm 1/40^{\text{th}}$ of a second is extremely short in musical time. Moreover, if we are using frequencies greater than 100 Hz, as for the vast majority of musical notes, then $\phi/(2\pi\nu)$ is no more than $\pm 1/200^{\text{th}}$ of a second. That brief amount of time is around the *threshold of human time discrimination,* as measured in controlled laboratory studies. This explains why we will be able to ignore phase differences and yet still do significant musical analysis. It is typically sufficient to find amplitudes and frequencies when analyzing musical recordings.

Synthesizing music, however, is another matter. We will revisit the question of phase when we discuss musical synthesis in Chapter 8.

Example 4.5.2. Estimates for another sum of harmonic oscillators. The previous example showed how we were able to exactly determine amplitudes and frequencies. In that example, however, the frequencies of the harmonic oscillators were members of our set of test frequencies. Unfortunately, this circumstance does not always apply. As an example, here is a function for a waveform that contains the fundamentals of a first inversion A-minor chord (C_4-E_4-A_4):

$$\begin{aligned} f(t) = 50\cos(2\pi \cdot 261t + 0.4) + 100\cos(2\pi \cdot 330t) \\ + 200\cos(2\pi \cdot 440t - 0.2). \end{aligned} \tag{4.20}$$

We again compute the amplitude estimators $\mathcal{A}(8k)$ by using values of $f(t)$ over the time-interval $[0, 0.125]$, with $N = 4096$ points. In this case, the frequencies 330 and 261 are not multiples of 8 so they do not belong to the set of testing frequencies. On the right of Figure 4.17, we have graphed the amplitude estimator $\mathcal{A}(8k)$. The graph does not locate a frequency as precisely as in the previous example. However, there are three peaks located at the frequencies 264 Hz, 328 Hz, and 440 Hz, which are approximately the values of the frequencies used in the model (actually 440 Hz is exact).

These approximate frequencies are certainly accurate enough for correct note identification. The heights of those peaks, however, are not exactly equal to the amplitudes for the frequencies in (4.20). Table 4.1 shows the heights of these peaks, in comparison with the exact values of the amplitudes for the waveform in (4.20).

Table 4.1   ESTIMATION ACCURACIES FOR EXAMPLE 4.5.2

Frequency	Amplitude	Estimations	% Error
261	50	$\mathcal{A}(264) = 39.6$	20.8
330	100	$\mathcal{A}(328) = 89.9$	10.1
440	200	$\mathcal{A}(440) = 200$	0

Although our estimators did not perform perfectly in this example, it is important to keep in mind that for musical analysis we do not need perfect estimates. The estimates are close enough to identify fundamentals and other harmonics, as well as giving rough estimates of the amplitudes of the harmonics. We now turn to a more realistic example, illustrating how our estimators perform on a musical recording.

Example 4.5.3. Estimates for a musical recording. Our method of analysis will be to examine a musical waveform within small time intervals in order to get accurate estimates of changing frequencies and amplitudes. As a first illustration of this idea, we look at a localized analysis of a recording of four piano notes played in succession. The recording was done at a rate of $22,050$ digitally recorded values per second. For each of our estimators, we used $N = 4096$ points over a large number of time intervals. These time intervals were **not** taken as adjacent time intervals. Instead, they were *slid* along, by just slightly moving the next time interval forward in time from the previous one. In fact, there is about 95% overlap from one interval to the next. This is done in order to smoothly depict the evolution of the sound. Within each time interval, the amplitude estimators $\mathcal{A}(k\nu_o)$ are computed for each frequency $k\nu_o$, and the results are plotted along a *vertical* direction, with relative brightness being used to indicate the relative magnitude of the values of $\mathcal{A}(k\nu_o)$. The vertical plots for all these estimators $\mathcal{A}(k\nu_o)$ are positioned at the initial time-values for each of the small time intervals. The result is shown in Figure 4.18.

This figure provides a wealth of information, not always immediately evident to our ears, about this musical recording. Some of the information that it tells us includes:

1. It displays the harmonics for the four tones and the time intervals over which they are sounding. The display is accurate enough to estimate the frequencies for these harmonics as well.

2. The timings of the harmonics for the tones significantly overlap. For example, the harmonics for the E_4 tone can be seen to extend into the time interval for the F_4 tone. This overlapping of tones is difficult for our ears to detect with the recording used for this example, but the display of the amplitude estimators clearly shows it.

3. The third harmonic for the G_4 tone is much weaker than for the other tones, and its fundamental also appears relatively more intense (more loud). This explains the slightly unusual sound of this tone, compared to the other three tones.

Although this last example was fairly effective in analyzing the four piano tones, the graph in Figure 4.18 suffers from pronounced vertical smearing, and from "interference patterns" which resemble interference patterns from the bright light of the white bars. In more complex musical recordings, this smearing and interference would significantly impede our ability to analyze the harmonics and other features in the sound. In the next section, we will show how some simple processing will remove these smearing and interference effects, and yet retain all of the good features listed in items **1** through **3**.

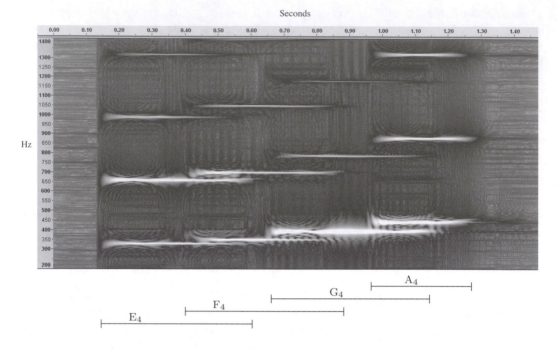

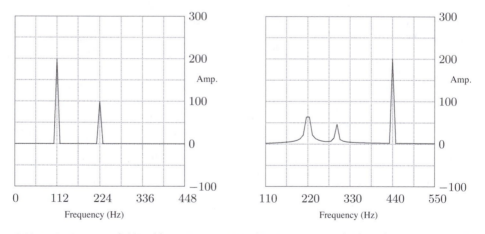

Figure 4.18. Multiple graphs of the amplitude estimators $\mathcal{A}(k\nu_o)$, over sliding time intervals, for four piano tones (E_4, F_4, G_4, A_4). The first 4 harmonics of E_4 are contained within the white bars above the line segment labeled E_4. Likewise, the first 3 harmonics of F_4, G_4, and A_4 are contained within the white bars above the line segments marked F_4, G_4, and A_4. In each case, the bars are centered vertically on the harmonics of these notes.

Figure 4.19. Left: Graph of $\{\mathcal{A}(k\nu_o)\}$ for Exercise 4.5.1. Right: Graph of $\{\mathcal{A}(k\nu_o)\}$ for Exercise 4.5.2.

Exercises

4.5.1. On the left of Figure 4.19 we show a graph of the amplitude estimators $\{\mathcal{A}(k\nu_o)\}$ for a waveform $f(t)$ defined over the time interval $[0, 0.125]$. **(a)** Estimate the significant frequencies, and their pitches, for this waveform. **(b)** Estimate the amplitudes for these frequencies.

4.5.2. On the right of Figure 4.19 we show a graph of the amplitude estimators $\{\mathcal{A}(k\nu_o)\}$ for a waveform $f(t)$ defined over the time interval $[0, 0.125]$. **(a)** Estimate the significant frequencies, and their pitches, for this waveform. **(b)** Estimate the amplitudes for these frequencies.

4.5.3. On the left of Figure 4.20 we show a graph of the amplitude estimators $\{\mathcal{A}(k\nu_o)\}$ for a waveform $f(t)$ defined over the time interval $[0, 0.125]$. **(a)** Estimate the significant frequencies, and their pitches, for this waveform. [Hint: For estimating pitches, use Table 1.1 for just tuning on p. 11.] **(b)** Estimate the amplitudes for these frequencies.

4.5.4. On the right of Figure 4.20 we show a graph of the amplitude estimators $\{\mathcal{A}(k\nu_o)\}$ for a waveform defined over the time interval $[0, 0.125]$. **(a)** Estimate the significant frequencies, and their pitches, for this waveform. **(b)** Estimate the amplitudes for these frequencies.

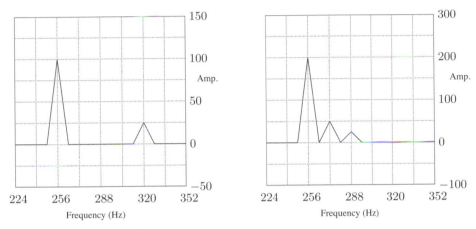

Figure 4.20. Left: Graph of $\{\mathcal{A}(k\nu_o)\}$ for Exercise 4.5.3. Right: Graph of $\{\mathcal{A}(k\nu_o)\}$ for Exercise 4.5.4.

4.5.5. On the left of Figure 4.21 we show a graph of the amplitude estimators $\{\mathcal{A}(k\nu_o)\}$ of a recording of a piano tone for a short time interval. Estimate the significant frequencies that are displayed in this graph. What note is being played?

4.5.6. On the right of Figure 4.21 we show a graph of the amplitude estimators $\{\mathcal{A}(k\nu_o)\}$ of a recording of a piano tone for a short time interval. Estimate the significant frequencies, and their pitches, that are displayed in this graph. What note is being played?

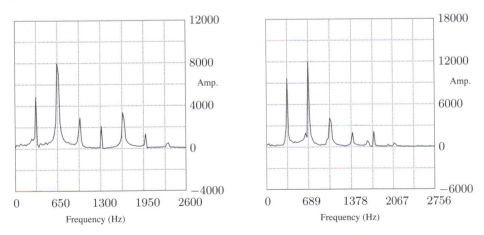

Figure 4.21. Left: Graph of $\{\mathcal{A}(k\nu_o)\}$ for Exercise 4.5.5. Right: Graph of $\{\mathcal{A}(k\nu_o)\}$ for Exercise 4.5.6.

4.5.7. The amplitude estimators graphed in Exercises 4.5.5 and 4.5.6 were obtained from the spectrogram in Figure 4.18. They are graphs of amplitude estimators for time intervals centered on time positions $t = 0.3$ and $t = 0.6$, respectively. Do the notes you found in those exercises match the notes indicated in Figure 4.18?

4.6   Windowing the Waveform: Spectrograms

In the previous section we saw how multiple computations of the amplitude estimators $\mathcal{A}(k\nu_o)$ over a sliding time interval produced an informative time-frequency description of a musical waveform (as shown in Figure 4.18). However, this time-frequency description suffers from artifacts which impede its analysis. In this section we explain where these artifacts come from, and the standard method for eliminating them. We will illustrate this elimination of artifacts through examples. A more general discussion will be given in the next section.

First, we look at what is causing the artifacts. When we only analyze a waveform over a short time interval, we are in effect cutting it off to zero values outside of the interval of analysis. As shown on the left side of Figure 4.22, this looks like we are viewing the signal through a *rectangular window*. This creates artificial discontinuities in the waveform, where it meets the left and right edges of the window at non-zero values, but then jumps to zero values on either side of the window. These artificial discontinuities in each of the sliding intervals (sliding rectangular windowing) is the source of the artifacts in Figure 4.18. By eliminating these artificial discontinuities, we will eliminate the artifacts.

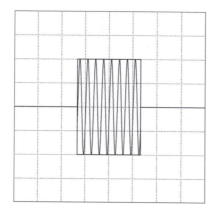 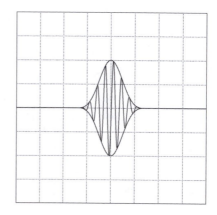

Figure 4.22. Left: Waveform over a short interval (rectangular windowing). Right: Blackman-windowed waveform over same short interval.

The graph on the right of Figure 4.22 shows the signal viewed through a window that decays gently to zero on each side. This gentle decay to zero does not create any discontinuities. Although this windowing distorts the waveform, we shall see that it does not distort it enough to cause difficulties for musical analysis, *and* it will eliminate the unpleasant artifacts in Figure 4.18.

The windowing of a waveform over a short time-interval $[0, \Omega]$ is accomplished by using this function

$$w(t) = \frac{0.42 + 0.5 \cos \left(\frac{2\pi(t - \Omega/2)}{\Omega} \right) + 0.08 \cos \left(\frac{4\pi(t - \Omega/2)}{\Omega} \right)}{0.42}. \tag{4.21}$$

A graph of $y = w(t)$ is shown in Figure 4.23. It is called the *Blackman window*.[7] Many other functions are used for windowing, but they all perform similarly. We shall just use the Blackman window. An example of Blackman windowing is shown on the right of Figure 4.22. At each time-value t, the waveform value $f(t)$ is multiplied by the window value $w(t)$. The windowed waveform is then written as $w(t)f(t)$.

---

[7]The numerator of $w(t)$ is also commonly referred to as the Blackman window.

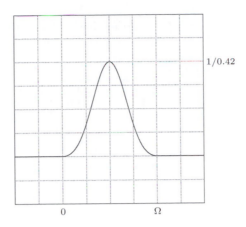

Figure 4.23. Graph of the Blackman window function $w(t)$.

When performing the computations for producing $\mathcal{C}(k\nu_o)$, $\mathcal{S}(k\nu_o)$, and $\mathcal{A}(k\nu_o)$ in Equations (4.17) and (4.18), we substitute values of $w(t)f(t)$ in place of the values of $f(t)$ used in those equations. This gives us the ***windowed estimators***, $\mathcal{C}_w(k\nu_o)$, $\mathcal{S}_w(k\nu_o)$, and $\mathcal{A}_w(k\nu_o)$, defined by

$$
\mathcal{C}_w(k\nu_o) = \frac{2}{N} \sum_{m=1}^{N} w(t_m)f(t_m)\cos(2\pi k\nu_o t_m)
$$

$$
\mathcal{S}_w(k\nu_o) = \frac{2}{N} \sum_{m=1}^{N} w(t_m)f(t_m)\sin(2\pi k\nu_o t_m) \tag{4.22}
$$

$$
\mathcal{A}_w(k\nu_o) = \sqrt{[\mathcal{C}_w(k\nu_o)]^2 + [\mathcal{S}_w(k\nu_o)]^2}.
$$

We now look at some examples that show we can accurately detect frequencies and amplitudes with these windowed estimators.

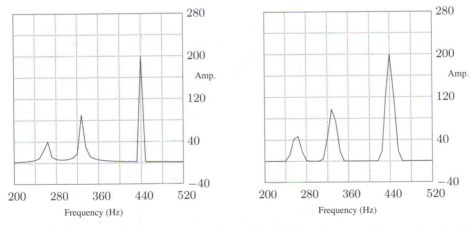

Figure 4.24. Amplitude estimators for a waveform containing fundamentals for a first inversion A-minor chord. Left: Rectangular windowing. Right: Blackman windowing. In each case, the peaks are located at 264 Hz, 330 Hz, and 440 Hz, which are sufficiently accurate to identify the notes.

Example 4.6.1. Estimates for a sum of harmonic oscillators. We examine how windowing affects the estimates for amplitudes and frequencies of the sum of harmonic oscillators considered in

Example 4.5.2. See the left of Figure 4.24. The waveform $f(t)$ used in that example was given in Equation (4.20) on p. 114. For comparison with that previous example, we will compute $\mathcal{A}_w(k\nu_o)$ by using values of $w(t)f(t)$ over the same time-interval $[0, \Omega] = [0, 0.125]$, and using the same number of points $N = 4096$. On the right of Figure 4.24, we have graphed the Blackman-windowed amplitude estimator $\mathcal{A}_w(8k)$. The graph locates the frequencies in the waveform just as precisely as in Example 4.5.2. There are again three peaks centered on the frequencies 264 Hz, 328 Hz, and 440 Hz, and those frequencies are accurate enough for correct note identification. In Table 4.2 we show that the estimates of amplitude are also fairly accurate. In fact, they are better than the estimates given in Table 4.1 for rectangular windowing.

Table 4.2   ESTIMATION ACCURACIES FOR EXAMPLE 4.6.1

Frequency	Amplitude	Estimations	% Error
261	50	$\mathcal{A}_w(264) = 46.6$	6.8
330	100	$\mathcal{A}_w(328) = 96.9$	3.1
440	200	$\mathcal{A}_w(440) = 200$	0

Just as importantly, the Blackman windowing eliminates the artifacts described for a sliding window analysis. In this example, the graph on the left in Figure 4.24 for the rectangularly windowed case, has "tails" extending away from the peaks. It is those tails that cause the vertical smearing artifacts when sliding rectangular windows are used to analyze waveforms. The graph on the right in this figure, for the Blackman windowed case, shows that those tails have been greatly reduced in extent. This eliminates the smearing artifact, as our next example illustrates.

Example 4.6.2. Estimates for a musical recording. For this example we compute sliding Blackman-windowed estimates for the musical recording considered in Example 4.5.3. The result is shown in Figure 4.25. We have previously referred to this type of time-frequency plot as a ***spectrogram***. The spectrogram in Figure 4.25 has the same three advantages described in Example 4.5.3:

1. It displays the harmonics for the four tones and the time intervals over which they are sounding. The display is accurate enough to estimate the frequencies for these

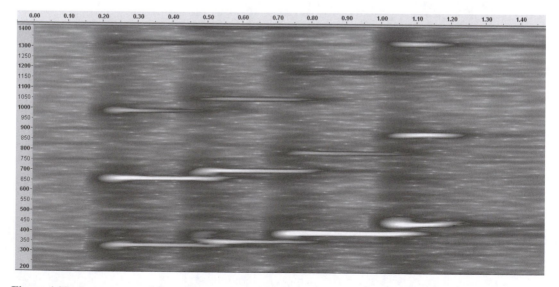

Figure 4.25. Spectrogram of four piano tones (E_4, F_4, G_4, A_4). Harmonics of tones are clearly located, both in time and frequency, and there are no smearing or interference effects as in Figure 4.18. Dark vertical bands at beginning of each tone are not interference effects. They are caused by percussive striking of piano strings.

harmonics.

2. The timings of the harmonics for the notes significantly overlap.

3. The third harmonic for the G_4 tone is much weaker than for the other tones, and its fundamental also appears relatively more intense (more loud).

and it has two additional advantages as well:

4. In contrast to the plot in Figure 4.18, the spectrogram in Figure 4.25 does not suffer from any smearing or interference artifacts.

5. Because there are no smearing artifacts, it is much easier to see the *percussive* onsets of the piano notes. These percussive onsets appear as dark vertical bands. They are about 0.1 seconds in duration, occurring at the beginning of each tone, and extending over the entire frequency range from 200 to 1400 Hz.

This example has illustrated that spectrograms are free of the artifacts found with rectangularly windowing, and thereby provide an enhanced portrait of the time-frequency content of musical sound. Therefore, we will use spectrograms exclusively from now on.

Remark 4.6.1. Percussive strikes in spectrograms. We have already mentioned several times that percussive strikes appear as vertical bands in spectrograms. To see why, we use an elementary model for a percussive strike. Suppose that the strike is described by the function

$$f(t_m) = \begin{cases} B, & \text{when } t_m = t_j \\ 0, & \text{when } t_m \neq t_j. \end{cases}$$

The constant B is assumed to be a relatively large positive number, and t_j is the single time value when the strike occurs. An actual percussive strike will last for a small amount of time, rather than a single instant of time, but this function $f(t_m)$ provides a first approximation to the more complicated function that describes an actual strike. If we compute a spectrogram for $f(t_m)$, then whenever t_j lies within the non-zero values of one of the windows, the estimator $\mathcal{C}_w(k\nu_o)$ will satisfy for each frequency $k\nu_o$:

$$\mathcal{C}_w(k\nu_o) = Bw(t_j)\cos(2\pi k\nu_o t_j)$$

because only the one term with $t_m = t_j$ will contribute to its sum. Likewise, the estimator $\mathcal{S}(k\nu_o)$ will satisfy for each frequency $k\nu_o$:

$$\mathcal{S}_w(k\nu_o) = Bw(t_j)\sin(2\pi k\nu_o t_j).$$

Therefore, the amplitude estimator $\mathcal{A}_w(k\nu_o)$ will satisfy for each frequency $k\nu_o$:

$$\begin{aligned} \mathcal{A}_w(k\nu_o) &= \sqrt{\left[Bw(t_j)\cos(2\pi k\nu_o t_j)\right]^2 + \left[Bw(t_j)\sin(2\pi k\nu_o t_j)\right]^2} \\ &= \sqrt{\left[Bw(t_j)\right]^2\left(\left[\cos(2\pi k\nu_o t_j)\right]^2 + \left[\sin(2\pi k\nu_o t_j)\right]^2\right)} \\ &= Bw(t_j)\sqrt{1}. \end{aligned}$$

This calculation shows that $\mathcal{A}_w(k\nu_o)$ equals a constant for **all** frequencies $k\nu_o$. Consequently, this amplitude estimator will appear as a vertical line segment within the spectrogram. Since we get various constant values, depending on the value of $w(t_j)$ for the particular windowing that contains the point t_j, we will get a collection of adjacent vertical line segments of varying brightness. This collection of vertical line segments is the vertical band that appears in the spectrogram for the percussive strike. It will have a width not greater than the time interval for the window w, since a windowing must overlap with t_j in order for the calculation above to hold. This provides some explanation for why vertical bands appear in spectrograms when percussive strikes occur.

We have now essentially completed our description of how spectrograms are produced. In the next section, we provide more general analysis of amplitude estimation, including mathematical proofs of some of the key ideas. This discussion, however, may be too technical for some readers. It could be safely skipped on a first reading. In the next chapter, we will discuss a wide variety of examples in music, and show how spectrograms provide an important tool for musical analysis.

Exercises

4.6.1. On the left of Figure 4.26 we show a graph of the Blackman-windowed amplitude estimators $\{\mathcal{A}_w(k\nu_o)\}$ for a waveform $f(t)$ defined over the time interval $[0, 0.125]$. **(a)** Estimate the significant frequencies, and their pitches, for the waveform $f(t)$. **(b)** Estimate the amplitudes for these frequencies.

4.6.2. On the right of Figure 4.26 we show a graph of the Blackman-windowed amplitude estimators $\{\mathcal{A}_w(k\nu_o)\}$ for a waveform $f(t)$ defined over the time interval $[0, 0.125]$. **(a)** Estimate the significant frequencies, and their pitches, for this waveform $f(t)$. **(b)** Estimate the amplitudes for these frequencies.

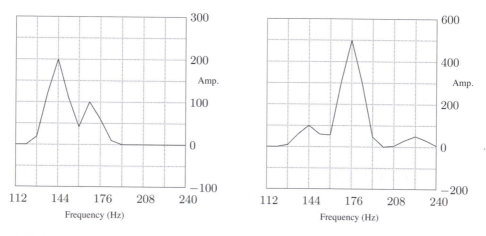

Figure 4.26. Left: Graph of $\{\mathcal{A}_w(k\nu_o)\}$ for Exercise 4.6.1. Right: Graph of $\{\mathcal{A}_w(k\nu_o)\}$ for Exercise 4.6.2.

4.6.3. On the left of Figure 4.27 we show a graph of the Blackman-windowed amplitude estimators $\{\mathcal{A}_w(k\nu_o)\}$ for a waveform $f(t)$ defined over the time interval $[0, 0.125]$. **(a)** Estimate the significant frequencies, and their pitches, for this waveform $f(t)$. **(b)** Estimate the amplitudes for these frequencies.

4.6.4. On the right of Figure 4.27 we show a graph of the Blackman-windowed amplitude estimators $\{\mathcal{A}_w(k\nu_o)\}$ for a waveform $f(t)$ defined over the time interval $[0, 0.125]$. **(a)** Estimate the significant frequencies, and their pitches, for this waveform $f(t)$. [Hint: For estimating pitches, use Table 1.1 for just tuning on p. 11.] **(b)** Estimate the amplitudes for these frequencies.

4.6.5. At the top of Figure 4.28 we show a spectrogram of four flute tones. On the bottom left of Figure 4.28 we show a graph of the amplitude estimators $\{\mathcal{A}_w(k\nu_o)\}$ from this spectrogram for the time $t = 0.35$ seconds. Estimate the significant frequencies that are displayed in the graph of the amplitude estimators. What note is being played at time $t = 0.35$ seconds?

4.6.6. At the top of Figure 4.28 we show a spectrogram of four flute tones. On the bottom right of Figure 4.28 we show a graph of the amplitude estimators $\{\mathcal{A}_w(k\nu_o)\}$ from this spectrogram for the time $t = 0.6$ seconds. Estimate the significant frequencies that are displayed in the graph of these amplitude estimators. What note is being played at time $t = 0.6$ seconds?

4.7   A Deeper Study of Amplitude Estimation

In this section we will discuss the theoretical rationale for the coefficient estimators $\mathcal{C}(k\nu_o)$ and

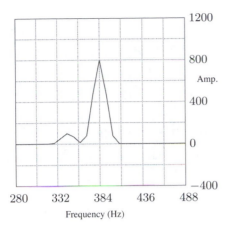
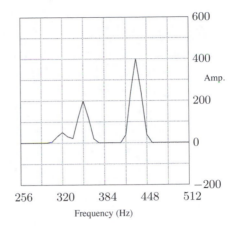

Figure 4.27. Left: Graph of $\{\mathcal{A}_w(k\nu_o)\}$ for Exercise 4.6.3. Right: Graph of $\{\mathcal{A}_w(k\nu_o)\}$ for Exercise 4.6.4.

$\mathcal{S}(k\nu_o)$ given in (4.17). We shall also provide more details as to what causes the smearing and interference artifacts for rectangular windowing, and why those artifacts are eliminated by Blackman windowing.

The theoretical rationale for the coefficient estimators is based on the following theorem.

Theorem 4.7.1. *Let $\nu_o = \frac{1}{\Omega}$. For $\ell\nu_o = \frac{\ell}{\Omega}$, with $\ell = 0, \pm 1, \pm 2, \ldots, \pm(N-1)$, and points $t_m = \frac{(m-1)\Omega}{N}$ defined in (4.15), we have*

$$\frac{1}{N} \sum_{m=1}^{N} \sin(2\pi\ell\nu_o t_m) = 0 \tag{4.23}$$

and

$$\frac{1}{N} \sum_{m=1}^{N} \cos(2\pi\ell\nu_o t_m) = \begin{cases} 0, & \text{when } \ell \neq 0 \\ 1, & \text{when } \ell = 0. \end{cases} \tag{4.24}$$

Sketch of Proof. The proof of this theorem is elementary. However, it is rather long and involves multiple applications of trigonometric identities. Therefore, in this section we will only sketch the idea behind why the theorem is true. A complete proof is given in Appendix C.

To see why (4.23) holds, we show on the left of Figure 4.29 a graph of $\sin(2\pi\ell\nu_o t_m)$ for $\ell = 3$. The symmetry of this graph implies that the values in adjacent lobes of this sine graph will cancel each other out, due to opposite signs. Therefore the sum in (4.23) must be 0. Similarly, for each $\ell \neq 0$, the sum in (4.24) must be 0. (See the graph on the right of Figure 4.29 for the case of $\ell = 3$.) Finally, when $\ell = 0$, we have

$$\frac{1}{N} \sum_{m=1}^{N} \cos(2\pi\ell\nu_o t_m) = \frac{1}{N} \sum_{m=1}^{N} \cos(0)$$

$$= \frac{1}{N} \sum_{m=1}^{N} 1.$$

The last sum consists of adding N terms that are all 1. Therefore, it equals N, and so we have $\frac{1}{N} N = 1$. That proves (4.24) for the case of $\ell = 0$, and completes our sketch of the proof. $\square$

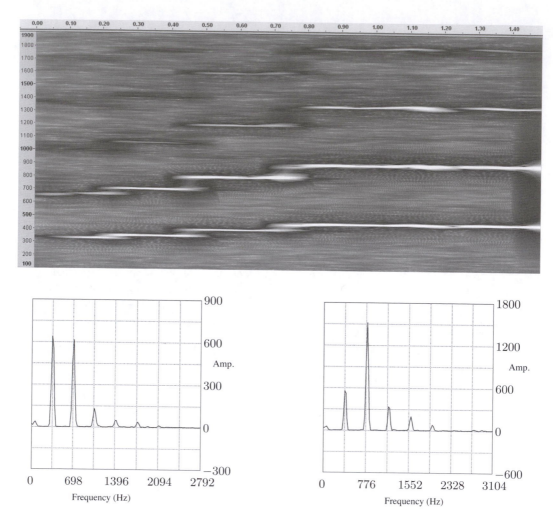

Figure 4.28. Top: Spectrogram of four flute tones. Bottom Left: Amplitude estimators $\{\mathcal{A}_w(k\nu_o)\}$ from the spectrogram at time position $t = 0.35$ seconds. Bottom Right: Amplitude estimators $\{\mathcal{A}_w(k\nu_o)\}$ from the spectrogram at time position $t = 0.6$ seconds.

Using Theorem 4.7.1, we can show why we obtain exact estimations of frequency and amplitude for those waveforms whose frequencies all belong to our set of testing frequencies. As we found in Example 4.5.1, for instance. We begin with the simplest case, a single harmonic oscillator:

$$f(t) = \alpha_j \cos(2\pi j \nu_0 t). \tag{4.25}$$

where $0 < j < N/2$. For $k = 1, 2, 3, \ldots, N/2 - 1$, we find that:

$$\mathcal{C}(k\nu_o) = \frac{2}{N} \sum_{m=1}^{N} f(t_m) \cos(2\pi k \nu_o t_m)$$

$$= \alpha_j \frac{2}{N} \sum_{m=1}^{N} \cos(2\pi j \nu_0 t) \cos(2\pi k \nu_o t_m)$$

$$= \alpha_j \frac{1}{N} \sum_{m=1}^{N} \left\{ \cos\big(2\pi[j - k]\nu_0 t_m\big) + \cos\big(2\pi[j + k]\nu_0 t_m\big) \right\}.$$

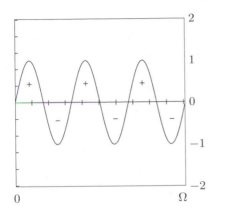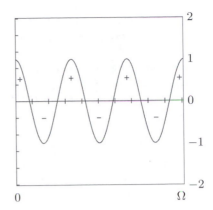

Figure 4.29. Graphs illustrating Theorem 4.7.1. Left: Graph of $\sin(2\pi \cdot 3\nu_o t_m)$ over $[0, \Omega]$. Because of the symmetry of the graph, values from adjacent lobes will cancel out, due to opposite signs. Right: Graph of $\cos(2\pi \cdot 3\nu_o t_m)$ over $[0, \Omega]$. A similar cancellation occurs, but the first and last half-lobes must be combined together into one lobe.

We then have

$$\mathcal{C}(k\nu_o) = \alpha_j \frac{1}{N} \sum_{m=1}^{N} \cos\big(2\pi[j-k]\nu_0 t_m\big) + \alpha_j \frac{1}{N} \sum_{m=1}^{N} \cos\big(2\pi[j+k]\nu_0 t_m\big). \tag{4.26}$$

Using Theorem 4.7.1, alternately for $\ell = j - k$ and $\ell = j + k$, we find that both of the sums in (4.26) are 0 when $k \neq j$. However, when $k = j$, the first sum equals $\alpha_j \cdot 1 = \alpha_j$. Thus, we have found that

$$\mathcal{C}(k\nu_o) = \begin{cases} \alpha_j, & \text{when } k = j \\ 0, & \text{when } k \neq j. \end{cases} \tag{4.27}$$

A similar calculation shows that

$$\mathcal{S}(k\nu_o) = 0, \quad \text{for each } k. \tag{4.28}$$

We have shown that the estimators in (4.17) give us exact values for this single harmonic oscillator. By similar reasoning they will also give us exact values for this harmonic oscillator:

$$f(t) = \beta_j \sin(2\pi j \nu_0 t). \tag{4.29}$$

In fact, they give these results:

$$\mathcal{S}(k\nu_o) = \begin{cases} \beta_j, & \text{when } k = j \\ 0, & \text{when } k \neq j \end{cases} \tag{4.30}$$

and

$$\mathcal{C}(k\nu_o) = 0, \quad \text{for each } k \tag{4.31}$$

which are exact.

We showed previously that every harmonic oscillator, $f(t) = A_j \cos(2\pi j \nu_0 t - \phi_j)$ can be expressed in the form

$$f(t) = \alpha_j \cos(2\pi j \nu_o t) + \beta_j \cos(2\pi j \nu_0 t)$$

with $\alpha_j = A_j \cos(\phi_j)$ and $\beta_j = A_j \sin(\phi_j)$. Therefore, when we apply our estimators to this harmonic oscillator, they will give us the exact values of $\alpha_j = A_j \cos(\phi_j)$ and $\beta_j = A_j \sin(\phi_j)$, and the exact value of the amplitude $A_j = \sqrt{\alpha_j^2 + \beta_j^2}$.

It is not difficult to extend this argument to all waveforms that are finite sums of harmonic oscillators, all having frequencies less than $N/2$. The details, however, are somewhat lengthy. Moreover, this is not the most important case, because it is rare that all of the frequencies for such waveforms will belong to the set of testing frequencies. For these reasons, we have put this extension to a finite sum of harmonic oscillators in Appendix C. The main idea is that, because our estimators involve multiplication and summation, we can use the distributive law to get exact estimates for each term (each harmonic oscillator) in such waveforms. For those readers who are interested, the rigorous details are in Appendix C.

Theorem 4.7.1 is the key ingredient for proving that our estimators give exact values for amplitudes. It requires, however, that the oscillators in the waveform being analyzed all have their frequencies belonging to the set of test frequencies used by our estimators. In Examples 4.5.2 and 4.5.3 we saw how our estimators performed reasonably well, although with some prominent artifacts for Example 4.5.3. We now provide more discussion of why our estimators performed in this way.

4.7.1   More on Rectangular Windowing

In Theorem 4.7.1, we proved identities for the two sums:

$$\frac{1}{N} \sum_{m=1}^{N} \sin(2\pi \ell \nu_o t_m) \quad \text{and} \quad \frac{1}{N} \sum_{m=1}^{N} \cos(2\pi \ell \nu_o t_m).$$

To examine what happens when a waveform contains harmonic oscillator terms having frequencies that do not belong to the set of testing frequencies, we examine analogues of these two sums of sines and cosines. That is, we examine the following two functions of all frequencies x:

$$\mathsf{S}(x) = \frac{1}{N} \sum_{m=1}^{N} \sin(2\pi x t_m) \quad \text{and} \quad \mathsf{C}(x) = \frac{1}{N} \sum_{m=1}^{N} \cos(2\pi x t_m). \tag{4.32}$$

These sums will occur in the analysis of harmonic oscillators when our sine and cosine estimators are applied. For instance, suppose we are analyzing a single harmonic oscillator of the form:

$$f(t) = \alpha \cos(2\pi \nu t).$$

Then (4.26) is modified as follows

$$\mathcal{C}(k\nu_o) = \alpha \frac{1}{N} \sum_{m=1}^{N} \cos\big(2\pi(\nu - k\nu_o)t_m\big) + \alpha \sum_{m=1}^{N} \frac{1}{N} \cos\big(2\pi(\nu + k\nu_o)t_m\big). \tag{4.33}$$

Using the definition of $\mathsf{C}(x)$, we can rewrite this formula as:

$$\mathcal{C}(k\nu_o) = \alpha \,\mathsf{C}(\nu - k\nu_o) + \alpha \,\mathsf{C}(\nu + k\nu_o). \tag{4.34}$$

Similarly, the sine estimator in this case satisfies

$$\mathcal{S}(k\nu_o) = \alpha \,\mathsf{S}(\nu - k\nu_o) + \alpha \,\mathsf{S}(\nu + k\nu_o). \tag{4.35}$$

In lieu of a complete analysis of this case, which is quite complex, we will content ourselves with showing how these functions S and C provide us with a clear explanation for the results we found in Examples 4.5.2 and 4.5.3. In those examples, we shall see that it is justified to make the following approximations:

$$\mathcal{C}(k\nu_o) \approx \alpha \,\mathsf{C}(\nu - k\nu_o) \quad \text{and} \quad \mathcal{S}(k\nu_o) \approx \alpha \,\mathsf{S}(\nu - k\nu_o) \tag{4.36}$$

which are based on the assumption that the frequencies $\nu - k\nu_o$ provide the dominant terms in Equations (4.34) and (4.35). In Examples 4.5.2 and 4.5.3, we computed the amplitude estimator

$$\mathcal{A}(k\nu_o) = \sqrt{\big[\mathcal{C}(k\nu_o)\big]^2 + \big[\mathcal{S}(k\nu_o)\big]^2}.$$

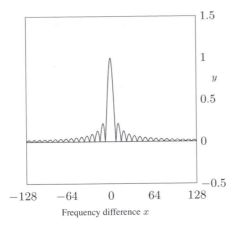
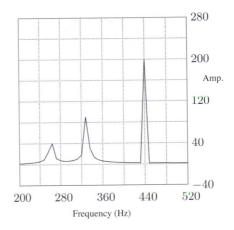

Figure 4.30. Left: Amplitude multiplier $y = A(x)$ for rectangular windowing. Right: Amplitude estimator found in Example 4.5.2.

Since the harmonic oscillators involved in those examples had frequencies that were not in the set of testing frequencies, we replace this amplitude estimator by the following one, for all frequencies $x = \nu - k\nu_o$:

$$A(x) = \sqrt{[C(x)]^2 + [S(x)]^2}. \tag{4.37}$$

To see how this function $A(x)$ applies to our examples, we show the graph of $y = A(x)$ on the left of Figure 4.30. On the right of Figure 4.30 we show the amplitude estimator $\mathcal{A}(k\nu_o)$ that we computed for Example 4.5.2. The relationship between these two graphs is the following. For a given value of $k\nu_o$, the amplitude estimator $\mathcal{A}(k\nu_o)$ will have a value that is closely approximated by $A_k A(\nu - k\nu_o)$, where ν is the closest frequency in the waveform to $k\nu_o$. For example, in Table 4.3 we show data for the rectangular windowing case in Example 4.5.2. The data shows that our analysis gives a good explanation for the estimation results in that example.

Table 4.3 AMPLITUDE MULTIPLIER COMPARISONS

			Rectangular Windowing, Example 4.5.2		
ν	$k\nu_o$	A_k	$A(\nu - k\nu_o)$	$A_k A(\nu - k\nu_o)$	$\mathcal{A}(k\nu_o)$
261	264	50	0.78	39.6	39.6
330	328	100	0.9	90	89.9
440	440	200	1	200	200
			Blackman Windowing, Example 4.6.1		
ν	$k\nu_o$	A_k	$A_w(\nu - k\nu_o)$	$A_k A_w(\nu - k\nu_o)$	$\mathcal{A}_w(k\nu_o)$
261	264	50	0.932	46.6	46.6
330	328	100	0.969	96.9	96.9
440	440	200	1	200	200

Examining the form of the function $A(x)$, graphed on the left of Figure 4.30, we can see that the general falloff in size of its y-values accounts for the tails connecting the spikes located at the peaks in the graph of $\mathcal{A}(k\nu_o)$. These tails are the source of the vertical smearing artifacts we observe in sliding rectangularly windowed computations of estimators, like the one shown in Figure 4.18. The interference effects in Figure 4.18 are a result of values of $x = \nu - k\nu_o$ that are near to the zeroes of $A(x)$, alternating with values of $x = \nu - k\nu_o$ that are near to where the peaks of $A(x)$ are located. This gives rise to an alternation of dark and light regions for the estimators $\mathcal{A}(k\nu_o)$ plotted in Figure 4.18.

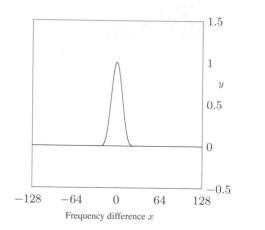 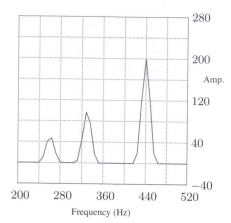

Figure 4.31. Left: Amplitude multiplier $y = \mathsf{A}_w(x)$ for Blackman windowing. Right: Amplitude estimator found in Example 4.6.1.

4.7.2 More on Blackman Windowing

The analysis of Blackman-windowed amplitude estimation is quite similar to the rectangular case. The only difference is that for Blackman windowing we replace the functions S and C by the windowed functions S_w and C_w:

$$\mathsf{S}_w(x) = \frac{1}{N}\sum_{m=1}^{N} w(t_m)\sin(2\pi x t_m) \quad \text{and} \quad \mathsf{C}_w(x) = \frac{1}{N}\sum_{m=1}^{N} w(t_m)\cos(2\pi x t_m). \qquad (4.38)$$

in Formulas (4.33) through (4.37). Doing this replacement produces the amplitude multiplier for Blackman windowing, $\mathsf{A}_w(x)$. On the left of Figure 4.31 we have graphed $\mathsf{A}_w(x)$. In Table 4.3 we show how this amplitude multiplier gives a precise explanation for the estimation results from Blackman windowing in Example 4.6.1. The tails for the Blackman windowing amplitude multiplier $\mathsf{A}_w(x)$ are much narrower in extent, compared to the tails for the rectangular windowing amplitude multiplier $\mathsf{A}(x)$. These narrower tails account for the lack of smearing for the Blackman windowed estimates on the right of Figure 4.31, and especially for the spectrogram in Figure 4.25. Furthermore, the absence of the alternation of zeroes with peaks in the Blackman windowing amplitude multiplier explains the lack of interference effects in the spectrogram.

Exercises

4.7.1. Show that (4.30) is valid.

4.7.2. Show that (4.28) is valid.

4.7.3. Show that (4.31) is valid.

4.7.4. Explain why the approximations in (4.36) are more accurate for Blackman windowing than rectangular windowing. Also explain how this greater accuracy for Blackman windowing relates to the results shown in Table 4.3.

4.7.5. For the waveform

$$f(t) = 100\cos(2\pi \cdot 260t) + 50\cos(2\pi \cdot 326t - 0.1)$$

we show the Blackman windowed amplitude estimator $\mathcal{A}_w(k\nu_o)$ on the right of Figure 4.32. For computing this estimator, we used $\Omega = 0.125$ and $N = 4096$. Use the information shown in the table on the left of Figure 4.32

to complete the following table:

ν	$k\nu_o$	A_k	$\mathsf{A}_w(\nu - k\nu_o)$	$A_k\mathsf{A}_w(\nu - k\nu_o)$	$\mathcal{A}_w(k\nu_o)$
260	256				88
260	264				88
326	320				37.5
326	328				48.4

Also explain why there are two equal heights for the peaks at 256 Hz and 264 Hz.

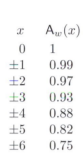

x	$\mathsf{A}_w(x)$
0	1
± 1	0.99
± 2	0.97
± 3	0.93
± 4	0.88
± 5	0.82
± 6	0.75

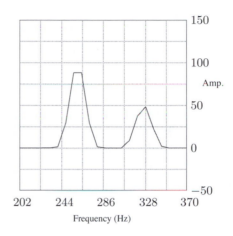

Figure 4.32. Left: Table of values for amplitude multiplier $\mathsf{A}_w(x)$ for Blackman windowing. Right: Amplitude estimator for waveform in Exercise 4.7.5.

Chapter 5

Spectrograms and Music

We live in a time I think not of mainstream, but of many streams, or even, if you insist upon a river of time, that we have come to a delta, maybe even beyond delta to an ocean which is going back to the skies.

—John Cage

In this chapter we apply spectrograms to musical recordings in order to better understand the music that we hear. Spectrograms are a tool for aiding our listening comprehension. They certainly cannot substitute for our hearing, which is primary. However, like musical scores, spectrograms allow us to better interpret the music as we listen to it. Some important aspects of music that are displayed well in spectrograms are (1) timbre, (2) vibrato, (3) tonal harmonics, (4) dynamics (loudness), and (5) short-term pattern and structure. The first four of these aspects—timbre, vibrato, tonal harmonics, dynamics—relate to the performance of the music, and how we perceive it as we listen to it. The fifth aspect, short-term pattern and structure, is an aspect of music that is typically analyzed using musical scores. Spectrograms provide an additional tool for perceiving pattern and structure, at least over a time span of a few minutes, since they reveal the full range of harmonics of notes, not just the fundamental pitches described by scores. They also are valuable in analyzing improvisational music, or music from other parts of the world, where scores are simply not available.

We will use spectrograms to analyze several components of musical performance. First, we look at singing, as it is perhaps the oldest and most universal form of music. Spectrograms have long been used in the field of linguistics for analyzing vocal production, and the techniques developed in that field can be applied to singing as well. Second, we look at characteristics of musical instruments, which typically amplify and diversify the musical qualities of the human voice. Third, we show how spectrograms help us to analyze the performance of musical compositions, revealing aspects of pattern and structure.

5.1 Singing

Our understanding of singing is greatly enhanced by spectrogram analysis. With spectrograms we can observe vibrato, harmonic emphasis, and texture of vocal sound. Our examples are chosen from different kinds of performance styles: operatic, blues, and gospel. These examples all confirm what we know intuitively, that the human voice is a remarkably flexible and expressive musical instrument. Perhaps it was our first musical instrument, and so it is fitting to look at it first.

5.1.1 An Operatic Performance by Renée Fleming

Our first example is an excerpt from a recording of a live performance of the aria, *O Mio Babbino Caro* from Puccini's opera, *Gianni Schicchi*. The soprano, Renée Fleming, is singing the aria. A spectrogram of a brief passage from this performance is shown in Figure 5.1. There are several points to make about this spectrogram and how it relates to the sound of the passage.

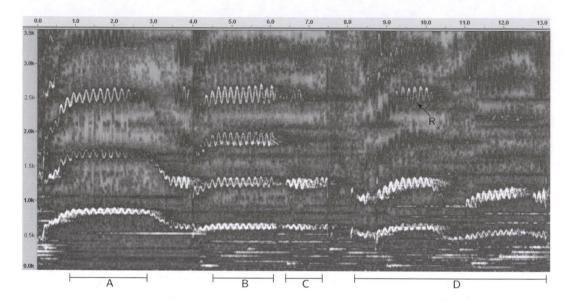

Figure 5.1. Spectrogram from recording of *O Mio Babbino Caro* with Renée Fleming. Labels explained in text.

First, Fleming is maintaining her vibrato throughout. Throughout the passage, she maintains a frequency spread in her vibrato of about 100 Hz for her fundamentals. She even maintains her vibrato when she is bending the pitch to move from one note to the next. These gliding pitch transitions are called *portamento*. A good example of Fleming's vibratoed portamento is shown in Figure 5.1 during the time interval 3.0 to 3.5 seconds. The gliding, vibratoed fundamental lies within the frequency band from 500 Hz to 700 Hz. Fleming also changes the rate of her vibrato. For example, the vibrato rate of 5 oscillations/sec for the tone in time interval A changes to a rate of 6 oscillations/sec for the lower pitched tone in time interval B. This change of rate can be heard in the recording. A change in vibrato rate is used to enhance expressiveness or artistry in singing.[1] In this case, the increase in vibrato rate provides a slight emphasis on the move to the lower pitched tone. *Second*, Fleming's tones in the passage fall into two categories. The first category consists of tones within the time intervals A and B. For these tones, the 3rd and 4th harmonics have a fairly high intensity. This creates a *ringing timbre*, which is audible at a higher pitch than the main pitch determined by the fundamental. This pronounced ringing quality is highly prized in operatic style. Its quintessential expression occurs in what is called *chiaroscuro* timbre. We will discuss *chiaroscuro* timbre at the end of this section. The second category of tones lie within the time intervals C and D. In contrast to the first category, these tones have the most intensity within just their first two harmonics, with slightly more intensity on the 2nd harmonic. Consequently, there is less ringing for these notes. An exception is the brief ringing corresponding to the heightened intensity harmonic pointed to by arrow R in the spectrogram. *Third*, in between the time intervals C and D, there is a brief moment where Fleming breathes in an enormous quantity of air, prior to the next set of vocals in interval D. This inhalation is marked by a dark vertical band on the spectrogram. Fleming uses it to good effect as part of her character's sobbing within the aria. *Fourth*, in the time interval D, Fleming maintains her vibrato while altering its frequency spread and curving the pitches of her tones. In this same time interval, she maintains an unbroken gently varying tone, a *legato*. The spectrogram provides a clear visual record of the quality of her legato. *Fifth*, the horizontal bands are the string background for Fleming's singing. Several places on the spectrogram show that Fleming's harmonics are centered on harmonics of the string notes, which is part of the total harmony of the sound.

---

[1]Richard Miller, *On the Art of Singing*, Oxford (1996), p. 304.

Formants

The differing qualities of Fleming's tones are examples of the importance of *formants* in singing. Formants are frequency bands that are differentially amplified by the human vocal system. The regions surrounding the vocal chords—throat passage, mouth cavity, nasal passages, sinus cavities, chest cavity, bone structures—work together to form a human *graphic equalizer* that differentially amplifies specific frequency bands. This differential amplification changes the timbre of the sound emerging from the vocal chords.

In Figure 5.2 we show schematic diagrams of amplification functions for formants. There are two classes of formants. The first class creates amplification of frequencies centered on 500 Hz, 1500 Hz, 2500 Hz, and 3500 Hz. These are formants found most often in female voices. The second class of formants, found typically in male voices, creates amplification of frequencies centered on 250 Hz, 1500 Hz, 2500 Hz, and 3500 Hz. There is also a fifth formant, centered on a higher frequency, which we did not show in Figure 5.2.

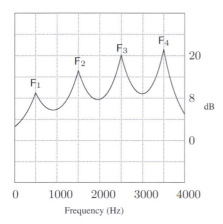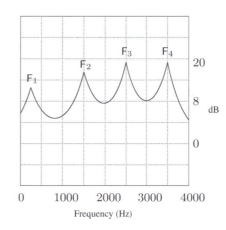

Figure 5.2. Schematic graphs of formant amplification functions. Size of amplification measured using standard sound volume units of *decibels* (dB). First formant denoted by F_1, higher formants denoted by F_2, F_3, F_4. Left: Formants at 500 Hz, 1500 Hz, 2500 Hz, and 3500 Hz, for typical female voice. Right: Formants at 250 Hz, 1500 Hz, 2500 Hz, and 3500 Hz, for typical (adult) male voice.

In Figure 5.3 we show schematically how formant amplification functions are applied to the harmonics for the sound coming from a person's vocal chords. The graph on the top left of the figure shows a sequence of equal-amplitude harmonics for a tone with fundamental ν_1. When each harmonic's amplitude is multiplied by the corresponding amplification value of the formant function shown on the middle left of the figure, then that alters the amplitude for each harmonic. To be precise, if A_k is the amplitude for a vocal chord harmonic at frequency $k\nu_1$, and $F(k\nu_1)$ is the value of the formant amplifier at that same frequency $k\nu_1$, then the value of the amplitude $\widetilde{A}_k$ of the voice harmonic at $k\nu_1$ is given by

$$\widetilde{A}_k = F(k\nu_1) \cdot A_k. \tag{5.1}$$

Equation 5.1 holds for each harmonic $k\nu_1$, $k = 1, 2, 3, \ldots$. We show these voice harmonics in the graph at the top right of Figure 5.3. The heights of the harmonics now follow the profile for the formant, because the initial vocal chord harmonics were closely spaced and of equal height. See Exercise 5.1.7. As another example, we show a set of vocal chord harmonics with a larger fundamental ν_2 on the bottom left of the figure. Since these harmonics are more widely spaced the resulting voice harmonics are not so clearly related to the formant amplifier graph.

The diagrams in Figures 5.2 and 5.3 are only general schematic diagrams. The relative heights of the formant peaks F_1 through F_4 will vary depending upon the vowel being sung. For example,

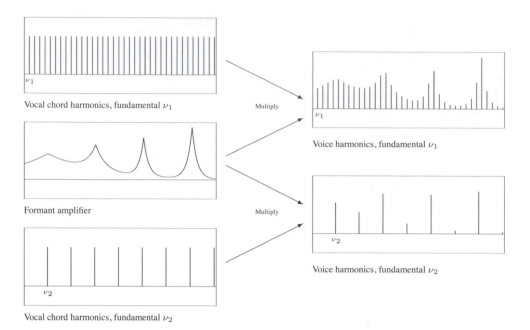

Figure 5.3. Schematic diagram of formant amplification. Formant function graphed on middle left is used to multiply amplitudes from vocal chord harmonics on top left, producing voice harmonics on top right. Same formant is used to multiply amplitudes from vocal chord harmonics at bottom left, producing vocal harmonics at bottom right.

in the passage sung by Fleming, each of the different timbres referred to above have different ratios of relative heights of formant peaks, which causes us to hear them as different vowels. Moreover, the harmonics for vocal chord sounds will not generally have constant amplitude. We have a rather complex situation: the amplitudes of the initial vocal sounds can vary in magnitude for each pitch, and the formant amplifier profile can vary for different vowel sounds. But it is precisely this complexity that allows for the wide range of complex tones achievable with the human voice. It is beyond the scope of this book to describe formants in detail. A good reference is Scott McCoy's book.[2]

Chiaroscuro timbre

We previously mentioned *chiaroscuro* timbre. This is a timbre to the singing voice that is highly prized in operatic style. The word originates in painting, where it refers to balancing of bright and dark regions. In opera it refers metaphorically to balancing bright and dark sound. Bright sound comes from higher frequency harmonics, and dark sound from lower frequency harmonics. The higher frequency harmonics are the ones emphasized by what is known as the *singer's formant*. McCoy provides a good description of the singer's formant (S. McCoy, ibid., p. 47):

> the singer's formant...is created by clustering the third, fourth, and fifth vocal tract formants tightly together within a narrow frequency range. This new combined formant—a sort of "super" formant— gives extra amplification to harmonics in the frequency range of approximately $2,400$ Hz–$3,200$ Hz.

The *chiaroscuro* timbre consists of balancing the intensities of the harmonics within the singer's formant region and the first formant region, with typically less emphasis on the second formant region. The spectrogram in Figure 5.1 shows the *chiaroscuro* timbre to Fleming's voice during the time interval A, and also for a brief moment during the time that the harmonics marked by R are being sung. Another singer, famous for his *chiaroscuro* timbre, was Luciano Pavarotti.

---

[2]S. McCoy, *Your Voice: An Inside View. Multimedia Voice Science and Pedagogy.* Inside View Press, 2004.

5.1.2 An Operatic Performance by Luciano Pavarotti

Our second example of singing is a passage from the aria, "Nessun Dorma," in Puccini's opera *Turandot*. It is taken from the last 21 seconds of a live recording featuring the tenor, Luciano Pavarotti. (See Figure 5.4.) The spectrogram in the figure reveals many of the same qualities that we saw in our previous analysis of Renée Fleming's vocals. For example, a formant structure of differing emphases on various harmonics is clearly visible. Pavarotti changes the formants for his two vocalizations of *vincerò*, and this is clearly audible as a change of timbre. The spectrogram also verifies that Pavarotti's

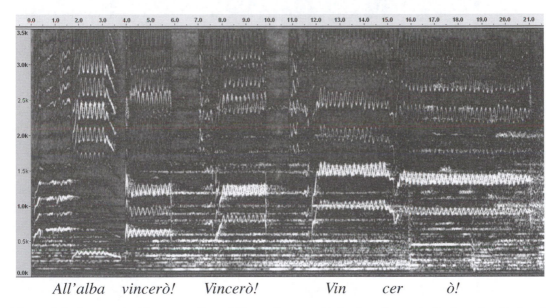

Figure 5.4. Spectrogram from recording of *Nessun Dorma* with Luciano Pavarotti. Lyrics printed below it in their approximate time positions. Pavarotti's vibrato has large frequency spread. Changes in formants for different vocalizations of the lyrics are clearly displayed, including *chiaroscuro* timbre throughout.

singing has *chiaroscuro* timbre. The super-formant amplification of higher harmonics, along with an amplification of lower harmonics and less emphasis on mid-frequency harmonics, are all clearly evident in the spectrogram. They are especially noticeable in the lyric sung between 2.0 and 3.0 seconds. An interesting contrast of Pavarotti's singing with Fleming's is how Pavarotti has extremely rapid stops to some of his tones. The spectrogram reveals these as narrow vertical bands at about 5.8 seconds, 9.8 seconds, and 21.0 seconds. These vocal stops are so rapid that they are registered as percussive sound in the spectrogram. The aria concludes with Pavarotti singing a five-second-long, constant vibratoed tone. It is remarkable how long Pavarotti sustains this vibrato with such a large frequency spread.

5.1.3 A Blues Performance by Alicia Keys

Our third example involves singing from a different genre, the blues. In Figure 5.5 we show a spectrogram of Alicia Keys singing along with her own piano playing in the opening of her song, *If I Ain't Got You*. The harmonics of her vocal tones can be seen in the curved bands. These curved bands form an interesting contrast with the straight, horizontal bands that are the harmonics of the piano tones. Unlike the operatic styles of the previous two examples, we see here that Keys is not consistently producing vibrato. Rather her vibrato is appearing in short bursts, as she bends her pitches around the harmonics from the piano tones. This shows that her singing is in a typical blues style. As we have analyzed previously in Example 3.3.5, the piano notes are following a classical pattern. Keys's singing provides an interesting contrast: a blues style of singing, contrasted with her classical style of piano playing. Also, unlike the opera singers discussed above, Keys does not sing with *chiaroscuro*

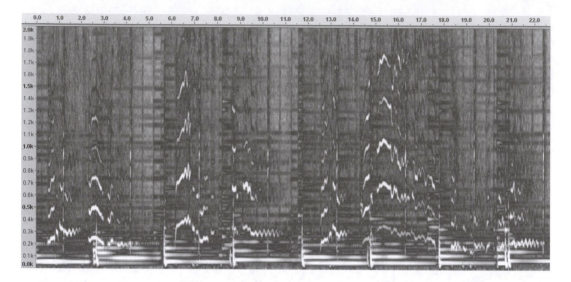

Figure 5.5. Spectrogram of Alicia Keys singing in blues style, with classical style piano accompaniment.

timbre. Her highest intensity harmonics are at lower frequencies at or near the fundamentals. This provides a "husky" timbre to her voice, which is part of the ambience of the song. In this song, an operatic *chiaroscuro* timbre would probably not be artistically satisfying.

5.1.4   A Choral Performance by *Sweet Honey in the Rock*

For our final example of singing, we turn to a recording of choral voices. In Figure 5.6 we show a spectrogram of the beginning of a recording of the song, *Sylvie,* by the all female chorus, *Sweet Honey in the Rock*. In the spectrogram we have indicated four time intervals, marked A through D, corresponding to important passages. Here is an analysis of each of these passages:

A: In this passage, the baritone lead singer introduces the song. Her vocals are in the blues style of bending the pitches in her tones. She also has some vibrato in her tones. Her 2nd harmonics are the most intense harmonics throughout the first half of this passage. They also stand out prominently in the entire spectrogram, reflecting her status as the lead singer. In the second half of this passage, her 2nd harmonics diminish somewhat in intensity, and more vibrato is introduced, which softens the sound texture at the end of the passage.

B: This passage is divided into five sub-passages, reflecting a series of choral harmonies of increasing complexity. In the first sub-passage, the baritone lead singer reprises her introductory passage A, although with a slight variation at the end, where the switch of the most intense harmonic from the 2nd to the 3rd provides a sense of anticipation for the next sub-passage. The second sub-passage marks the introduction of the bass singer of the chorus. Her fundamentals are so low in pitch that she is actually in the male vocal range. In the third sub-passage the baritone singer joins with the bass singer, to begin the choral harmony of the piece. The fourth sub-passage brings in additional members of the chorus. They are now singing the chords of the song, with each individual voice contributing its distinctive timbre. The lead singer's 2nd harmonic is the most prominent harmonic at the beginning of this sub-passage. She then drops out briefly at the end of the sub-passage. The fifth sub-passage is a more extended version of the fourth sub-passage. It begins with a chordal harmony featuring the lead singer and ending with

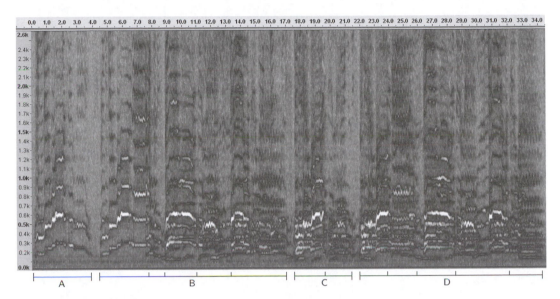

Figure 5.6. Spectrogram from recording of *Sylvie* by the choral group, *Sweet Honey in the Rock*. Passage A: baritone lead singer introduces song. Passage B: series of harmonies of increasing complexity, marked off as five sub-passages. Passage C: swelling, full texture of entire chorus in its first half, followed by thinner texture in second half. Passage D: complex sequence of shifting harmonies, marked as five sub-passages.

the lead singer fading to the background.

C: This passage divides into two halves. In the first half, there is a swelling, full texture of sound resulting from the entire chorus singing. The spectrogram corresponds to this full texture sound by displaying harmonics that essentially fill the entire frequency band from 100 Hz to 600 Hz. Again, the 2nd harmonic of the lead singer is the most intense harmonic in this first half of the passage. In the second half of the passage, the lead singer's second harmonic is less prominent (except for a brief moment in the very middle of this second half). The frequency band from 100 Hz to 600 Hz is less filled than the first half, and consequently the sound has a thinner texture.

D: This concluding passage has five sub-passages. These sub-passages form a complex sequence of shifting harmonies. The first sub-passage provides another swelling full-textured sound from the whole chorus, with the lead singer's 2nd harmonic being the most intense. As we have seen before, the lead singer then fades slightly, with a transition to her 3rd harmonic as the most intense harmonic. In the second sub-passage, the rest of the chorus is singing with a greater frequency spread of their vibrato, centered on fixed pitches, which produces a gentle contrast to the first sub-passage. The third, fourth, and fifth sub-passages are essentially reprises of the third, fourth, and fifth sub-passages of passage B. The fifth sub-passage finishes the introduction of the song with the full chorus singing a vibratoed chord. All of the tones in this chord have equal intensity, with a very brief pitch variation at the start that emphasizes the first phoneme "wh" of the final lyric "while."

This choral example illustrates how different voices, which because of their different timbres are akin to distinct instruments, can be effectively combined and contrasted.

5.1.5 Summary

In this section we have illustrated the value of spectrograms for analyzing vocal technique and artistry. The many advantages of spectrograms are summarized by the vocalist and singing teacher, Richard

Miller, as follows:[3]

1. Through visual and audio feedback, the singer's awareness of the quality of sound can be heightened.

2. Spectrograms provide audio and visual feedback, for a particular singing voice, in the following ways:

   (a) adaption of formants for particular vocal timbres. Including, within operatic style, the *chiaroscuro* timbre,

   (b) accuracy of vocal onset and release,

   (c) accuracy of fundamental frequency (pitch) targeting, and consequent accuracy of overtones,

   (d) vowel tracking or targeting (vowel definition) and vowel modification,

   (e) vibrato rate and its variation for artistic and expressive purposes,

   (f) presence or absence of legato,

   (g) effect of voiced and unvoiced consonants on the vocal line,

   (h) artistic maneuvers, including vocal coloration, dynamic variation, and *portamento*.

Exercises

5.1.1. In Figure 5.7 we show a spectrogram of Renée Fleming's vocals from another passage of *O Mio Babbino Caro*. Analyze this spectrogram in relation to her singing.

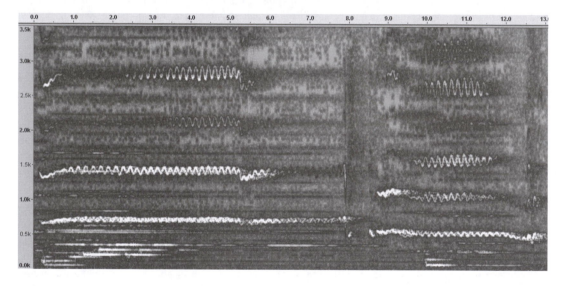

Figure 5.7. Spectrogram from a recording of *O Mio Babbino Caro* with Renée Fleming.

5.1.2. In Figure 5.8 we show a spectrogram of a brief portion of vocals of Sarah Brightman, from a recording of the composition by Lucio Quarantotto and Francesco Sartoni, *Time to Say Goodbye*. During most of the passage she is singing a single note with vibrato. Just above her fundamental for this note there is a less intense, smudged, vibratoed tone (centered on a frequency just below 600 Hz). Based on this spectrogram, answer the following four questions:

   (a) What is producing this other tone?

   (b) What is the note for this other tone?

   (c) Brightman's vibratoed fundamental is centered on a frequency just below 500 Hz. Identify the note she is singing.

---

[3]Adapted from R. Miller, *On the Art of Singing*, Oxford (1996), pp. 304–305.

(d) Right underneath Brightman's vibratoed fundamental there is an intermittent straight tone harmonic with frequency just below 400 Hz. Identify the note for this harmonic. Considering all three of the notes identified here, what musical chord do they form?

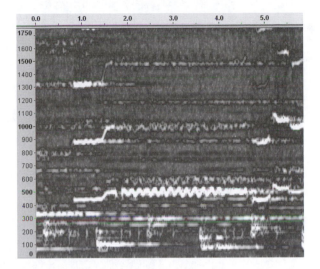

Figure 5.8. Spectrogram of vocals of Brightman for Exercise 5.1.2.

5.1.3. Read the following quote from Renée Fleming, and then answer the questions that follow it:

> We use the word *passaggio,* which is Italian for "passage," to describe usually two transition points in the voice. A singer must make sure the passage is a smooth and seamless one. Within a range of anywhere from one and a half to three octaves, a classical singer, unlike a pop singer, needs to have a sound that is homogeneous throughout, without any breaks the audience can hear... for me [the passaggio] consists of the tones between E-flat and F-sharp at the top of the staff... It's in the dangerous straits of the passaggio that many singers come to grief.[4]

(a) Can you find any evidence of Fleming negotiating the "dangerous straits" of her *passaggio* in the spectrogram in Figure 5.1?

(b) What might be the underlying source of *passaggio* in the voice system, based on terms that were discussed in this section?

5.1.4. In Figure 5.9 we show a spectrogram of the conclusion of the song, *Time to Say Goodbye*. There are two singers, the tenor Andrea Bocelli and the soprano Sarah Brightman, performing in this passage. Based on this spectrogram, answer these questions:

(a) Identify the parts of the spectrogram that correspond to their fundamentals.

(b) Identify parts of the spectrogram that correspond to the orchestral background, and any relevant features for the sound of those instruments that is revealed by the spectrogram.

5.1.5. In Figure 5.10 we show a spectrogram from a recording of the *Beatles'* song, *Dear Prudence*. Identify the frequency range in this spectrogram that contains the vocal harmonics. Are the fundamentals shown? How would you describe the vocal effect that is occurring here, and how is it reflected in the spectrogram?

5.1.6. A spectrogram of a clip from the song, *Bohemian Rhapsody,* is shown in Figure 5.11. The arrows point to fundamentals for the singers' vocals, with frequencies indicated in the caption for the figure. Using Table 1.2 on page 13, identify the notes that are being sung. Describe what is being done musically.

---

[4]R. Fleming, *The Inner Voice: The Making of a Singer,* Viking (2004), pp. 44–45.

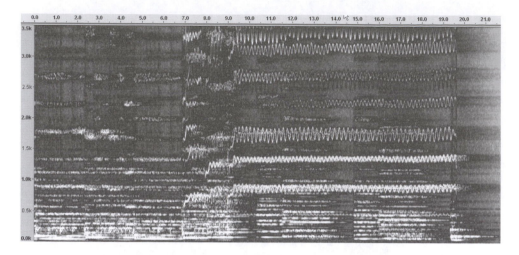

Figure 5.9. Spectrogram of vocals of Brightman and Bocelli at the conclusion of *Time to Say Goodbye*.

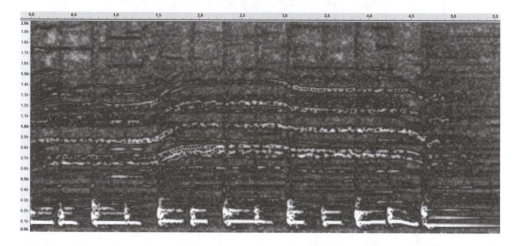

Figure 5.10. Spectrogram from a recording of the *Beatles'* song, *Dear Prudence*.

5.1.7. Assume that the amplitude A_k for each vocal chord harmonic is equal to a positive constant c. Use Equation (5.1) to explain why the graph of the voice harmonics in Figure 5.3, for the fundamental ν_1, follows the profile of the formant amplifier function.

5.1.8. Here are three interesting facts: (**1**) In the example from *Sylvie,* one singer's tones were in the range of a male bass singer, and yet our ears can distinguish that it is a female singer. (**2**) Young male speakers are often confused on the telephone with young females. (**3**) For formants of an adult male voice, the position of F_1 is significantly lower than for a female's voice. Discuss how these three facts are related and explained by the theory of formants described in this section.

5.2   Instrumentals

The lessons we drew about the human voice in the preceding section apply just as well to manufactured instruments. Many of the same features of spectrograms—such as displaying pitch and harmonics, or displaying vibrato and legato—apply to instruments such as trumpets, clarinets, guitars, and violins, for these instruments magnify and extend the human voice.

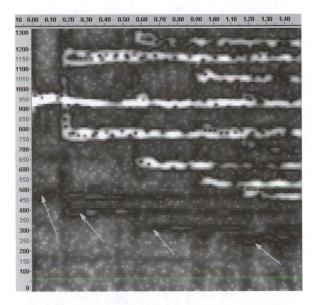

Figure 5.11. Spectrogram for a clip from *Bohemian Rhapsody*. Arrows point to fundamentals for the singers' vocals. From left to right they point to fundamentals centered on these frequencies: 466, 392, 311, and 233 Hz.

5.2.1 Jazz Trumpet: Louis Armstrong

Our first example is from the trumpet master, Louis Armstrong. In Figure 5.12, we show a spectrogram of Armstrong's trumpet playing from his recording of the song, *La Vie En Rose*. The spectrogram shows that the fundamentals of Armstrong's trumpet tones are in the pitch range of 350 Hz to 500 Hz, with relatively high amplitude overtones extending all through the displayed frequencies (harmonics 2 through 6 are all equally bright). Just as with many vocal tones, Armstrong's trumpet tones have their most intense harmonics in overtones rather than fundamentals.

Armstrong's trumpet tones tend to fall into three categories. The first category consists of tones that begin at a constant pitch, with little or no vibrato, and then vibrato is increased significantly as

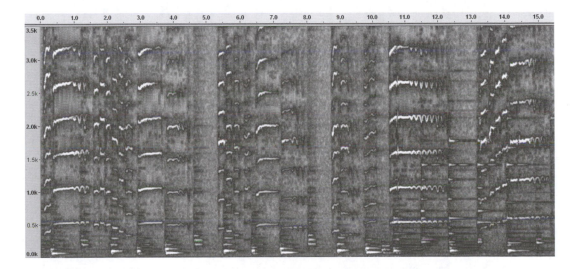

Figure 5.12. Spectrogram from a Louis Armstrong recording of *La Vie En Rose*.

the note is played. These tones we shall refer to as ***rising vibrato tones***. Examples of rising vibrato tones appear between 0.4 seconds to 1.4 seconds, between 10.5 seconds to 12.3 seconds, and between 14.2 seconds to 15.2 seconds. The steady increase of vibrato produces a progressive softening of the sound of these rising vibrato tones. The second category of tones consists of tones that remain at a constant pitch without any vibrato. These are called ***straight tones***. One of these straight tones occurs between 3.0 seconds to 3.5 seconds. Compared to the rising vibrato tones, these straight tones have a somewhat harsher sonority. The third category consists of ***constant vibratoed tones***. Armstrong ends the recording with a constant vibratoed tone that he holds for several seconds. (See Figure 5.13.) These tones have a constant frequency spread, oscillating around a fixed pitch, similar to the tone that Pavarotti ends his *Nessun Dorma* aria with in Figure 5.4, although Armstrong's trumpet tones have much less frequency spread than Pavarotti's sung tones.

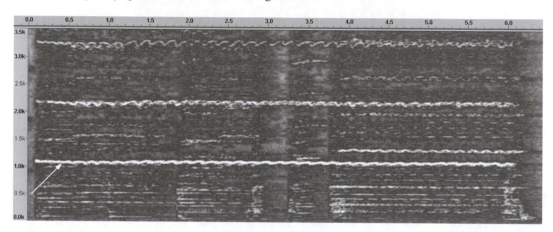

Figure 5.13. Spectrogram of ending of Louis Armstrong recording of *La Vie En Rose*. Arrow points to constant vibratoed tone held by Armstrong for entire 6.0 seconds of passage.

By varying the presence of vibrato in his tones, Armstrong is able to vary their emphasis in a way that differs from their loudness. The jazz composer, performer, and theorist, André Hodeir, has described very well the musical use that Armstrong makes of vibrato:

> ...the vibrato he uses on certain sounds, [gives] them an expressive density that makes each completely different than the others. (A. Hodeir, *Jazz: Its Evolution and Essence,* Da Capo Press, 1975, p. 159.)

Hodeir goes on to emphasize an important point about the rhythmic importance of vibrato:

> The esthetic importance of vibrato is evident; its rhythmic value seems less well known. Nevertheless, it appears certain that vibrato can frequently reinforce the feeling of the tempo. One of our best drummers used to claim that a single note, played by a good musician, was enough to get him started. This would be possible only if the note had a movement of its own; and how could it have one except by virtue of its vibrato? (Ibid., p. 226.)

In regards to Armstrong again, Hodeir is pointing out how he can vary the rhythm of a melody by varying the vibrato within the tones as he plays them. The example of *La Vie En Rose* is a good one. When Armstrong uses his three different ways of playing notes—rising vibrato, straight tone, constant vibrato—he is able to vary the basic melody of the song *without changing the pitches of the notes*. In this way, he inflects the melody with his own emotion.

As we have seen before with jazz singing, Armstrong also bends the pitches of his notes. Each of these effects, pitch bending and vibrato, provide a sharp contrast with the simple strumming of rhythm guitar tones in the passage. These guitar tones appear as horizontal line segments at lower pitches than the trumpet tones. For the guitar harmonics, the fundamentals tend to be the most intense, another contrast with the trumpet tones.

In sum, Armstrong has produced a passage of trumpet playing that builds on characteristics of singing. That his trumpet playing is an amplification of his voice has been noted by several commentators. The great conductor, composer, and music educator, Leonard Bernstein wrote the following:

> ...when Louis Armstrong plays his trumpet, he is only doing another version of his own voice. Listen to an Armstrong record, like "I Can't Give You Anything but Love," and compare the trumpet solo with the vocal solo. You can't miss the fact that they are by the same fellow. (L. Bernstein, *The Joy of Music*, Amadeus Press, 2004, p. 118.)

For another example illustrating Bernstein's point, see Exercise 5.2.2.

5.2.2 Beethoven, Goodman, and Hendrix

We now compare and contrast three legendary giants from the worlds of classical, jazz, and rock music: Ludwig van Beethoven, Benny Goodman, and Jimi Hendrix. The instruments used in our examples, piano in a sonata by Beethoven, clarinet in a jazz piece by Goodman, and electric guitar in a rock song by Hendrix, are all different. Spectrograms provide visual representations of the differences between these instruments, and they will help in analyzing performance technique.

First, we begin with a short passage from a Beethoven composition. In Figure 5.14 we show a spectrogram of a passage from Beethoven's *Piano Sonata in E* (Opus 109, Movement 1, Measures 16 and 17). This spectrogram shows the typical horizontal line segments for the harmonics of a piano.

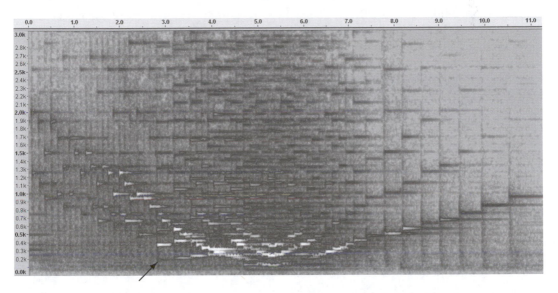

Figure 5.14. Spectrogram from recording of Beethoven's *Piano Sonata in E* (Opus 109). Arrow points to an ascending bass scale, entering in contrast to descending treble scale.

In the spectrogram, we see a descending series of tones followed by an ascending series, reflecting an approximate mirror symmetry. The symmetry is broken, however, by an ascending set of lower pitch tones (indicated by the arrow on the spectrogram). The mirror symmetric pattern is also broken by gently rising higher pitch tones trailing off to the right.

It is interesting to compare the spectrogram with the score for the passage. (See Figure 5.15.) In the score, we can see the symmetry in the set of notes that flow from the upper staff into the lower staff and then back into the upper staff. If we exclude the very first note in the upper staff, there is a palindrome symmetry for this set of notes. We can also see the bass notes that counteract this symmetry. They are the four notes immediately after the bass clef in the lower staff. The trailing off in the second half of the passage occurs because Beethoven indicates that its notes are to be played

Figure 5.15. Passage from score of Beethoven's *Piano Sonata in E* (Opus 109) corresponding to spectrogram in Figure 5.14.

ritardando. In other words, with a *reducing of tempo*. This slowing of tempo breaks the palindrome symmetry of the rhythm of the notes, and this symmetry breakage is displayed by the spectrogram. Individual performers will vary in how much they slow the tempo. The spectrogram provides a quantitative record of exactly how much the performer has reduced the tempo of the notes.

Another aspect of the spectrogram is that it provides a quantitative record of the varying dynamics, the different loudness levels, for the notes. As can be seen in the spectrogram in Figure 5.14, the notes are not all played with the same loudness. The score only provides some indications of such dynamics, their exact production is part of performance interpretation and will vary from performer to performer. The spectrogram provides us with a quantitative display of these sound dynamics. A display which, along with our ears, can guide us in evaluating a performer's interpretation.

The spectrogram of a Benny Goodman recording is shown in Figure 5.16. It features an improvised clarinet solo by Goodman. Like the Beethoven example, this spectrogram shows an approximate mirror symmetry from 3.5 seconds to 8.5 seconds, which is also broken by a gently rising scale trailing off to the right. In this case, Goodman is bending his pitches as is commonly done in jazz style. This pitch bending is clearly displayed in the spectrogram by the connected curved structures, *which possess a symmetry of their own*. This is a significant contrast to the discretely separated, constant harmonic, piano tones in the Beethoven passage. The spectrogram provides visual evidence of how closely matched these symmetric curved structures are. Goodman is reproducing them with almost identical form, especially the four symmetric structures between 10 seconds and 12 seconds, that are gliding gently upwards in pitch. From these exact reproductions of curved pitch structures, we can see that Goodman is sequencing transpositions (see Exercise 5.2.9). Unlike the transposition sequencing

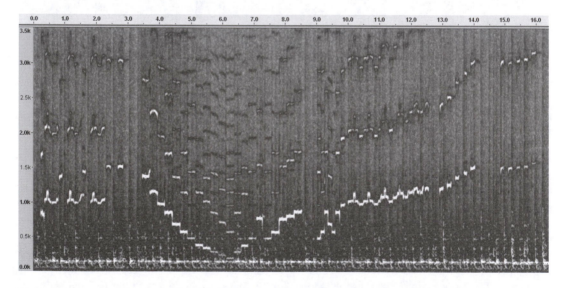

Figure 5.16. Spectrogram from recording of *Sing, Sing, Sing* by the Benny Goodman Orchestra.

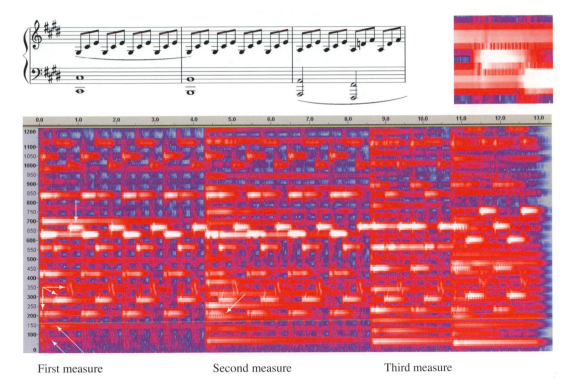

First measure   Second measure   Third measure

Color Figure 1. Top: Score of first three measures of Beethoven's *Moonlight Sonata*. Bottom: Spectrogram of a recording of these first three measures. Bottom two arrows point to fundamentals for $C_2^\sharp$ and $C_3^\sharp$. Triple arrow points to fundamentals for $G_3^\sharp$, $C_4^\sharp$ and E_4. Top arrow, pointing straight down, indicates beating between 2nd harmonic of E_4 and 5th harmonic of $C_3^\sharp$. Magnified view shown at top right. Arrow in second measure also points to beating between harmonics. (In color spectrograms: yellow-white is loudest, red is medium loud, purple is faint, blue is barely audible or inaudible. Within those color ranges, brighter is always louder.)

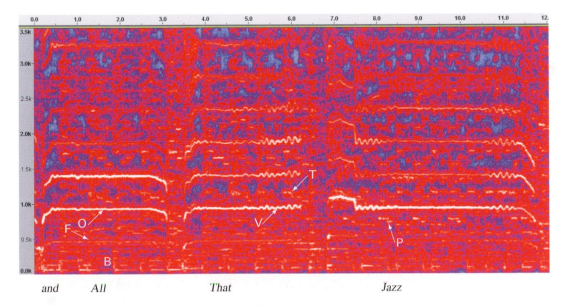

and   All   That   Jazz

Color Figure 2. Spectrogram from *And All That Jazz*. Arrow O points to 2nd harmonic of singer's note. Arrow F points to its fundamental, which is much fainter and corresponds to a $B_4^\flat$ note. Letter B marks some bass notes. Arrow V points to vibrato in singer's voice. Arrow T points to trumpet notes. Arrow P points to piano notes.

A PDF file for these color figures, ColorFigures.pdf, is available at the book's web site.

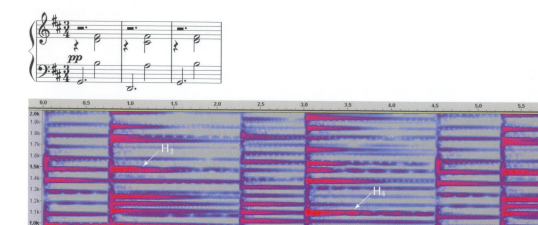

Color Figure 3. Top: Score of first three measures of Satie's *Gymnopedie I*. Bottom: Spectrogram of first three measures of *Gymnopedie I*. Fundamentals indicated with arrows labeled by the corresponding notes, G_2, B_3, $F_4^\sharp$. Arrow H_1 points to a near match of fundamental of D_4 with 3rd harmonic of G_2. Arrow H_2 points to 4th harmonic of G_2, not quite matching with fundamental of $F_4^\sharp$ lying just below it. Arrow H_3 points to approximate matching of overtones of G_2, B_3, D_4, and $F_4^\sharp$. Arrow H_4 points to another near match. Percussive striking of notes appear as vertical bands; one lies directly above letter P at time 4.5 seconds.

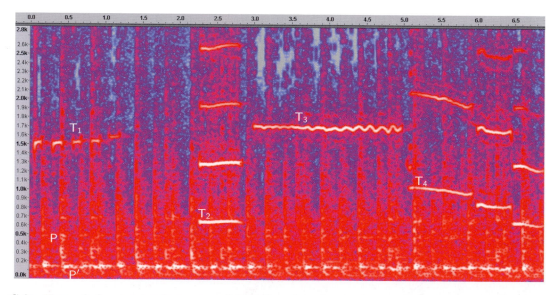

Color Figure 4. Spectrogram from a recording of *Sing, Sing, Sing* by the Benny Goodman Orchestra. P lies between two drum strikes. P' lies below a bass harmonic from a drum strike. T_1 lies near five brief clarinet notes, with pitch bending on the first two. T_2 lies above a mostly straight clarinet fundamental. T_3 lies above a clarinet fundamental with increasing vibrato. T_4 lies above a clarinet fundamental with descending pitch.

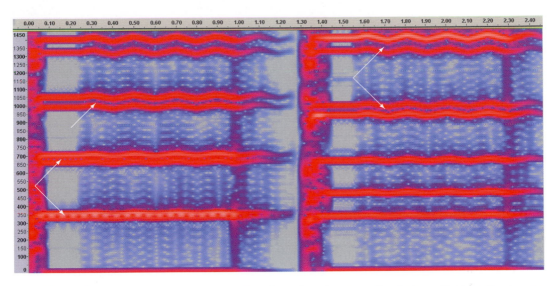

Color Figure 5. Dissonant harmonic intervals. Top: Score for minor second and tritone. Bottom: Spectrogram of these notes played on a violin. Arrows point to beating interference between harmonics.

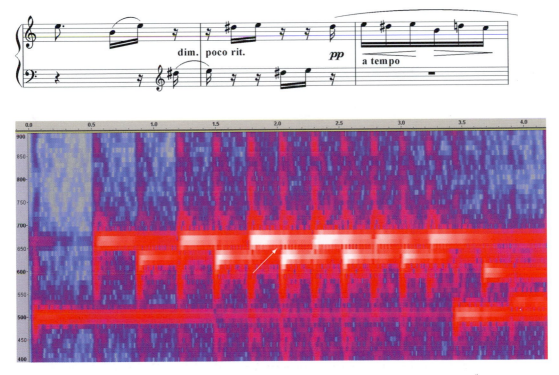

Color Figure 6. Passage from Beethoven's *Für Elise, WoO 59*. Top: Score shows repeated $D_5^\sharp$ and E_5 notes. Bottom: Arrow points to beating between fundamentals for these notes.

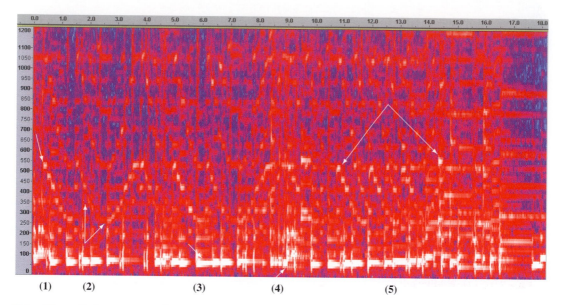

Color Figure 7. Spectrogram of portion of *Strasbourg/St. Denis*. Arrows above **(1)** through **(5)**: **(1)** descending saxophone tones; **(2)** descending trumpet tones followed by ascending saxophone tones; **(3)** fundamental of bass note; **(4)** brief burst of drumming; **(5)** between two arrows, a call and response between saxophone and trumpet.

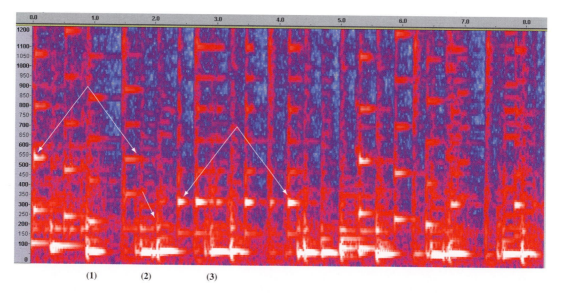

Color Figure 8. Spectrogram of portion of *Strasbourg/St. Denis*. Arrows above **(1)** through **(3)**: **(1)** between two arrows, piano notes played on keyboard; **(2)** plucked bass notes; **(3)** between two arrowheads, piano notes played by plucking the piano strings.

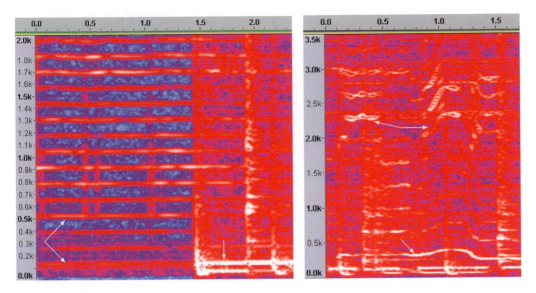

Color Figure 9. Spectrograms from the Beatles' song, *Tomorrow Never Knows*. Left: Introductory passage. Double arrow points to 1st and 4th harmonics for a drone of note C_3. Single arrow points to harmonic from drum strike matching with the drone's fundamental. Right: Clip showing vocal track. Lower arrow points to fundamental tone for vocals. Double arrow points to 7th and 6th harmonics, showing beating and fading.

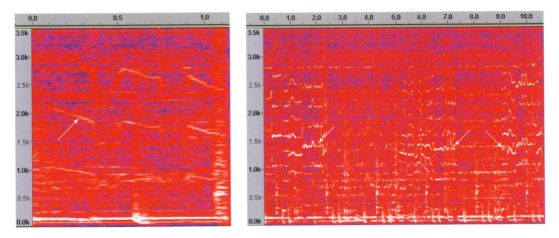

Color Figure 10. Spectrograms from the Beatles' song, *Tomorrow Never Knows*. Left: Three rapid tones resembling bird calls. The arrow points to 2nd harmonic of first of these tones. When the recording is slowed to half speed, these tones sound like someone laughing: "Ha, Ha, Ha." Right: Time-reversed guitar solo. Three arrows point to harmonics from these time-reversed tones.

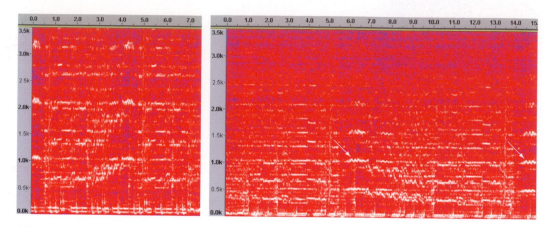

Color Figure 11. Multiple tape looping in the Beatles' song, *Tomorrow Never Knows*. Left: Spectrogram of clip from original recording. Right: Spectrogram of clip after time-reversal and slowing down to half speed. Two arrows point to vibrato in string instrument tones.

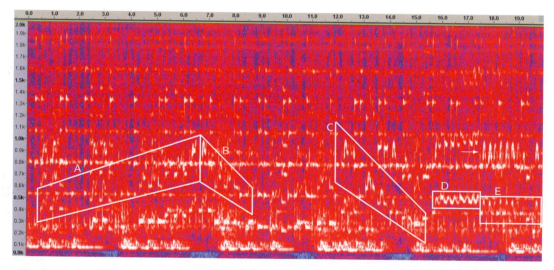

Color Figure 12. Spectrogram from Ravi Shankar composition, *Megh*. Quadrilaterals A through E contain harmonics for sequenced sitar tones: (A) ascending sequencing of a motif; (B) descending sequencing; (C) another descending sequencing; (D) rapid sequencing producing a vibrato-like sonority; (E) more vibrato-like sequencing, transposed down in pitch from D, but with greater frequency spread. Arrow at 17.0 sec, 900 Hz, points to bright 2nd harmonics for vibratoed sequencing of sitar tones with fundamentals shown in E. Ostinato of tabla drum strikes indicated by repetitive percussive and tonal structures (five and half in all), whose fundamentals lie at bottom of spectrogram, between about 80 to 180 Hz.

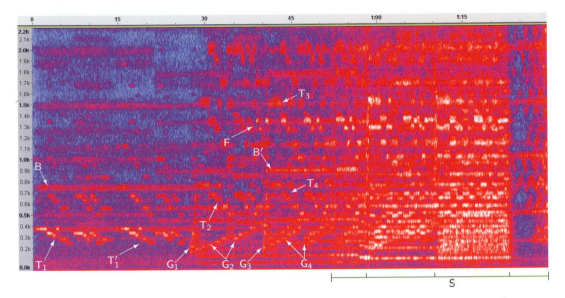

Color Figure 13. Spectrogram from finale of *Firebird Suite*. T_1: horn motif, main theme. B: constant pitch string background. T_1': variation of horn motif. G_1: fast string glissando. T_2: strings playing transposition upwards of main theme. G_2: two slow string glissandos. F: bird-like flute motif. G_3: fast string glissando. B′: constant pitch string background. T_3: flutes playing transposition upwards of main theme. T_4: strings repeating transposition of main theme. G_4: overlayed slow string glissandos. S: consecutive strata, containing multiple glissando overlays and repetitions of main theme by flutes and strings, strata boundaries marked by fast string glissandos, final stratum contains "swirling" string motif.

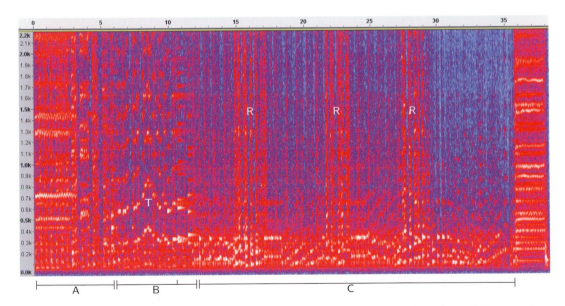

Color Figure 14. Spectrogram of finale of *Jack the Bear*. A: bass note fundamentals at about 73 Hz, saxophone fundamentals at about 725 Hz with significant vibrato, brass fundamentals at about 440 Hz and 523 Hz with less vibrato. B: first part contains main theme, indicated by saxophone 2nd harmonics labeled T. Second part contains piano notes forming accented completion of saxophone theme. C: extended melodic section of bass notes, punctuated by three rhythmic bursts of orchestra in vertically striped regions labeled R. After C, finale ends with a combination of orchestral notes of varying vibrato.

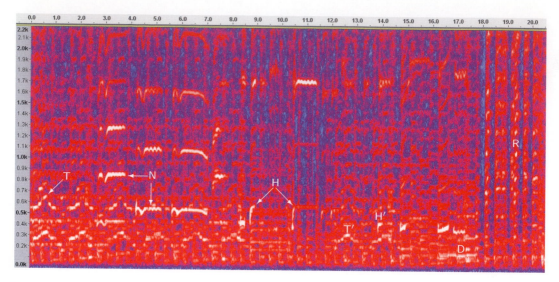

Color Figure 15. Passage from Duke Ellington's *Jack the Bear*. Arrow T points to 2nd harmonics of main theme, played by saxophone section. Fundamentals of this theme repeat six times from 0 to 8.5 sec. Second harmonics of trombone notes are indicated by arrow pointing from left of N. An arrow points down from N to fundamentals of these trombone notes. Arrows from H point to two fundamentals of horn tones, with very rapid attacks and differing amounts of energy in their harmonics (the 3rd harmonic is very bright for second horn tone, creating a ringing timbre). T′ marks repetition of main theme by saxophone section. H′ marks fundamentals for more horn tones. D marks low pitch saxophone notes with beating effects (a growling "low down" sound). R marks rhythmic ostinato provided by rapid playing of orchestra.

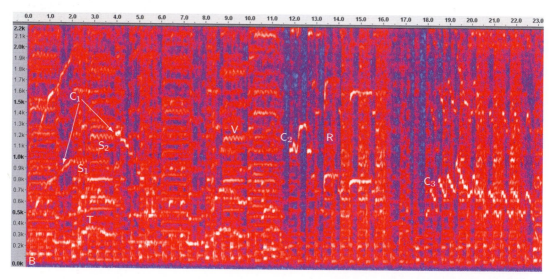

Color Figure 16. Another passage from Duke Ellington's *Jack the Bear*. Label B at lower left corner indicates bass note fundamentals. Arrows from C$_1$ point to clarinet solo. First arrow points to upward pitch excursion, which cuts off abruptly and is followed by vibratoed saxophone note indicated by S$_1$. Immediately following S$_1$, there is a higher pitch fundamental for saxophone note indicated by S$_2$, cutting off abruptly with entry of clarinet (pointed to by right arrow from C$_1$). Below this sequence of clarinet, saxophone, clarinet, there is theme T played by saxophone section. Label V marks third harmonic of vibratoed saxophone note. Clarinet begins again with high bent pitches marked by C$_2$. Label R marks vertical bands for the rhythmic playing of orchestra. C$_3$ marks rapid succession of clarinet descensions, punctuated by rhythmic orchestral tones.

that we considered in Chapter 3, which dealt with transposing groups of discrete, straight tones (the played notes), here the transposed pitch structures are the connected pitch-bent tones.

The spectrogram also reveals how Goodman matches the durations of his tones with the steady drum rhythm in the background. In many instances we can verify that the endings of Goodman's tones precisely line up with the vertical bands marking the drum beats. This is an important aspect of the performance that is not so evident on first hearing, but the spectrogram helps us verify it.

The spectrogram of the Hendrix passage is shown in Figure 5.17. It is a clip from his recording of *All Along the Watchtower*, featuring his electric lead guitar solo. The spectrogram exhibits a similar pattern as the passages from Beethoven and Goodman, an approximately mirror symmetrical descension and ascension of pitch, followed by a gently rising trailing off of melodic contour. Hendrix, however, illustrates a unique aspect of his music. Rather than using discrete tones, he instead uses his electric guitar to generate a *continuous flow of chords*. The chord progression is continuous, rather than a discrete sequence. It is interesting that the electronic sound produced here is surprisingly warm and soft, especially in the later trailing off portion. Perhaps this is due to Hendrix synthesizing continuous chord transitions and vibrato within his blues style.

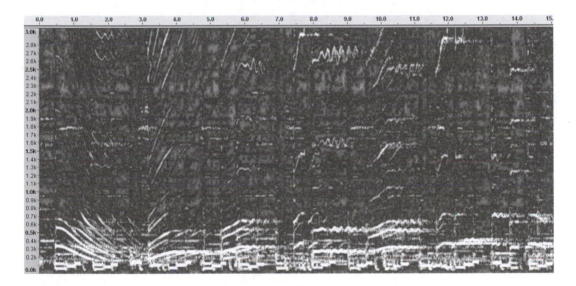

Figure 5.17. Spectrogram from a recording of *All Along the Watchtower* by Jimi Hendrix.

5.2.3 Harmonics in Stringed Instruments

Understanding the physics of how musical instruments produce their tones is a fascinating and complex area of study. Since the mathematics of this study is beyond the scope of this book, we shall limit ourselves to a brief summary of the basic facts. In particular, we describe how the harmonics in the tones arise from the motion of the strings being played.

The strings in musical instruments are typically held fixed at both ends with a fairly constant tension and material density over their length. We shall denote the length of a string by L, its density by ρ, and the force of tension over its length by T. There is a precise relationship between the fundamental frequency ν_o of the tone produced by a string and the values of these parameters. It is given by the following formula:

$$\nu_o = \frac{1}{2L} \sqrt{\frac{T}{\rho}}. \tag{5.2}$$

In this formula, for a particular string, the density ρ is a constant. But the tension T can be varied. Varying T is done when the string is tuned. On a guitar, for example, the knobs at the head of the guitar are turned in order to vary the tensions on the strings, and that varies the fundamental frequencies for each of the strings. Once the strings are in tune, then the length L of a string can usually be varied by pressing down on it to close it off. For example, a guitar has frets which the player presses the string down on in order to change the effective length of the string, thereby changing the fundamental frequency that the string plays.[5]

The discussion in the previous paragraph shows that a single string can be used to play various tones by changing its length. Its most basic tone is when it is played **open**, by not pressing down on it. Then its length, say $L = L_o$, determines its most basic tone, with fundamental ν_o. Suppose, however, the string is pressed down firmly in its middle, so that its length is halved to $L = \frac{1}{2}L_o$. Since ν_o is inversely related to L in formula (5.2), the fundamental will then be changed to $2\nu_o$. Similarly, if the string is pressed down firmly at a point where its length is cut in third to $L = \frac{1}{3}\nu_o$, then the fundamental is changed to $3\nu_o$. This shows that the string has the ability to play fundamentals that are positive integer multiples of ν_o. Of course, the string can be pressed down firmly at other points along its length, and thereby play other tones as well. The tones that are played in this manner, with a string either open or closed, are called **natural harmonics**. We have already used the term, harmonics, to describe the components of the tones from musical instruments. If the context does not make it clear, we shall refer to these components of musical tones as **tonal harmonics**. We shall now describe the physical source of the tonal harmonics of vibrating strings.

It has been known for over two centuries that a string vibrates in a complex motion that is a sum, or superposition, of simpler vibratory motions. These simpler vibratory motions are called the **characteristic modes** for the string's vibrations. The 1st characteristic mode of the string's vibration is shown on the left of Figure 5.18. Since the string is vibrating in an open position, the sound frequency

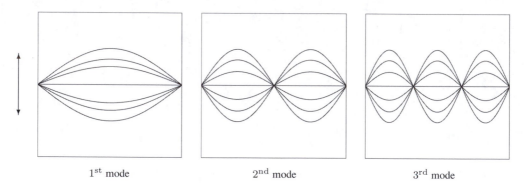

| 1st mode | 2nd mode | 3rd mode |

Figure 5.18. Vibrating String Modes. Double-arrow on left indicates up and down motion of string modes.

created by this motion is the open string's fundamental ν_o. The 2nd characteristic mode is illustrated in the middle of Figure 5.18. In this mode, the middle point of the string is **not** vibrating. This mode is analogous to when the string is firmly pressed in its middle. Following this analogy, we can see why the sound frequency created by this motion is $2\nu_o$. Furthermore, the 3rd characteristic mode is illustrated on the right of Figure 5.18. In this mode, the points at $\frac{1}{3}L$ and $\frac{2}{3}L$ do **not** vibrate. This mode is analogous to when the string is firmly pressed at one-third its length. The sound frequency created by this mode is $3\nu_o$. Continuing in this way, for each positive integer n there is an nth characteristic mode, which generates the sound frequency $n\nu_o$. In this mode, the points $\frac{k}{n}L$, for $k = 1, 2, \ldots, n-1$, do **not** vibrate. These non-vibrating positions are called **nodes**. The motion of the string is a sum of all of these characteristic modes of vibration. Consequently, the sound from the string is a combination of the harmonics with frequencies $\nu_o, 2\nu_o, 3\nu_o, \ldots$. In other words, **the tonal harmonics from the**

---

[5]The mathematical details of the placement of guitar frets are described in an article by Ian Stewart, *Faggot's fretful fiasco*. See Chapter 4, pp. 61–76, of *Music and Mathematics: From Pythagoras to Fractals*, Oxford, 2003.

instrument's string are the result of the vibratory motions for each of the string's characteristic modes. We now give an example of how this physics can be put to use in musical performance. The key idea is to make use of the nodes for the string's characteristic modes of vibration.

Example 5.2.1. Playing harmonics. The physics of string harmonics can be put to use in musical performance. An nth characteristic mode, for $n > 1$, will have nodes that are not vibrating. For example, the 3rd characteristic mode will have nodes at $\frac{1}{3}L$ and $\frac{2}{3}L$, where L is the length of the string. In fact, each nth characteristic mode, for n a multiple of 3, will have nodes at these points as well. The other characteristic modes, where n is not a multiple of 3, will ***not*** have nodes at these points. These other characteristic modes will be *vibrating* at the positions $\frac{1}{3}L$ and $\frac{2}{3}L$. If the player *gently touches* one of these positions, then these other characteristic modes will be suppressed. The tonal harmonics from these other characteristic modes will then also be suppressed. Consequently, the tonal harmonics from the nth characteristic modes, where n is a multiple of 3, will be made relatively more prominent. That is, the tonal harmonics $3\nu_o, 6\nu_o, 9\nu_o, \ldots$, will be *amplified* relative to the other tonal harmonics. There are two ways to play harmonics. One way is for an open string, and the other way is for a closed string. For the open string, just one finger is used to gently touch a node position. This is called a *normal harmonic*. For the closed string, one finger is needed to firmly press down on the string in order to close it off, while a second finger is used to gently touch a node position for the closed string. This is called an *artificial harmonic*.

In Figure 5.19, we show a spectrogram illustrating this technique, which is known as *playing harmonics*. The spectrogram shows the effect on the tonal harmonics for the guitar tones. The first

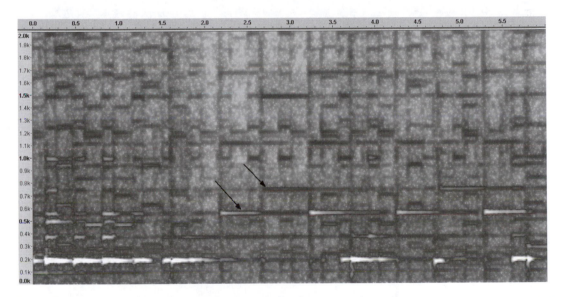

Figure 5.19. Spectrogram from recording of *Black Rabbit* by Peter Mulvey. He is playing harmonics from 2.2 seconds onward. First arrow points to harmonic created from 1/3rd the string length. Second arrow points to harmonic created from 1/4th the string length.

arrow in the figure points to the relative enhancement of the tonal harmonic $3\nu_o$, which is the first of the tonal harmonics $3\nu_o, 6\nu_o, 9\nu_o, \ldots$ that are enhanced relative to the other tonal harmonics. Since the other harmonics are still present, the sound is not quite the same as the normal method of playing a fundamental of $3\nu_o$. (Normally, to produce a fundamental of $3\nu_o$, the player *firmly presses down* on the position $\frac{1}{3}L$.) Moreover, the amplitudes for the harmonics at multiples of $3\nu_o$ are created from the string's length L (open or closed), so they are also ***not*** the same as when the string is played by firmly pressing at $\frac{1}{3}L$. The sonority for a harmonic is a delicate higher pitched tone. The figure also shows an arrow pointing to the playing of another harmonic, in this case a relative enhancement of $4\nu_o$. See Exercise 5.2.13.

Exercises

5.2.1. In Figure 5.20 we show spectrograms of trumpet notes played with a cup mute, and without a mute. Compare and contrast the timbre of the sound in these two cases and its relation to these spectrograms.

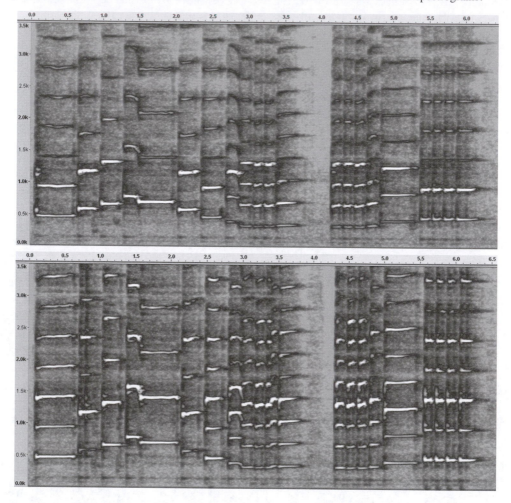

Figure 5.20. Top: Notes played by trumpet without mute. Bottom: Same notes using cup mute.

5.2.2. In Figure 5.21 there is a spectrogram of Louis Armstrong's singing. Armstrong's singing has frequently been compared with his trumpet playing. For example, the classical pianist Edna Stern makes this comparison:

> The way he plays trumpet is very similar to his singing, it comes through in the way he is vibrating or marking the expressive moments in the music. Also the timing and the way he is building his phrases are both done in the same declamatory way as his singing. When he is singing, his vibratos are very similar to the trumpet vibratos and I think that there is a clarity in the way he punctuates his syllables, using the exploding ones in a sort of trumpet way.[6]

Find examples of what Stern is talking about in the spectrograms shown in Figures 5.12 and 5.21.

5.2.3. In Figure 5.22 we show spectrograms of two versions of another passage from Mussorgsky's *Pictures at an Exhibition*. One by the pianist, Evgeny Kissin, playing the original piano score by Mussorgsky. The other by the Chicago Symphony Orchestra performing the Ravel orchestration of the piece. Compare and contrast these

---

[6]E. Stern, from *Pandolous* web site, **www.pandalous.com**, with search topic: Louis Armstrong.

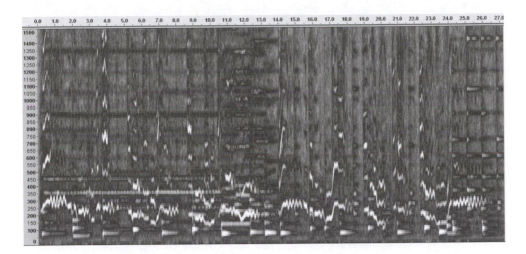

Figure 5.21. Spectrogram of Louis Armstrong's singing in *La Vie En Rose*.

two performances, making reference to their spectrograms. In particular, describe how Ravel makes use of the different spectral characteristics (timbres) of the orchestral instruments to change the aural effect of the passage.

5.2.4. Suppose an open string produces the note C_4. **(a)** If the string is firmly pressed down at a position that is one-third from its end, then what note will the string produce when its longer part is played? **(b)** If the string is firmly pressed down at a position that is one-fourth from its end, then what note will the string produce when its longer part is played?

5.2.5. Suppose an open string produces the note C_4. **(a)** If the string is firmly pressed down at a position that is one-fifth from its end, then what note will the string produce when its longer part is played? **(b)** If the string is firmly pressed down at a position that is one-ninth from its end, then what note will the string produce when its longer part is played?

5.2.6. In Figure 5.23 we show spectrograms of portions of two trumpet performances. One by a musician first learning the piece, with only two days practice. The other by a different musician performing in concert. Compare and contrast these two performances, making reference to their spectrograms.

5.2.7. On the top of Figure 5.24 is a spectrogram from a recording of *Gymnopedie I* by Erik Satie. On the bottom of Figure 5.24 is a spectrogram from the recording, *Variations On A Theme By Erik Satie* by the group, *Blood, Sweat & Tears*. Compare and contrast the sound from these two recordings as represented in the spectrograms, including the relation between Satie's piano score (shown in Figure 5.25) and the instruments used in the *Blood, Sweat & Tears* recording.

5.2.8. Identify in the spectrogram of the Beethoven passage in Figure 5.14 where there is beating interference within the time interval 4.0 to 5.0 seconds. Identify the notes in the score for the passage, shown in Figure 5.15, that are creating this beating interference.

5.2.9. Explain why a transposition corresponds to a shifting, in time and frequency, of some discrete structure in the time-frequency plane. In particular, for the following score

determine the transposition used to map the notes in the first measure to those in the second measure, and then determine the amount of shifting in time and frequency that this transposition is performing. If the instrument used to play the notes in the score were a piano, then describe the appearance in the spectrogram of the discrete structure that is shifted.

5.2.10. Using Formula (5.2), explain the following phenomena:

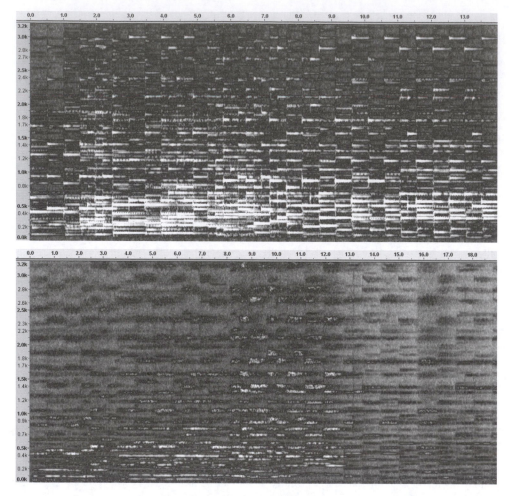

Figure 5.22. Spectrograms from two different versions of Mussorgsky's *Pictures at an Exhibition*. Top: Mussorgsky's original piano version. Bottom: Ravel's orchestral version.

(a) If a string is stretched tighter, tension T increased, then the tones from the string are higher in pitch.

(b) If a string is effectively shortened, by firm pressure from a finger or bridge or some such device, then the tones from the string are higher in pitch.

(c) If all other variables are equal, then decreasing the density of a string makes a higher pitch tone.

5.2.11. (a) Give at least one example from playing technique for a stringed instrument, say a violin or electric guitar, that uses the phenomenon described in Exercise 5.2.10(a). (b) Describe how the phenomenon described in Exercise 5.2.10(c) relates to the tuning of a piano. [Hint: it involves humidity.]

5.2.12. By what factor would you need to increase the tension T on a string in order to double the fundamental frequency of the string?

5.2.13. Discuss how the harmonics $4\nu_o, 8\nu_o, 12\nu_o, \ldots$, are amplified in the spectrogram shown in Figure 5.19. Where does the player touch the string?

5.2.14. Suppose a violin player is bowing the open string for G_3. While bowing, the player very gently touches the string at a node position of $3/4$ its length. What is the pitch of the resulting tone? What are the frequencies of its harmonics (list the first four)?

5.2.15. Suppose a violin player is bowing the open string for G_3 and gently touches it at a node position of $2/3$ its length. Suppose that the player also, a moment later, simultaneously touches the front third of the string (the

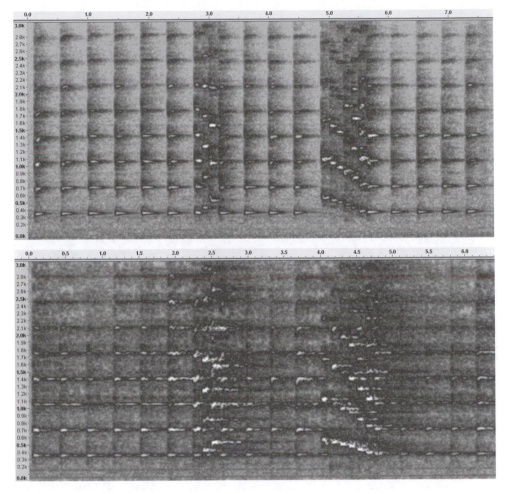

Figure 5.23. Top: Notes played by a musician with two days practice. Bottom: Same notes played by a different musician in a concert.

part closest to the pegs) at a position of $1/5$ its length. What tone will the vibration of that front third of the string now produce? What tone would it produce if the player touched it at $1/3$ its length? Considering the three tones produced here—one from the original touch at the $2/3$ length of the open string, and the two others just described—what chord is arpeggiated? [Note: This may not actually be playable by a human being, but it is possible in theory.]

5.2.16. This exercise describes a mathematical formulation for the superposition of string harmonics.

(a) Consider the following equation

$$y = B_1 \cos(2\pi\nu_o t - \phi_1) \sin \frac{\pi x}{L}$$

where B_1 is an amplitude, ν_o is the fundamental frequency as given in (5.2), and ϕ_1 is a phase constant. Explain why this equation describes a motion similar to the one shown for the 1st characteristic mode in Figure 5.18. In particular, for $\phi_1 = 0$, find all times t for which the string has the form described by $y = B_1 \sin \frac{\pi x}{L}$, and all times t for which it has the form $y = -B_1 \sin \frac{\pi x}{L}$.

(b) Consider the following equation

$$y = B_2 \cos(2\pi\, 2\nu_o t - \phi_2) \sin \frac{2\pi x}{L}$$

where B_2 is an amplitude, ν_o is the fundamental frequency as given in (5.2), and ϕ_2 is a phase constant. Explain why this equation describes a motion similar to the one shown for the 2nd characteristic mode in

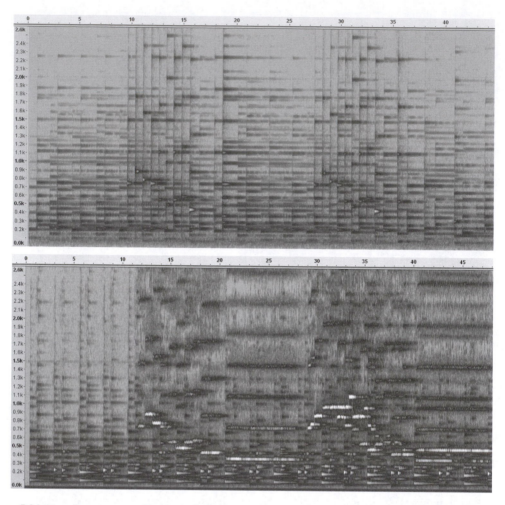

Figure 5.24. Top: Spectrogram of beginning of *Gymnopedie I*. Bottom: Spectrogram of beginning of *Variations On A Theme By Erik Satie*.

Figure 5.25. Score of beginning of *Gymnopedie I* by Erik Satie.

Figure 5.18. In particular, if $\phi_2 = 0$, then find all times t for which the string has the form described by $y = B_2 \sin \frac{2\pi x}{L}$, and all times t for which it has the form $y = -B_2 \sin \frac{2\pi x}{L}$.

(c) Consider the following equation

$$y = B_3 \cos(2\pi\, 3\nu_o t - \phi_3) \sin \frac{3\pi x}{L}$$

where B_3 is an amplitude, ν_o is the fundamental frequency as given in (5.2), and ϕ_3 is a phase constant. Explain why this equation describes a motion similar to the one shown for the 3rd characteristic mode in Figure 5.18. In particular, if $\phi_3 = 0$, then find all times t for which the string has the form described by $y = B_3 \sin \frac{3\pi x}{L}$, and all times t for which it has form $y = -B_3 \sin \frac{3\pi x}{L}$.

(d) Explain why the following formula, where M is some large positive integer, is a reasonable model for the superposition of characteristic modes described in the text:

$$y = \sum_{k=1}^{M} B_k \cos(2\pi\, k\nu_o t - \phi_k) \sin \frac{k\pi x}{L}.$$

Describe how this formula relates to the basic model for a sound waveform in formula (4.6) on p. 103.

5.3 Compositions

We now turn to using spectrograms to analyze musical compositions, particularly their performance aspects. We show how spectrograms help us identify important performance qualities such as timbre, tonal harmonics, dynamics, and vibrato. The compositions we look at cover a wide range of musical styles, including jazz, electronic, symphonic, and music from non-Western cultures.

5.3.1 Roy Hargrove's *Strasbourg/St. Denis*

Our first example is a modern jazz composition by the Roy Hargrove quintet, *Strasbourg/St. Denis*. This piece, like many jazz compositions, contains a combination of virtuoso performances by individual players along with tightly organized combined playing. We shall look at just two passages from this piece. They contain some interesting examples from all of the musicians in the quintet. Two spectrograms for these passages are shown in Color Figures 7 and 8. The spectrogram in Color Figure 7 shows a sequence of ascending and descending harmonics in the frequency range 250 Hz to 550 Hz. These harmonics are the fundamentals of saxophone and trumpet tones. These tones alternate in groups, each an ascending or descending group, played as a call-response sequence by the saxophone and trumpet players. The first arrow on the left points to a descending set of saxophone tones, the call. The double arrow points to the response of the trumpet, followed by a retrograde of the saxophone tones. This call and response pattern is repeated for about 10 seconds. It is then followed by a different pattern, shown in the time interval marked off by the double arrow on the right. This latter pattern begins by the saxophone playing a descending set of notes, followed by a repetition of most of those notes by the trumpet. That basic motif is then rapidly repeated in a sequence. These call and response patterns are good examples of sequencing, carried out by two different instruments. There are also bass drum tones indicated by the thick horizontal bars in the frequency range 50 to 100 Hz. An example is indicated by the third arrow in the figure. The drummer also uses his drums in a brief burst indicated by the fourth arrow in the figure.

The spectrogram for the second passage is shown in Color Figure 8. It consists of a short piano and bass duet with some drum accompaniment. The piano tones are particularly interesting here because they are created in two ways. The first four piano notes, within the time span indicated by the double-arrow on the left, are played in the standard way using the keyboard. We shall refer to these as *struck piano notes*. Right after those notes are played, there are several notes played in a different way, by *plucking the strings of the piano*. These plucked piano notes occur within the time span indicated by the second double-arrow. Plucked piano notes sound a bit like the tones from a bass, due to the percussive onset from the plucking. They conclude with a piano sonority, due to the distinctive harmonics created from the piano string vibrations. The relative amplitudes in those harmonics are different than for the struck piano notes. The plucked piano notes have mostly one prominent harmonic, much like the plucked bass tones, while the struck piano tones have several prominent harmonics. In this passage, the piano player alternates these two methods of playing. Moreover, when the plucked piano notes are played, they are interspersed with the plucked bass notes. The single arrow, in between the two double arrows, points toward some of these plucked bass notes. The two different methods of plucking string notes makes for a delightful contrast.

5.3.2 The Beatles' *Tomorrow Never Knows*

The Beatles' music went way beyond the limits of music in the popular realm, extending across boundaries into a wide variety of musical genres. We will examine just one of their songs, *Tomorrow Never Knows,* which lends itself well to analysis with spectrograms. *Tomorrow Never Knows* is a perfect example of music as defined by the composer Edgard Varèse. He said that "music is organized sound." The Beatles used a wealth of sound production techniques, some invented specifically for this song, to create the unusual aural landscape of *Tomorrow Never Knows.* These techniques include: (**1**) use of a *drone,* a single pitch held throughout the composition, providing a single harmonic foundation for other pitches, (**2**) extensive processing of recorded sounds, using fuzz boxes, dynamic compression, and amplifying John Lennon's vocals through a rotating speaker, and (**3**) multiple dubbing of prerecorded taped loops, often time-reversed and at altered speeds. Using spectrograms, we can reverse engineer some of these production techniques.

Tomorrow Never Knows begins with a drone, an Indian tambura playing the single note C_3. This is shown in the spectrogram on the left side of Color Figure 9. The harmonization with this drone actually begins with the drum strikes that break the monotony of the drone sound. The drum strikes have a tonal quality consisting of two harmonics, the second harmonic matching the fundamental for the drone. These drum strikes were *compressed* when they were recorded. Compression involves mapping the dynamic range of the sound, the sound loudness values, into a smaller set of values. This reduces the extreme contrast of the drum strikes with other sounds in the recording, such as the tambura's drone in this case. Most listeners find that compression improves overall sound quality.

In addition to compression, the Beatles did several other kinds of audio processing. For example, the recording of guitar notes was often done by sending them through a *fuzz box.* The fuzz box clips off the high and low values of the guitar's sound waveform, thereby creating additional harmonics or even a noisy distortion in the sound. Fuzz boxes were commonly used at the time in rock music. The engineers working with the Beatles, however, created some entirely new techniques in producing this song. One of their new techniques was *automatic double tracking.* They would send the vocal signal through two lines, one slightly delayed, and then dub them together. Automatic double tracking created an automatic doubling of voices on the final recording. See the right side of Color Figure 9. This voice doubling is similar to a reverberation, or even a choral effect, and thereby enhances the music. The beating that often results with this doubling is illustrated in the spectrogram on the right of Color Figure 9. Automatic double tracking was subsequently used extensively by the Beatles, and is now commonly employed in all musical recording. Another technique that originated with *Tomorrow Never Knows* is the feeding of John Lennon's vocals through a Leslie cabinet. A Leslie cabinet is an enclosure containing a spinning loudspeaker, used for amplifying the sound of an electronic Hammond organ. Some of the unusual contrast of intensities in the vocal harmonics in this spectrogram may be a result of their processing through the Leslie cabinet.

One of the most strikingly innovative aspects of *Tomorrow Never Knows* is the Beatles' extensive dubbing of pre-recorded tape loops. We will look at a few notable examples. First, we show on the left side of Color Figure 10 a spectrogram of a clip that seems to contain three rapid bird calls. However, when this clip is played at half speed, it sounds like someone laughing ("Ha, Ha, Ha").

It is known that several of the bird call sounds in the song were, in fact, created by speeding up tapes of Paul McCartney laughing. This may be one of those instances. Second, on the right of Color Figure 10, we show a spectrogram of one of the most famous tape effects used in the recording. George Harrison's lead guitar solo was played in reverse, and randomly cut into smaller parts, which were then dubbed onto the rest of the recording. By playing this recording in reverse, thereby un-reversing the reversal, the different parts of the original solo can be heard clearly. This solo was later used in recording another song on the same album, George Harrison's *Taxman*.

Finally, we show in Color Figure 11 an example of tape looping that involves both time-reversal and speed change. The spectrogram on the left of the figure was created from a clip from the recording. There is an array of many harmonics appearing in the spectrogram, due to simultaneous dubbing of several tapes. The sound resembles a stringed instrument playing a rising scale of notes with a very grating timbre. However, if this recording is reversed and played at half speed, one can hear multiple riffs of a guitar or sitar, possibly both. The bell-like timbre and the vibrato of many of these riffs can also be clearly heard.

A good summary of the musical importance of *Tomorrow Never Knows* was given by the composer, Howard Goodall, in a television documentary he did on the Beatles:[7]

> ...there was a big difference between what the Beatles were up to and what avant-garde composers were doing. For the classical experimentalist, it was enough simply to do the looping and listen to that as a performance on its own. For the Beatles, however, it was a *means* to create a nightmarish or hallucinogenic *soundscape* within their highly imaginative songs ...*nothing like the aural landscape of this song had ever been heard before.*

Another good description of *Tomorrow Never Knows* can be found in Walter Everett's magisterial treatment, *The Beatles As Musicians: Revolver through the Anthology* (Oxford University Press, pp. 34–38).

5.3.3 A Portion of Morton Feldman's *The Rothko Chapel*

Our next example is a brief analysis of a composition that draws a contrast between more traditional melodic sounds and more experimental aspects of musical production. In Figure 5.26 we show a spectrogram of a passage from Morton Feldman's *The Rothko Chapel*.

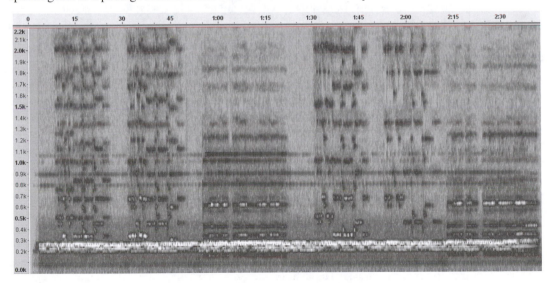

Figure 5.26. Spectrogram from a recording of Morton Feldman's *The Rothko Chapel*.

---

[7]H. Goodall, *20th Century Greats — The Beatles*, Channel Four Television, 2004.

The spectrogram provides us with a clear display of the structure of the music in this passage. We can see that there are just a few basic patterns that combine to create the sound. They consist of the following:

(1) There is a thick horizontal band near the bottom of the spectrogram, which is a combination of pedaled notes played near C_4 on a keyboard instrument known as a *celesta* (or *celeste*). It provides a lower pitch continuous sound, which is slightly dissonant due to the beating that we can see in the spectrogram between closely spaced harmonics. Second, there are thinner horizontal bands at higher frequencies (around 800 Hz and 900 Hz, for example), which are actually composed of a large number of short duration notes. These provide an *ostinato* consisting of sustained repetition of a small set of higher pitched notes over a long time interval. Together, this low pitch, continuous sound with the contrasting higher pitched *ostinato*, provide an ambience for the passage.

(2) In the time interval from 0 to 50 seconds, there are two sections of melodic passages played by a viola. The fundamentals for this melody are clearly discernable in the frequency range of 300 Hz to 500 Hz, located just above the thick horizontal band. The overtones of the viola notes are also clearly displayed. The spectrogram confirms what we hear, that the second section is a repetition with slight variations of the first section. Some of this variation consists of transposition to higher pitch of parts of the first section.

(3) In the time interval from 55 seconds to 1 min : 20 sec, there is a completely different style of viola playing. Here, rather than a time sequence of varying pitches in a melody, there is instead a sequence of changes of timbre of a fixed pitched tone. In other words, during the whole time interval the viola player is holding a tone with a fixed pitch while varying its timbre. We shall call this *timbre sequencing*. The timbre sequencing is revealed in the spectrogram by the variations in the intensity of the harmonics.

(4) From 1 min : 30 sec to the end of the passage, the contrast between melody and timbre sequencing is repeated.

In this example, the spectrogram method has illustrated the contrasts that Feldman creates between different aspects of music. Such as the contrasts between high and low pitch ostinatos, and between melody and timbre sequencing.

5.3.4   A Portion of a Ravi Shankar Composition

In this example we shall briefly illustrate how spectrograms provide us with a tool for analyzing the music of other cultures. For other cultures, scores may not be available or they may be difficult to interpret due to our lack of knowledge of a given culture's musical tradition. Nevertheless, spectrograms provide a way of visualizing tonal relationships and thereby help us make sense of what we hear in the music. Our case in point here is a short passage from some Indian music by Ravi Shankar, a composition entitled *Megh*. We show a spectrogram of this passage in Color Figure 12.

This spectrogram reveals several features of the music. First, it shows structures reflective of the two main instruments in the passage: a sitar and a tabla. The tabla's drumming is revealed by the series of structures aligned in a row within the frequency range of about 80 Hz to 180 Hz. The tabla is clearly being used to create a rhythm for the passage in a bass register. Second, it shows the curved, connected harmonics of the sitar. The sitar is the famous Indian stringed instrument, for which Ravi Shankar was known as the world's most accomplished player. The quadrilaterals marked A through E contain harmonics for the sitar's tones. These harmonics, similar to the clarinet tones in the Goodman example we discussed, form bent pitch structures that connect notes on the scale used for the music. In quadrilateral A, Shankar is sequencing one basic motif in an ascension, followed by a descension in quadrilateral B. This basic motif—which lies in a single octave between about 300 Hz and 600 Hz, and in the time interval 0.5 seconds to 1.5 seconds—also both ascends and descends. Quadrilateral C contains another example of this sequencing of one motif by descension of pitch. In this case, the connection of the sequencing of the basic motif to a descension through a musical scale can be heard quite clearly. The passage concludes with two sequencings, shown in rectangles D and E. In both cases, the sequencing approximates a slow-rate vibrato.

Although there are important differences between the Western music that we have concentrated on, compared with this example from Indian music, there are also some similarities. The differences are profound, including: (1) using the tabla for tonal drumming, (2) using the sitar to produce elaborate bent-pitch tonality, and (3) using an Indian musical scale. However, the spectrogram enabled us to identify some similarities to Western music, including (1) using sequencing, and (2) using the tabla's bass tones as counterpoint to the much more elaborate melodic tones of the sitar.

5.3.5 Musical Illusions: *Little Boy* and *The Devil's Staircase*

We now turn to an intriguing aspect of musical composition: the use of auditory illusions. Almost 50 years ago, Roger Shepard discovered an auditory illusion in our pitch perception.[8] Shepard used electronic tones that sound as if they are rising endlessly in pitch, even though they in fact stay within just one octave. The tones can also be arranged to create an illusion of endless descent. A nice demonstration of Shepard tones can be accessed from the link, Shepard Tones Demo, at the book's web site (see Videos/Chapter 5). We shall explain his illusion and show how two composers have used it to create musical compositions.

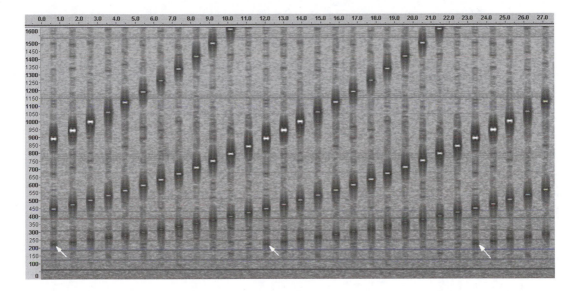

Figure 5.27. Spectrogram of Shepard's rising pitch illusion. Arrows point to fundamentals of disguised tones.

Shepard's illusion can be explained using spectrograms and the chromatic clock. (See Figure 5.27.) The arrows in that figure point to the tones where the illusion occurs. Our brains are tracking the pitch contour of the fundamentals for the tones and expecting the next fundamental to be exactly one octave higher than where it occurs. The harmonics of these electronic tones are all designed in an octave series—the fundamental multiplied by two, four, eight, etc.—to hide the drop in pitch. The series of tones shown in the figure begin with a fundamental of 220 Hz, A_3, which is hour 9 on the chromatic clock. Then they rise one half step at a time through the chromatic scale, ($A_3^\sharp$, B_3, C_4, $C_4^\sharp$, ...), moving clockwise one hour at a time around the chromatic clock. When $G_4^\sharp$ is reached, which

---

[8]R.N. Shepard, Circularity in judgments of relative pitch. *J. Acoustical Society of America*, **36** (1964), 2346–53.

is hour 8, then the next tone will be at hour 9 but with a fundamental of pitch A_3. However, that is a drop in pitch of almost one octave, a drop of 11 half steps. Dropping in pitch by 11 half steps is a rather unlikely event in music. As the spectrogram shows, the fundamentals are tracking towards A_4, and A_4 would represent just a single half step up from $G_4^\sharp$. Going up one half step is a minor second interval, a much more common musical event than the actual change of down eleven half steps. Consequently, our brains tend to hear the much more likely occurring tone following $G_4^\sharp$, the tone of A_4 rather than the actual fundamental of A_3. Once you are fooled, then you are hearing tones as if their fundamentals are tracking along the upper curve of 2nd harmonics shown in the spectrogram. The tones are heard *as if* they are A_4, $A_4^\sharp$, B_4, C_5, $C_5^\sharp$, …. In terms of the chromatic clock, we are again circling clockwise around from hour 9. So, when hour 9 is reached this time, you may be tricked into hearing A_5 instead of the actual tone A_3. The abstract for Shepard's paper summarizes our discussion quite well:

> A special set of computer-generated complex tones is shown to lead to a complete breakdown of transitivity in judgments of relative pitch. Indeed, the tones can be represented as equally spaced points around a circle in such a way that the clockwise neighbor of each tone is judged higher in pitch while the counter-clockwise neighbor is judged lower in pitch. Diametrically opposed tones—though clearly different in pitch—are quite ambiguous as to the direction of the difference.

The connection with the chromatic clock is quite clear in this description. The diametrically opposed tones are the tritones, tones separated by 6 half steps on the chromatic clock. We shall say more about tritones later.

Shepard's illusion is well documented in the psychological literature. Like a magician's illusion, if one is aware that the illusion is coming, then it is sometimes possible to perceive the external reality. In this case, to hear the lower pitch fundamental disguised by the contrived tones. However, when one is not told in advance that there is an illusion—or the illusion is disguised further by other tones within a musical composition—then listeners will generally hear an endlessly rising or descending pitch.

Just a few years after Shepard's paper on his illusion was published, Jean-Claude Risset used the illusion in his electronic composition, *Little Boy*. A spectrogram of the beginning of the second movement of *Little Boy* is shown in Figure 5.28. The spectrogram shows how Risset is using Shepard's method in order to create an illusion of sound descending endlessly in pitch.[9] The spectrogram also shows several steeply sloped pitch descents in the higher frequencies. Although these are part of the artistry of the composition, they also may serve to distract the listener from exclusive concentration on the descending Shepard tones and thereby heighten the illusion. Later in the piece, Risset gives musical credit to Shepard by including some of Shepard's original tones.

In addition to the pitch illusion, *Little Boy* also includes tones that create a sound similar to blowing wind, and tones that sound like breaking glass. The whole piece is a dramatic evocation of the dropping of the first atomic bomb, nicknamed "Little Boy." In that context, Risset's comments on his work are quite incisive:

> I have used these paradoxical sounds in compositions such as *Little Boy*, …They are not mere tricks: as Purkinje wrote, illusions are errors of the senses but truths of perception.[10]

Both Shepard and Risset utilized electronic tones. György Ligeti, however, wrote a piano etude entitled *The Devil's Staircase* (*L'Escalier du Diable*, Études pour piano, Book II, No. 13) which also contains Shepard's pitch illusion. This composition is remarkable because it uses an acoustic instrument. In Figure 5.29, we show a spectrogram of the beginning of *The Devil's Staircase*. The

---

[9]The explanation for this endless descent illusion is essentially the same as the one we gave above for endless ascent. The only differences are that, for endless descent, the motion on the chromatic clock is counter-clockwise and the signs of the half-step changes are reversed.

[10]In article by J.C. Risset, Chap. 13, of *Mathematics and music: a Diderot Mathematical Forum*, G. Assayag, H. Feichtinger, J.-F. Rodrigues, (Eds.), p. 225.

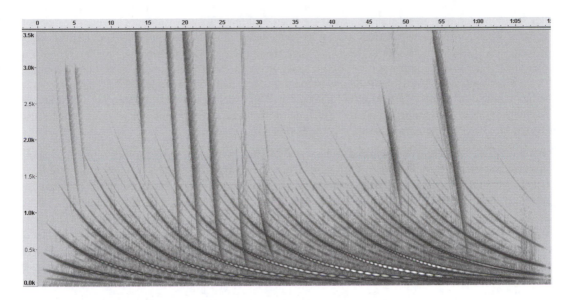

Figure 5.28. Spectrogram from J.-C. Risset's composition, *Little Boy*.

connection between the multiple rising piano glissandos in this spectrogram and the Shepard tone spectrogram is quite evident. Listening to the piece as the spectrogram traces out, we can hear the pitch illusion occur in some places but not in others. This is certainly in keeping with the title's figurative allusion to a staircase in hell. The multiple piano glissandos in this piece are a good example of Ligeti's technique of *glissando clusters*, which is one of his innovative compositional techniques.

The Devil's Staircase contains a couple of other noteworthy musical effects. First, about halfway through the piece there is a glissando cluster that culminates in a very high-pitched, rapid ostinato of C_8, the highest pitched key on the piano. The extremely high pitch, along with its rapid emphasis, creates considerable tension. This tension is released by an abrupt transition to a very low pitch dissonant chord. See the spectrogram on the top of Figure 5.30. The acoustic dissonance in the chord is evident from the beating between harmonics. This beating is heard as a low pitch rumbling, which

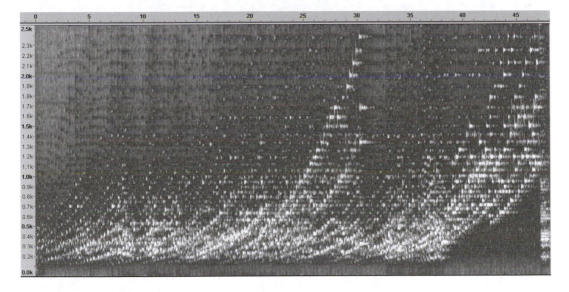

Figure 5.29. Spectrogram from Ligeti's *The Devil's Staircase*.

after the considerable tension created by the preceding high pitches, comes as a welcome release. So, here we have dissonance being used to relieve rather than create tension. Second, the piece finishes with a single chord that is held for over half a minute. See the spectrogram on the bottom of Figure 5.30. It is fascinating how long a fine piano can hold these notes. The chord contains the harmonic intervals, A_1-$E_2^{\flat}$ and F_2-B_2, which are both tritones. Tritones were characterized in medieval times as "diabolus in musica" ("the devil in music"). So it is fitting that *The Devil's Staircase* ends with a chord containing tritones. This is the most obvious occurrence of tritones. Ligeti also uses tritones in several other parts of the composition.

Remark 5.3.1. The Tritone Paradox. The tritone interval is the source of another musical illusion, known as the ***tritone paradox***. Recall the quote from Shepard above, which ends with this statement: "Diametrically opposed tones—though clearly different in pitch—are quite ambiguous as to the direction of the difference." Diametrically opposed tones on the chromatic clock are separated by $+6$ hours clockwise and -6 hours counter-clockwise. Such pairs of tones constitute the tritones. Shepard says that a tritone interval of tones is quite ambiguous as to whether one tone is heard as higher or lower in pitch. The psychologist, Diana Deutsch, verified Shepard's idea by constructing sequences of tritones that would be perceived in precisely the opposite pitch relations to the ones that were played (e.g., C_3 followed by $F_3^{\sharp}$ is heard as C_3 followed by $F_2^{\sharp}$, or vice versa). To hear examples of Deutsch's

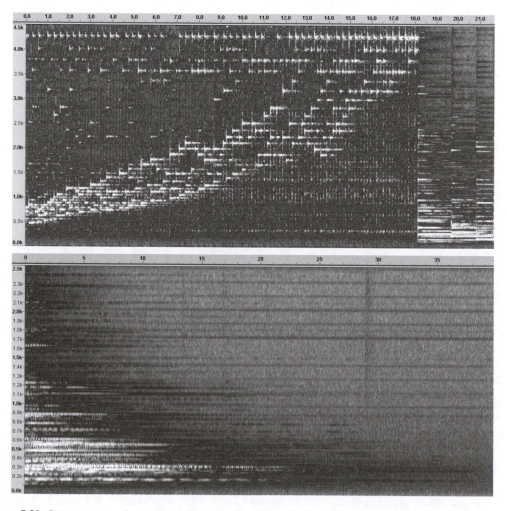

Figure 5.30. Spectrograms of two parts of Ligeti's *The Devil's Staircase*. Top: Major transition from extremely high register to extremely low register. Bottom: Final, extremely dissonant chord.

tritone illusion, go to the following link:

Tritone Illusion

at the book's web site. It is an interesting question as to whether this illusion has played any role in musical composition.

5.3.6 The Finale of Stravinsky's *Firebird Suite*

Our next example is a passage from the first half of the finale of Stravinsky's *Firebird Suite*. Here we shall find that spectrograms reveal some of the underlying structure of the music and how Stravinsky makes use of the differing harmonic characteristics of the instruments in a symphony orchestra. We show the spectrogram for the passage in Color Figure 13. This spectrogram shows a number of ways in which Stravinsky packs a lot of musical material into the span of about one and a half minutes.

The passage begins with a haunting motif played by a horn. The fundamentals for this motif are indicated by the arrow T_1. Underlying this motif is a constant pitch string background, whose second harmonic is indicated by the arrow labeled B. The spectrogram shows that some of the notes from the motif T_1 are at the same pitch as the string background note, since they lie on its fundamental. The motif is then repeated, with a slight variation, by the horn. The fundamentals for this second version of the motif are indicated by the arrow T_1'. We shall see that Stravinsky performs multiple transpositions of the first motif T_1, using various instruments at different pitches.

The passage continues with a *glissando* of string tones. This *glissando* is indicated by the arrow G_1 in the spectrogram. After this *glissando*, the main motif is repeated by strings at an octave higher pitch than for the horn. The fundamentals for this transposition are indicated by the arrow T_2. Underlying this motif are a gently rising series of string tones, a very slowly rising *glissando*. Their fundamentals are indicated by the double-arrow G_2. Through the rest of this passage, Stravinsky repeatedly uses fast rising *glissandos* to mark sub-passages within the passage as a whole. They form a ***stratification*** of regions of increasing complexity within the passage.

The next rapid *glissando* is indicated by the arrow G_3. Right before this glissando there is a high pitched sequence of flute notes, whose fundamentals are indicated by the arrow F. This is a prelude to the transposition of the motif up another octave, played by flutes. The beginning of this transposition is indicated by the arrow T_3. Underlying this repetition of the motif is another repetition played simultaneously an octave lower by strings at a lower volume. The fundamentals for these string tones are indicated by the arrow T_4. There is also a constant pitch string background, indicated by the arrow B', at a higher pitch than the previous background and with significant vibrato. This constant pitch string background increases in pitch, with increasing vibrato, through each of the strata. This contributes to a heightening background tension as the complexity and volume of sound increase within each successive strata. The slow *glissando* is repeated in this sub-passage, but with increased complexity obtained through overlaying two halves of the previous slow glissando, and then repeating this overlaying. The fundamentals for these overlayed *glissandos* are indicated by the double-arrow G_4.

At this point, the spectrogram becomes too complex to label its parts. The whole time and frequency plane is almost completely filled with harmonics from various instruments. Nevertheless, the structure of the end of the passage is a repetition of the pattern described above. There are successive strata marked off by four sub-intervals indicated within the time interval labeled S. The first three of these strata, lying within the first three of the sub-intervals of S, begin with fast *glissandos* and have these features in common:

(1) There are transpositions to higher pitch ranges of the motif, played by various instruments. For example, the stratum which begins at the point marked by the left side of the line segment S, has the motif played by strings in a higher pitch range than the previous flute notes. The variance of tonal color (timbre) of the different instruments plays an important role in the repetition of the motif in different pitch ranges.

(2) Underlying repetitions of the motif at previously lower pitches.

(3) Increasingly complex overlays of the slow *glissando* structures.

The final stratum of the passage is a relaxation of tension. It is shown above the last sub-interval of S. It consists of just a succession of string notes, alternating up and down in staggered linear melodic contours, producing a "swirling" sonority. This relaxed passage is an interlude that preludes the second, very dramatic, half of the finale. This second half is also organized in strata, using powerful strikes of tympani notes rather than *glissandos*. We leave its analysis to the reader as an exercise.

Remark 5.3.2. In our discussion of this passage of Stravinsky's we found some important aspects of his music that have been identified within several other of his compositions. The eminent musicologist, Robert Cogan, has described these features as follows:

> No composer possesses a more distinctive sonic fingerprint than Igor Stravinsky. Certain textures have become instantly recognizable as Stravinskian: a bright, clear, high sonority, often sustained; an array of sharp pulsations, often (but not always) dense and dark-hued; the built-up attack of a single sound, whose continuation then hangs "disembodied" after the pronounced attack.[11]

5.3.7   Duke Ellington's *Jack the Bear*

Our final example is an analysis of the recording of *Jack the Bear,* by the Duke Ellington orchestra. Like the Stravinsky piece, we shall find that spectrograms reveal some of the underlying structure of the music and how Ellington makes use of the differing harmonic characteristics of the instruments in his orchestra. As with many of Ellington's compositions, *Jack the Bear* concentrates an extraordinary amount of musical material into a short time span. It is known as one of Ellington's greatest recordings, and contains the following essential features of Ellington's style:

1. Masterful short solos by members of Ellington's orchestra. In this case, solos by Jimmy Blanton, Joe Nanton, and Barney Bigard are particularly noteworthy. Ellington designed his compositions to highlight the strengths of members of his orchestra. He would often involve his whole orchestra in late-night composition sessions, incorporating ideas from his performers. One of his important observations on composition is the following:

   > Sometimes I think a piano limits you when you are writing. You employ what you can play when you use a piano. Otherwise, your mind is much broader.[12]

   The recording of *Jack the Bear* features two extraordinary solos by Joe Nanton, on trombone, and Barney Bigard, on clarinet, employing those instruments for producing sounds that would be simply impossible with a piano.

2. The use of the orchestra as a rhythmic instrument. Ellington's son noted that his father's "band is used as a rhythm instrument to allow the bass to make a statement."[13] We will see an example of this in *Jack the Bear*. In order for the band to perform as a rhythm instrument, it must play in a highly synchronized way. One of Ellington's trumpet players noted that they

   > ...worked hard to get a good section sound, and paid particular attention to maintaining matched vibrato, whether fast, medium, or slow, that was appropriate to the different numbers and tempos.[14]

---

[11] Robert Cogan, *New Images of Musical Sound,* Harvard Univ. Press, 1984, p. 56.

[12] Duke Ellington, quoted in Stanley Dance, *The World of Duke Ellington,* Scribner, 1970, pp. 273-4.

[13] Mercer Ellington, quoted in Stanley Dance, *The World of Duke Ellington,* Scribner, 1970, pp. 39.

[14] Shorty Baker, quoted in Stanley Dance, *The World of Duke Ellington,* Scribner, 1970, pp. 166.

They could match not only their pitches and note lengths, but even their vibrato rates, so as to produce a matched sound, as if from one instrument. There are good examples of this in the *Jack the Bear* recording.

We begin our analysis of the *Jack the Bear* recording with a spectrogram of the finale of the piece. (See Color Figure 14.) This spectrogram shows a number of the features we have noted, and helps us understand the rest of the recording. The passage begins with the time interval A, which contains an orchestral sound produced by brass and saxophones playing with different degrees of frequency spread in their vibrato. There are saxophone fundamentals at about 725 Hz with significant vibrato, and brass fundamentals at about 440 Hz and 523 Hz with less vibrato. The spectrogram also shows low-pitch bass notes at fundamental frequency about 73 Hz. One of the most important features in the passage occurs in time interval B, where the main motif for the composition occurs. This main motif is a series of notes played by the saxophone section. In the spectrogram we have indicated this motif by marking the 2nd harmonics of the saxophone tones with the label T. The saxophone tones rise and then descend, with a jump to another descension at the end. Ellington accents the conclusion of this motif by some brief piano notes. Ellington would typically play in this reserved style, only sporadically emphasizing particular passages in the music, mostly remaining in the background as the conductor of the orchestra.[15] Most of the finale is taken up by an extended melodic line of bass notes played by Blanton, which is shown at the bottom of the spectrogram during time interval C. The bass melody is punctuated by three rhythmic orchestral bursts, shown in the spectrogram as three dark vertically striped regions labeled by R. The vertical stripes appear because the notes occur so rapidly that there is a percussive effect to the sharp breaths of the players during the attack phases of the notes. As Blanton's bass line culminates, the orchestra plays several constant pitched notes in a chord that holds for the last two seconds of the finale. These chordal tones have varying degrees of frequency spread in their vibrato.

During the finale, the main motif of the composition as a whole was played. We see this motif used repeatedly throughout the composition. For example, in Color Figure 15 we show a passage from earlier in the composition, illustrating the use of this motif. The passage is introduced by six repetitions of the motif, played by the saxophone section. The first of these is indicated by the arrow T, pointing to the second harmonics for the saxophone tones. The fundamentals for these saxophone tones can be seen repeating six times within the time interval from 0 to 8.5 seconds, a very rapid sequencing. During this part, the trombonist Joe Nanton plays several distinctive tones. The second harmonics of these tones are indicated by the arrow pointing from the left of N. Pointing down from N is an arrow indicating fundamentals for these tones. The contours of these tones are very similar to human vocals, because of their rapid pitch excursions with sharp bending attacks. The sound from Nanton's trombone imitates, sometimes with an extraordinary exactness, the sound of a human cry. Joe Nanton, also known as "Tricky Sam" Nanton, is famous for this unique style. Following Nanton's trombone tones, there is a short sequence of horn tones of varying timbre indicated by the arrows marked by H. These arrows point to two fundamentals of horn tones, with very rapid attacks and differing amounts of energy in their harmonics. The spectrogram clearly shows the varying attacks and distributions of energy between harmonics that creates the varying timbre of these horn tones. In particular, for the horn tone pointed to by the right arrow, the 3rd harmonic is very intense, and this creates a ringing timbre.

The passage then contains some further repetitions of the motif, indicated by T′. Horn tones, with fundamentals labeled by H′, punctuate these repetitions of the motif. Just as the horn tones finish, there is a short sequence of alto saxophone notes played at low pitches, marked by D in the spectrogram. These saxophone tones are closely spaced in frequency, creating beating effects that produce a growling sound, a "low-down" sonority. The end of the passage consists of a succession of orchestral rhythm tones, labeled R.

We conclude our analysis of *Jack the Bear* with a passage that contains a virtuoso performance on clarinet by Barney Bigard. A spectrogram of this passage is shown in Color Figure 16. The

---

[15] See, however, Exercise 5.3.3 for an interesting example of an Ellington piano solo.

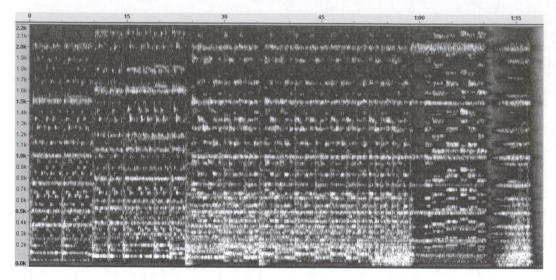

Figure 5.31. Spectrogram from the second half of the finale of Stravinsky's *Firebird Suite*.

passage begins with a rapidly rising ascension by Bigard's clarinet, pointed to by the left-hand arrow under the label C_1. Underlying it are a rhythmic ostinato of bass tones played by Blanton. The fundamentals for these tones are marked by the label B at the lower left corner. Following the cutoff of the clarinet ascension, right at its crest at about 980 Hz, there is a continuation by a saxophone with significant frequency spread in its vibrato. These saxophone tones are labeled by S_1. When the saxophone cuts off (see the vibratoed harmonic above the label S_2), Bigard's clarinet resumes with a rapid descension (pointed to by the right-hand arrow under the label C_1). When Bigard's solo resumes, a fascinating effect occurs. We perceive that the clarinet sound rose into inaudibility, and now has returned to audibility, even though the spectrogram shows no trace of a sound being played by the clarinet between the ascension and descension. This perception of inaudible music has been described incisively by the musicologist and concert pianist, Charles Rosen:

> Inaudible music may seem an odd notion, even a foolishly Romantic one—although it is partly the Romantic prejudice in favor of sensuous experience that makes it seem odd. Still, there are details of music which cannot be heard but only imagined, and even certain aspects of musical form which cannot be realized even by the imagination. (C. Rosen, *The Romantic Generation*, p. 1.)

This whole sequence of clarinet, saxophone, clarinet, is a large-scale version of the motif for the composition. As if to emphasize this, the motif (labeled by T) is repeated by the saxophone section as a whole during the sequence, and immediately following the sequence. During the second repetition of the motif there is a ringing vibratoed tone played by the saxophone section, labeled by V in the spectrogram, at about 1140 Hz from 8.5 seconds to 9.5 seconds. The spectrogram focuses our attention on this tone, played in synchronous vibrato by the saxophone section as an accent on the main motif. It would be easy to miss just by listening alone. Bigard then returns on clarinet with a couple of quick bent pitches, labeled by C_2. There is then a sequence of rhythmic playing of the orchestra, vertical striped bands starting with the one labeled R. Here the orchestral rhythm serves as counterpoint to several of Bigard's clarinet tones. The passage ends with an extremely rapid succession of clarinet descensions, labeled C_3, with the orchestra again providing rhythmic pulses in the background. The spectrogram shows that these clarinet tones are an example of sequencing, and it also shows the pitch accuracy and precise timing with which Bigard executes this sequencing.

Our analysis of *Jack the Bear* was based on spectrograms of a recording of the composition. The spectrograms reveal aspects of the musical performance that can be difficult to glean from analyzing the score for the piece. There is some good score analysis, however, in Gunther Schuller's monumental history of jazz.[16] In particular, Schuller points out an interesting doubling effect of Blanton's bass melody with the saxophone melody two octaves higher.

5.3.8 Concluding Remarks

We have discussed how spectrograms provide a visual depiction of a number of important qualities of the music that we hear performed. Spectrograms accurately portray harmonic relationships, timbre, and vibrato. They also provide at least a partial portrayal of dynamics—in this respect, not that much different from our hearing, which is not extraordinarily accurate in detecting volume levels. Finally, they provide an approach to detecting pattern and structure in musical composition. In this last aspect, they are not to be counterposed to score analysis. Instead, they should be used in conjunction with score analysis to provide further insight.

Exercises

5.3.1. In Figure 5.31 on p. 164, we show a spectrogram of the ending part of the finale of Stravinsky's *Firebird Suite*. Analyze this musical passage using this spectrogram.

5.3.2. In Figure 5.32 we show a spectrogram of a recording of the "Simple Gifts" section of Aaron Copland's *Appalachian Spring*. Analyze this musical passage using this spectrogram.

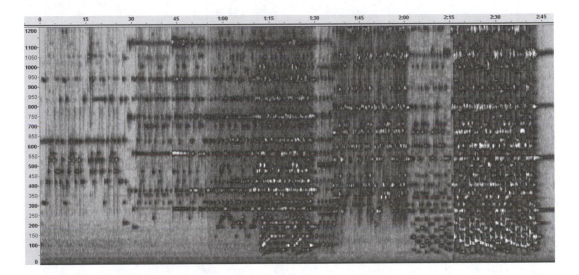

Figure 5.32. Spectrogram from a recording of Copland's *Appalachian Spring*.

5.3.3. In Figure 5.33 we show a spectrogram from *Jack the Bear* in which Duke Ellington plays a melody filled with blue notes. **(a)** Identify at least three occurrences of blue notes, what piano keys were played simultaneously to create them, and what the blue note frequencies are. **(b)** What characteristics from the analysis of *Jack the Bear* given in the text are also present in the music in this passage?

5.3.4. In Figure 5.34 we show a spectrogram of a recording of John Cage's *Piano Sonata II*. Analyze this musical passage using this spectrogram, include any uses of musical transformations and explain how the spectrogram relates to the timbre of the piano. (The piano used for the recording is called a *prepared piano*. It is a piano with bolts and other objects placed on the strings in order to change their harmonics.)

---

[16]G. Schuller. *The Swing Era: The Development of Jazz. 1930–1945*. Oxford, 1989, pp. 114–116.

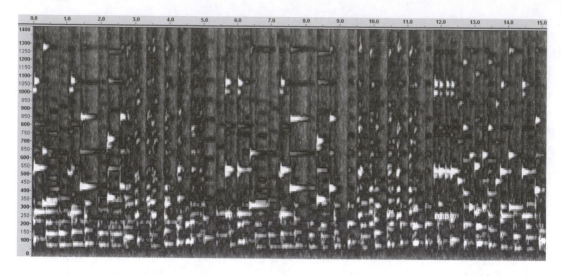

Figure 5.33. Spectrogram from *Jack the Bear* with Duke Ellington's piano melody.

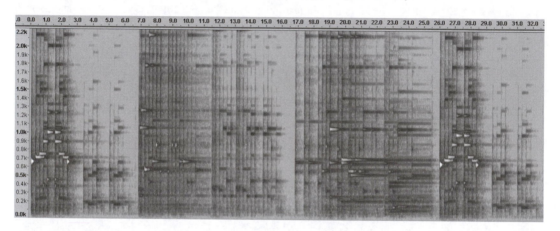

Figure 5.34. Spectrogram of passage from a recording of John Cage's *Piano Sonata II*.

Essay

Certainly the most exciting way to apply the ideas of this chapter is to perform a spectrogram analysis of a piece of music of your own choosing. Write an essay, of at least 5 pages in length, that analyzes a musical recording. Please include images of spectrograms to clarify your points. If possible, cite examples from a scoring of the music that relate to the spectrogram displays.

Chapter 6

Analyzing Pitch and Rhythm

...notice the close psychological parallel between scale degree and metric position. Both are schematic frameworks that allow us to hear events "as" something. We hear the pitch C *as the tonic. We hear a particular moment as the downbeat.*

—David Huron

In this chapter we further explore the geometric approach to pitch and musical scales introduced in Chapter 1. We also discuss how the same mathematics used for organizing pitches underlies periodically repeated musical rhythms (cyclic rhythms). This mathematics is a generalization of the clock arithmetic we used to organize pitches on the chromatic clock. In particular, the transpositions and chromatic inversions of pitches we discussed in Chapter 3 can also be applied to cyclic rhythms. To see this how this is done, we need to look at these transformations relative to the chromatic clock. This approach provides a deep analysis of the structure of harmony and the harmonic relations that govern melody in diatonic music. It also applies to other musical forms as well, such as music using pentatonic scales and the 12-tone chromatic scale. When applied to cyclic rhythms, it provides a key to understanding a variety of rhythms used throughout the world's music. It is quite satisfying to find that these two fundamental dimensions of music, pitch and rhythm, have some common logic underlying their organization.

6.1 Geometry of Pitch Organization and Transpositions

In this section we describe a geometric interpretation of how pitches are organized. We relate this geometry to transpositions and chromatic inversions.

6.1.1 Pitch Classes, Intervals, Chords, and Scales

In Chapter 1 when we introduced the chromatic clock, shown in Figure 1.16 on page 19, we pointed out that the note letters on the clock stand for collections of notes in different registers. For example, C is defined by

$$\mathbf{C} = \{C_1, C_2, C_3, C_4, C_5, C_6, C_7\}$$

and $\mathbf{C}^\sharp$ is defined by

$$\mathbf{C}^\sharp = \{C_1^\sharp, C_2^\sharp, C_3^\sharp, C_4^\sharp, C_5^\sharp, C_6^\sharp, C_7^\sharp\}$$

and so forth. These collections of notes, or pitches, are referred to as ***pitch classes***. In our discussion here, when we plot a point on the chromatic clock at a specific hour, then that point will stand for an entire pitch class. For example, if we plot a point at hour 2 of the chromatic clock, then it stands for the pitch class $\mathbf{D} = \{D_1, D_2, D_3, D_4, D_5, D_6, D_7\}$. In Chapter 1, we discussed the idea of octave equivalence. All of the notes in a pitch class are octave-equivalent. For instance, all of the notes $D_1, D_2, D_3, D_4, D_5, D_6, D_7$ that comprise the pitch class $\mathbf{D}$ are octave-equivalent. Because of the connection with octave equivalency, studying the relation between pitch classes is fundamental to understanding important musical ideas like chordal harmony and chordal progressions (both for harmony and melody).

We will now define a notion of distance between two pitch classes. This notion of distance is useful for comparing chords and musical scales.

Definition 6.1.1. *The* **chromatic distance** *between two pitch classes is the smallest number of half steps between two pitches selected from each of the two pitch classes.*

For example, on the left of Figure 6.1, we have plotted the pitch classes **C** and **E**, at hours 0 and 4, respectively. There are 4 half steps between C_1 and E_1, and also between C_2 and E_2, and other pairs as well. There is no other pair from **C** and **E** that has a smaller number of half steps separating its pitches. So the chromatic distance between **C** and **E** is 4. This is illustrated on the chromatic clock shown on the left side of Figure 6.1. We have drawn a line segment between the two points at hours 0 and 4, and have labeled that line segment with the number 4 to indicate the chromatic distance between the two pitch classes. Notice that 4 equals the smallest number of arc-length segments, hour increments, separating the two points on the clock.

As another example, suppose we have the pitch classes $\mathbf{C} = \{C_1, C_2, C_3, C_4, C_5, C_6, C_7\}$ and $\mathbf{G} = \{G_1, G_2, G_3, G_4, G_5, G_6, G_7\}$. On the second chromatic clock from the left in Figure 6.1 we have plotted points for the pitch classes **C** and **G** at hours 0 and 7, respectively. There are 5 half steps between G_1 and C_2 (and also between G_2 and C_3 and other choices as well). There is no other pair from **C** and **G** that has a smaller number of half steps separating its pitches. So the chromatic distance between **C** and **G** is 5. On the chromatic clock we have illustrated this by drawing a line segment between the two points at hours 0 and 7 and labeling this line segment with the number 5. We note that 5 equals the smallest number of arc-length segments, hour increments, separating the two points on the chromatic clock.

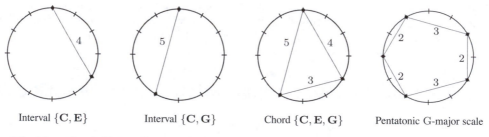

Interval {**C**, **E**}   Interval {**C**, **G**}   Chord {**C**, **E**, **G**}   Pentatonic G-major scale

Figure 6.1. Measuring half-step distances between pitch classes, represented as points on chromatic clock. Numbers next to line segments are smallest number of half steps between members of pitch classes connected by the line segments. These numbers also equal the smallest number of arc-length segments, hour increments, separating two points on the clock.

We have been careful to use the term, *line segment,* in the paragraphs above. The line segments connecting two points on the chromatic clock are usually called chords in geometry. Of course, we do not wish to confuse them with musical chords. Musical chords appear as polygons when graphed on the chromatic clock. For example, 3-note chords appear as triangles. On the chromatic clock graphed third from the left in Figure 6.1, we have graphed the pitch classes **C**, **E**, and **G**, comprising the pitch classes for the C-major chord. Connecting all the pairs of pitch classes from these three pitch classes produces a triangle. We have labeled the line segments of this triangle with the chromatic distances between the pitch classes that form their endpoints. This graphical depiction of the C-major chord in terms of pitch classes **C**, **E**, and **G** on the chromatic clock is important for the theory of harmony in music. It relates to the use of the C-major chord as part of the harmonic framework for some piece of music, not necessarily to an explicit chord on a score. If the C-major chord is the chord providing the harmonic framework for a musical passage, then notes will be (predominantly) selected from the pitch classes **C**, **E**, and **G**. The graph of these pitch classes on the chromatic clock in Figure 6.1

shows the smallest possible number of half steps between pairs of notes from distinct pitch classes.

We have seen that a 3-note chord corresponds to a triangle inscribed on the chromatic clock. One possible source of confusion, resulting from labeling the sides of this triangle by chromatic distance values, is that one might interpret those numbers as side lengths for the triangle as a geometric figure. For instance, the C-major chord triangle in Figure 6.1 is labeled with the numbers 3, 4, and 5. It is not a right triangle, however, as it would be if those numbers represented side lengths in geometry. In the hope of avoiding such confusion, we shall always refer to these numbers as ***chromatic distances***, and never use the unmodified term, distance.[1]

Besides chords, we can also plot musical scales on the chromatic clock. On the chromatic clock on the right side of Figure 6.1, we have plotted the 5 distinct pitch classes for the pentatonic G-major scale. These pitch classes are **G, A, B, D**, and **E**. For simplicity, we have only drawn the line segments connecting adjacent pitch classes and labeled them with their chromatic distances. Later in the chapter, we will consider all the possible line segments connecting pairs of distinct pitch classes and their chromatic distances. Nevertheless, even with this simplified diagram, we see an interesting symmetry for the pentatonic G-major scale. It has a mirror symmetry about a horizontal mirror line drawn through hours 3 and 9. We shall see later that this mirror symmetry shows that there is only one chromatic inversion that preserves all the notes of this scale. We now turn to the geometric interpretation of musical transformations such as chromatic inversions and transpositions.

6.1.2 Transpositions

In Section 3.5, we considered various chromatic transformations. These included transpositions and chromatic inversions. Transpositions and chromatic inversions have geometric interpretations as rotations and reflections of the chromatic clock. In this section, we look at transpositions. Chromatic inversions are examined in the next section.

In Section 3.5.1, we first defined transpositions as adding a specific number to each hour value of a note on the chromatic clock. For example, the transposition T_1 will add 1 to each hour value of a note. In Figure 6.2 we show the effect of applying T_1 to a B-minor chord. A B-minor chord consists

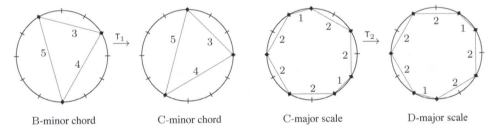

| B-minor chord | C-minor chord | C-major scale | D-major scale |

Figure 6.2. Transpositions as rotations. Transposition T_1 maps B-minor chord, **B-D-F$^\sharp$**, to C-minor chord, **C-E$^\flat$-G**. This appears on chromatic clock as clockwise rotation by 30° (or $2\pi/12$ radians) about circle's center. Transposition T_2 maps C-major scale to D-major scale. This appears on chromatic clock as clockwise rotation by $2 \cdot 30°$ (or $2 \cdot 2\pi/12$ radians) about circle's center.

of the pitch classes **B-D-F$^\sharp$**, which T_1 maps to **C-E$^\flat$-G**, the pitch classes for the C-minor chord. Since each hour value for the pitch classes in the B-minor chord is increased by 1, this corresponds to a clockwise rotation by the angle $2\pi/12$ radians (or $360°/12 = 30°$) about the origin. Moreover, the number of arc-length segments between each pitch class is preserved. This corresponds to the second way we handled transpositions: The transposition T_1 sends root note B to root note C, and preserves half-step changes (hour, or arc-length, increments). Consequently, chromatic distances between the pitch classes in the two chords are preserved.

---

[1]The measure of distance taught in high-school geometry is only one of many possible measures of distance. It is called *Euclidean* distance. The chromatic distance that we are using is a *non-Euclidean* distance.

Likewise, all other transpositions are rotations. For instance, T_2 is a clockwise rotation by $2 \cdot 2\pi/12$ radians about the origin. This follows from the fact that T_2 will add 2 to every hour for pitch classes, and that corresponds to a clockwise rotation by $2 \cdot 2\pi/12$ radians about the origin. In Figure 6.2 we show the effect of applying T_2 to the pitch classes in the C-major scale. Applying the transposition T_2 to the C-major scale produces the D-major scale.

We summarize our discussion of transpositions as rotations with the following theorem.

Theorem 6.1.1. *Every transposition corresponds to a rotation of the chromatic clock by a multiple of $2\pi/12$ radians about its center. In particular, for each integer k, the transposition T_k corresponds to a rotation by $k \cdot 2\pi/12$ radians about the origin. If k is positive, the rotation is clockwise. If k is negative, the rotation is counter-clockwise.*

6.1.3 Clock Arithmetic Formally Defined

Up till now we have been working with clock arithmetic based on telling time on a 12-hour clock. For some applications, however, it helps to have a more formal, purely arithmetic approach that does not rely on the clock metaphor. The basis of this formal approach is the following definition of equivalence, modulo 12, of two integers.

Definition 6.1.2. *Two whole numbers, m and n, are **equivalent**, **modulo** 12, if they differ by a whole number multiple of 12. When m and n are equivalent, modulo 12, we will write: $m \equiv n \bmod 12$. More generally, for a positive whole number $k > 1$, the whole numbers m and n are **equivalent**, **modulo** k, if they differ by a whole number multiple of k. When m and n are equivalent, modulo k, then we will write: $m \equiv n \bmod k$.*

For equivalence, modulo 12, the connection to the chromatic clock should be clear. When $m \equiv n \bmod 12$, then $m - n$ is a whole number multiple of 12. Consequently, by circling around the chromatic clock that same whole number of times, we can go from hour n to hour m, and these hours mark the same position on the chromatic clock.

Example 6.1.1. Equivalence modulo 12. (a) The whole numbers 16 and 4 are equivalent, modulo 12. Since $16 - 4 = 12$, they differ by a whole number multiple of 12. So, we can write $16 \equiv 4 \bmod 12$. Both 16 and 4 mark the position of the pitch class **E** on the chromatic clock. (b) We have $-18 \equiv 6 \bmod 12$, because $-18 - 6 = -24$, which is a whole number multiple of 12. Both -18 and 6 mark the position of the pitch class **F**$^\sharp$, or **G**$^\flat$, on the chromatic clock. (c) We have $11 \equiv -1 \bmod 12$, because $11 - (-1) = 12$, which is a whole number multiple of 12. Both 11 and -1 mark the position of the pitch class **B** on the chromatic clock.

Example 6.1.2. Equivalence modulo 7. (a) We have $14 \equiv 0 \bmod 7$, because $14 - 0 = 14$ which is a whole number multiple of 7. (b) We have $-8 \equiv 6 \bmod 7$, because $-8 - 6 = -14$ which is a whole number multiple of 7. (c) We have $94 \equiv 3 \bmod 7$, because $94 - 3 = 91$ and $91 = 13 \cdot 7$.

Equivalence modulo 7 is important musically because it relates to the cyclic repetition of notes on octave scales. In fact, as we explore in Exercise 6.1.9, the notes of octave scales can be placed on 7-hour clocks, in analogy with the 12-hour clock that we have been using for the chromatic scale.

Example 6.1.3. Equivalence modulo 24. (a) We have $36 \equiv 12 \bmod 24$, because $36 - 12 = 24$ which is a whole number multiple of 24. (b) We have $-23 \equiv 1 \bmod 24$, because $-23 - 1 = -24$ which is a whole number multiple of 24. (c) We have $74 \equiv 2 \bmod 24$, because $74 - 2 = 72$ and $72 = 3 \cdot 24$.

Equivalence modulo 24 is important musically because it relates to the placement of additional notes in between notes on the chromatic clock. In fact, as shown in Exercise 6.1.10, we can use a 24-hour clock to include blue notes and quarter-step notes in between chromatic scale notes. For this 24-hour clock, we have geometric interpretations of transpositions and inversions just like with a 12-hour clock.

Formulas for transpositions, mod 12

Using the idea of equivalence, modulo 12, there are arithmetic formulas for transpositions of pitch classes. Consider, for example, the transposition T_3. Since the transposition T_3 adds 3 hours to the hour of any given pitch class on the chromatic clock, we have the following description of T_3:

$$T_3: \quad m \longrightarrow 3 + m \mod 12 \qquad \text{(for each hour } m\text{)}. \tag{6.1}$$

So, for example, T_3 maps hour 0 to hour 3. It also maps hour 11 to hour 2, because $3 + 11 = 14$ and $14 \equiv 2 \mod 12$.

In general, for a given integer k, the transposition T_k is described by the following mapping:

$$T_k: \quad m \longrightarrow k + m \mod 12 \qquad \text{(for each hour } m\text{)}. \tag{6.2}$$

For example, T_7 is described by the mapping $m \longrightarrow 7 + m \mod 12$. Therefore, T_7 maps 0 to 7, and maps 8 to 3 (because $7 + 8 = 15$, and $15 \equiv 3 \mod 12$). As another example, T_{-4} is described by the mapping $m \longrightarrow -4 + m \mod 12$. Therefore, T_{-4} maps 9 to 5, and maps 2 to 10 (because $-4 + 2 = -2$ and $-2 \equiv 10 \mod 12$).

Because $12 \equiv 0 \mod 12$, there are only 12 distinct transpositions of pitch classes. The 12 distinct transpositions for pitch classes are

$$T_0, T_1, T_3, \ldots, T_{11}.$$

Other transpositions of pitch classes, such as T_{16}, are equal to one of these 12 transpositions. In the case of T_{16}, we have $16 \equiv 4 \mod 12$, so $T_{16} = T_4$ when applied to pitch classes.

Exercises

6.1.1. For the following graphs of musical intervals, label the pitch classes that are marked by dots, and find the chromatic distances between these pitch classes.

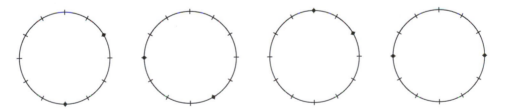

6.1.2. For the following graphs of musical intervals, label the pitch classes that are marked by dots, and find the chromatic distances between these pitch classes.

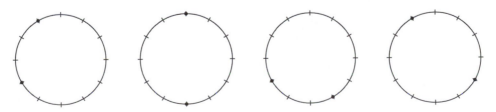

6.1.3. Find the angles of rotations of the chromatic clock about its center for the following transpositions: **(a)** T_4 **(b)** T_{-3} **(c)** T_{-5} **(d)** T_9.

6.1.4. Find the angles of rotations of the chromatic clock about its center for the following transpositions: **(a)** T_5 **(b)** T_{-4} **(c)** T_{-7} **(d)** T_6.

6.1.5. Label these chromatic clocks with dots for the pitch classes of each of the indicated chords, and draw line segments between pairs of these pitch classes with labels for the chromatic distances. The first chord is worked out for you.

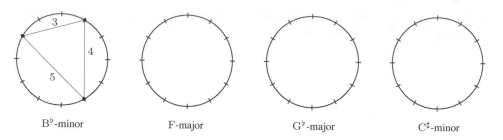

| $B^\flat$-minor | F-major | $G^\flat$-major | $C^\sharp$-minor |

6.1.6. Label these chromatic clocks with dots for the pitch classes of each of the indicated chords, draw line segments between pairs of these pitch classes with labels for the chromatic distances.

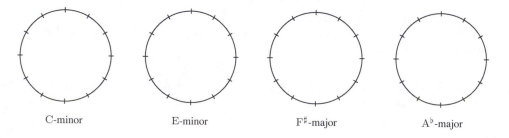

| C-minor | E-minor | $F^\sharp$-major | $A^\flat$-major |

6.1.7. For each of the items below, find the number from 0 to $k-1$ that is equivalent modulo k to the given number. The first item is done for you as an example.

(a) 44 mod 7 (Answer: $44 \equiv 2 \bmod 7$.)

(b) 83 mod 12

(c) 122 mod 24

(d) -63 mod 10

(e) 38 mod 16

(f) -12 mod 7

6.1.8. For each of the items below, find the number from 0 to $k-1$ that is equivalent modulo k to the given number.

(a) 64 mod 7

(b) -8 mod 12

(c) 93 mod 24

(d) 84 mod 10

(e) -34 mod 16

(f) 185 mod 7

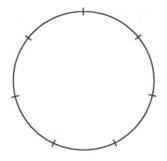 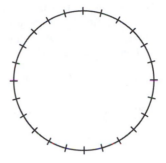

Figure 6.3. Left: A 7-hour clock. Right: A 24-hour clock.

6.1.9. On the left of Figure 6.3 there is a 7-hour clock. **(a)** What would be the best way to label the hours of this clock with hour numbers, and with note letters for displaying the C-major scale? **(b)** What would be the best way to label the hours of this clock with hour numbers, and with note letters for the G-major scale? **(c)** Suppose you have labeled this clock with hour numbers, and with letters for the C-major scale. What is the geometric interpretation of diatonic scale shifts? What would be a formula for a diatonic scale shift $\mathcal{S}_k$, using mod-arithmetic based on this 7-hour clock?

6.1.10. On the right of Figure 6.3 there is a 24-hour clock. **(a)** Label the hour marks on this clock with notes of the chromatic scale, and with blue notes and quarter-step notes in between the chromatic scale notes. Use notations such as $\mathbf{C}^{\sharp}$, and $\mathbf{D}^{\flat}$, for blue notes and quarter-step notes between $\mathbf{C}$ and $\mathbf{C}^{\sharp}$, and between $\mathbf{D}^{\flat}$ and $\mathbf{D}$, respectively. **(b)** Discuss the mathematics for transpositions involving pitch classes on this 24-hour clock. In particular, explain how these transpositions are described geometrically, and with mod-arithmetic.

6.2 Geometry of Chromatic Inversions

In the previous section we saw how transpositions of pitch classes can be thought of geometrically as rotations of the chromatic clock. In this section we shall see that chromatic inversions of pitch classes can be thought of geometrically as reflections of the chromatic clock through mirror lines passing through the center of the clock.

To keep things as simple as possible, and also because of their importance musically, we focus on chromatic inversions of chords of pitch classes. If the notes of chords are graphed by marking their pitch classes on the chromatic clock, then chromatic inversions can be accomplished by a reflection about a mirror line through the origin. The two types of mirror lines that are used are shown in Figure 6.4. For both types of mirrors, reflection will reverse the signs of all half-step changes between pitch classes. If, before reflection, half steps are being taken as +1 changes around the chromatic

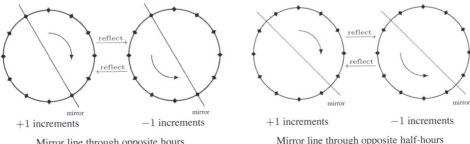

Figure 6.4. Chromatic inversion as reflection. First kind, shown on left, is reflection through mirror line passing through two opposite hours on chromatic clock. Those opposite hours are unchanged by reflection. Second kind, shown on right, is reflection through mirror line passing through two opposite half-hour positions on chromatic clock. All hours are changed by reflection.

clock, then reflection turns them into -1 changes. Likewise, if half steps are being taken as -1 changes around the chromatic clock before reflection, then reflection turns them into $+1$ changes. Therefore, any number of half steps between two pitch classes will reverse sign when either one of these types of reflection is performed.

Example 6.2.1. Chromatic inversion of a minor chord. As an example, consider a chromatic inversion IT_4 of the B-minor chord. Using the method described in Section 3.5.3, the following diagram shows how this chromatic inversion produces an $\mathrm{A}^\flat$-major chord:

$$
\begin{array}{ccc}
\mathrm{F}^\sharp & & \mathrm{E}^\flat \\
+4 \uparrow & & -3 \downarrow \\
\mathrm{D} & \quad T_4 & \mathrm{C} \\
+3 \uparrow & & -4 \downarrow \\
\mathrm{B} & & \mathrm{A}^\flat \\
\end{array}
$$

$$(\text{B-minor}) \xrightarrow{\ \mathsf{IT}_4\ } (\mathrm{A}^\flat\text{-major})$$

On the left side of Figure 6.5, we show that this application of IT_4 to the B-minor chord is the result of a mirror reflection on the chromatic clock. The mirror line passes through hours 1 and 7. For the B-minor chord, the half-step changes of $+3$ and $+4$ from its root **B** are indicated by arrows pointing from **B** to **D**, and from **D** to $\mathbf{F}^\sharp$. When the mirror reflection is performed, the root **B** is reflected to $\mathbf{E}^\flat$, and the arrow from $\mathbf{E}^\flat$ to **C** corresponds to a change by -3 half steps. The arrow from **C** to $\mathbf{A}^\flat$ corresponds to a change of -4 half steps. This produces the $\mathrm{A}^\flat$-major chord from the pitch class $\mathbf{E}^\flat$.

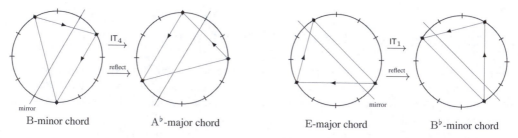

Figure 6.5. Geometric Interpretation of Chromatic Inversion. Left: B-minor chord marked on chromatic clock with mirror line. Reflecting through mirror line reverses direction of arrows, producing $\mathrm{A}^\flat$-major chord. This reflection is equivalent to performing chromatic inversion IT_4. Right: E-major chord marked on chromatic clock with mirror line. Reflecting through mirror line reverses direction of arrows, producing $\mathrm{B}^\flat$-minor chord. This reflection is equivalent to performing chromatic inversion IT_1.

Example 6.2.2. Chromatic inversion of a major chord. As a second example, let's examine a chromatic inversion IT_1 of the E-major chord. Using the method described in Section 3.5.3, the following diagram shows how this chromatic inversion produces a $\mathrm{B}^\flat$-minor chord:

$$
\begin{array}{ccc}
\mathrm{B} & & \mathrm{F} \\
+3 \uparrow & & -4 \downarrow \\
\mathrm{G}^\sharp & \quad T_1 & \mathrm{D}^\flat \\
+4 \uparrow & & -3 \downarrow \\
\mathrm{E} & & \mathrm{B}^\flat \\
\end{array}
$$

$$(\text{E-major}) \xrightarrow{\ \mathsf{IT}_1\ } (\mathrm{B}^\flat\text{-minor})$$

On the right side of Figure 6.5, we show that this application of IT_1 to the E-major chord is the result of doing a mirror reflection on the chromatic clock. The mirror line passes through the hours 4.5 and

10.5. The half-step changes of $+4$ and $+3$ from the chord's root **E** are indicated by arrows pointing from **E** to **G**$^\sharp$, and from **G**$^\sharp$ to **B**. When the mirror reflection is performed, **E** is reflected to **F**, and the directions of the arrows are reversed. The arrow from **F** to **D**$^\flat$ corresponds to a change by -4 half steps, and the arrow from **D**$^\flat$ to **B**$^\flat$ corresponds to a change of -3 half steps. This produces the B$^\flat$-minor chord from the pitch class **F**.

Formulas for chromatic inversions, mod 12

Using the idea of equivalence, mod 12, there are arithmetic formulas for chromatic inversions of pitch classes. Consider, for example, the chromatic inversion performed by reflection of the chromatic clock through a vertical mirror passing through the hours 0 and 6. This mirror inversion is described by the following mapping:

$$m \longrightarrow -m \mod 12.$$

Notice that 0 is mapped to 0, in accordance with its lying on the mirror. Also, 6 is mapped to $-6 \equiv 6$ mod 12. Therefore, 6 is mapped to 6 on the chromatic clock, in accordance with its lying on the mirror. Furthermore, 1 is mapped to $-1 \equiv 11$ mod 12, and that matches what the mirror reflection does. The reader may check that this mapping does indeed map all the hours on the chromatic clock to their mirror reflected positions. We will denote this mapping by $\mathcal{I}_0$:

$$\mathcal{I}_0 : \quad m \longrightarrow -m \mod 12 \qquad\qquad (6.3)$$

and refer to it as the ***pitch class chromatic inversion*** $\mathcal{I}_0$. We are using the modifier, *pitch class,* in referring to this chromatic inversion, because it is based on mod 12 arithmetic for the pitch classes on the chromatic clock. The chromatic inversions described in Chapter 3 are not limited to pitch classes; they can act on notes in different registers according to their score positions. For a pitch class chromatic inversion like $\mathcal{I}_0$, we are viewing the mapping on the chromatic clock solely in terms of pitch classes.

The other pitch class chromatic inversions are also reflections of the chromatic clock, either through mirrors passing through opposite hours on the clock, or through mirrors passing through opposite half-hours. There are 12 mirrors in all, 11 in addition to the vertical mirror used for $\mathcal{I}_0$. The 11 other mirrors are clockwise rotations of the vertical mirror by $k \cdot \pi/12$ radians ($k \cdot 15°$) for $k = 1, 2, \ldots, 11$. If the angle of clockwise rotation of the vertical mirror is $k \cdot \pi/12$ radians, then hour 0 is reflected to hour k. For example, if $k = 1$, then the angle of clockwise rotation of the vertical mirror is $\pi/12$ (or $15°$) and the mirror passes through the half-hours 0.5 and 6.5. Therefore, hour 0 is reflected to hour 1. Since all hour changes of $+1$ moving clockwise are changed to -1 moving counter-clockwise after reflection, the mapping

$$m \longrightarrow 1 - m \mod 12 \qquad \text{(for each hour } m\text{)}$$

describes this reflection. Notice that hour 1 is mapped to hour 0, as it is by the reflection through the mirror passing through hour 0.5. We leave it to the reader to check that the other hours on the clock are mapped to their reflected hours by this mapping. We will denote this mapping by $\mathcal{I}_1$. These examples of $\mathcal{I}_0$ and $\mathcal{I}_1$ are typical of all pitch class chromatic inversions, which we now define.

For each $k = 0, 1, 2, \ldots, 11$, we define a pitch class chromatic inversion $\mathcal{I}_k$ to be the mapping:

$$\mathcal{I}_k : \quad m \longrightarrow k - m \mod 12. \qquad\qquad (6.4)$$

For a given k, the pitch class chromatic inversion $\mathcal{I}_k$ describes the reflection of the chromatic clock through a mirror that is a clockwise rotation by $k \cdot \pi/12$ radians (or $k \cdot 15°$). This reflection maps hour 0 to hour k, and reverses signs for hour changes because of the $-m$ term.

Exercises

6.2.1. For the chromatic clocks below, draw the triangles for the chords indicated, and the positions of the mirrors for the reflections corresponding to the indicated chromatic inversions.

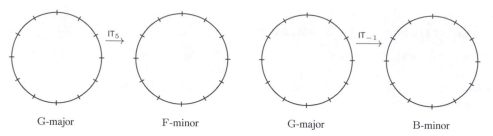

G-major · F-minor · G-major · B-minor

6.2.2. For the chromatic clocks below, draw the triangles for the chords indicated, and the positions of the mirrors for the reflections corresponding to the indicated chromatic inversions.

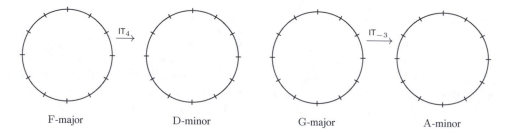

F-major · D-minor · G-major · A-minor

6.2.3. For the chromatic clocks below, draw the triangles for the chords indicated, and the positions of the mirrors for the reflections corresponding to the indicated chromatic inversions. Also draw the triangles for the chromatically inverted chords. Label the chromatic clocks for the chromatically inverted chords by the names for those chords.

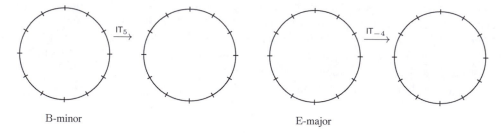

B-minor · E-major

6.2.4. For the chromatic clocks below, draw the triangles for the chords indicated, and the positions of the mirrors for the reflections corresponding to the indicated chromatic inversions. Also draw the triangles for the chromatically inverted chords. Label the chromatic clocks for the chromatically inverted chords by the names for those chords.

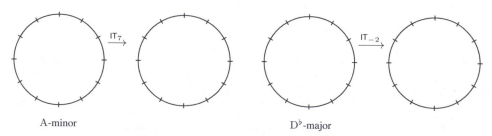

A-minor · D♭-major

6.2.5. In **(a)** through **(d)** below, we list a pair of hours on the chromatic clock. In each case, find the pitch class chromatic inversion $\mathcal{I}_k$ that has a mirror through those hours. (We have done part **(a)** for you.)

  (a) Hours $1.5, 7.5$ (Answer: $\mathcal{I}_3$)

  (b) Hours $1, 7$

  (c) Hours $2.5, 8.5$

  (d) Hours $4, 10$

There is a pattern that connects these pairs of hours, and their corresponding inversions $\mathcal{I}_k$; find a formula that encapsulates this pattern.

6.2.6. Find the pitch class chromatic inversions, $\mathcal{I}_k$, corresponding to the chromatic inversions in Figure 6.5.

6.2.7. Find the pitch class chromatic inversions, $\mathcal{I}_k$, corresponding to the chromatic inversions in Exercises 6.2.1 and 6.2.3.

6.2.8. Find the pitch class chromatic inversions, $\mathcal{I}_k$, corresponding to the chromatic inversions in Exercises 6.2.2 and 6.2.4.

6.2.9. On the left of Figure 6.3, on p. 173, there is the 7-hour clock we considered in Exercise 6.1.9. Suppose this clock has hour numbers 0 through 6, and these hours are labeled by note letters from the C-major scale. What is the geometric interpretation of diatonic scale inversions? How would you generalize the notion of pitch class chromatic inversions to this diatonic case?

6.2.10. On the right of Figure 6.3, there is the 24-hour clock we considered in Exercise 6.1.10. Suppose this clock has hour numbers 0 through 23, and these hours are labeled by note letters from the chromatic scale, and labeled with blue notes and quarter-step notes in between those chromatic notes. Provide a mathematical formula that generalizes the notion of pitch class chromatic inversion to this extended case.

6.3 Cyclic Rhythms

We now begin our analysis of mathematical patterns in musical rhythm. The process of musical rhythm has been succinctly described by Grosvenor Cooper and Leonard Meyer:

> To experience rhythm is to group separate sounds into structured patterns. Such grouping is the result of the interaction among the various aspects of the materials of music: pitch, intensity, timbre, texture, and harmony—as well as duration.[2]

Realizing a complete mathematical description of the multiple aspects of rhythm described by Cooper and Meyer is a major undertaking, one that has yet to be completed. We will begin our treatment with the last aspect that they mention: duration. The relative time duration of notes, and the timing pattern for the music, are important aspects of rhythm. One type of rhythmic organization in particular, *cyclic rhythm,* has been studied extensively in the last couple of decades. A cyclic rhythm is a sequence of note durations that repeats itself periodically. Cyclic rhythms can be described with the same mathematics that we have used for organizing pitch in the previous two sections. This allows for the same types of transformations, rotations and reflections, that we used on pitch classes to also be employed on cyclic rhythms.

6.3.1 Cyclic Rhythm in *Down in the Valley*

As a first example of a cyclic rhythm, we consider the traditional song, *Down in the Valley.* The entire melody for this song is on the bottom of Figure 6.6. The relative time durations (note values) of the notes in the first five measures are repeated sixteen times to form the whole song. In these first five measures—which provide the melody for the lyrics *"Down in the valley, Valley so low"*—the shortest

---

[2]G. Cooper and L. Meyer, *The Rhythmic Structure of Music,* p. 1.

Figure 6.6. Cyclic rhythm for *Down in the Valley*. Bottom: Melody for entire song. This melody is repeated 8 times. Top: Note onsets and values for Measures 1 through 5—for lyrics *"Down in the valley, Valley so low"*—indicated on 15-hour clock. This rhythm is repeated for measures 6 to 10, for lyrics *"Hang your head over, Hear the wind blow."* It repeats 14 more times for the entire song.

duration note is a quarter note. Since the time signature is $\frac{3}{4}$, there can be 3 quarter notes within each measure. Hence, there is sufficient time for 15 quarter notes within the first five measures. Therefore, we will mark the onset of each note as some hour on a 15-hour clock. This clock begins with hour 0 at the top, like all of our clocks in this book. On this 15-hour clock, each quarter note is thought of as occupying a duration of one hour. A measure will span 3 hours. A dotted half-note has a duration equal to 3 quarter notes, or one measure, so it occupies a duration of 3 hours. For each clockwise cycle around this clock, the rhythm repeats its five-measure duration.

This example from folk music has illustrated a couple of important features of rhythm. First, we can plot note onsets for a cyclic rhythm as hours on a clock. Second, we saw a connection between the rhythm of note onsets and the melodic structure of the music. This connection is commonly found in tonal music. Of course, there can also be pure rhythm, as found in many types of drumming.

6.3.2 Cyclic Rhythm in Drumming

As another case of cyclic rhythm, we consider an example of drumming. In Figure 6.7, we show a drum score for a cyclic rhythm, called *cinquillo*. The *cinquillo* rhythm is commonly used in Cuban drumming. The basic rhythm is laid down in the first measure of the score, and then repeats for the next three measures. The note onsets for this rhythm can be plotted on an 8-hour clock, as shown in the figure.

6.3.3 Time Transpositions of Cyclic Rhythms

We can plot the note onsets for a cyclic rhythm on a clock with N hours, for some positive integer N. Therefore, we can use the idea of transposition (or rotation) that we employed for pitch classes on the chromatic clock. For instance, the *cinquillo* cyclic rhythm uses an 8-hour clock. So there are 8

Figure 6.7. Top: four measures of *cinquillo* rhythm in drum score for conga. A conga drum can play a high note and a low note. Below: 8-hour clock with note onsets for single measure from drum score. These note onsets are repeated cyclically through four measures.

possible transpositions, T_k, for $k = 0, 1, 2, 3, 4, 5, 6, 7,$ defined by

$$\mathsf{T}_k : \quad m \longrightarrow k + m \mod 8 \qquad \text{(for each hour } m\text{)}.$$

On the left of Figure 6.8 we show the effect of the transposition T_3 on the *cinquillo* rhythm. Its note

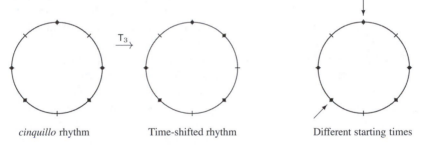

| *cinquillo* rhythm | Time-shifted rhythm | Different starting times |

Figure 6.8. Transposition (time-shifting) of *cinquillo* cyclic rhythm. Left: Time-shifting by T_3. Right: Different starting times for two rhythms. The arrow pointing to hour 0 marks starting time for *cinquillo* rhythm. Arrow pointing to hour $-3 \equiv 5 \mod 8$ marks starting time for time-shifted rhythm.

onsets are at the hours $0, 2, 3, 5, 6$. Applying T_3 to these hours produces the hours $3, 5, 6, 8, 9,$ which in mod 8 arithmetic are equivalent to $3, 5, 6, 0, 1$. Written in standard order, these hours are $0, 1, 3, 5, 6$ and those are the hours of the note onsets shown on the second clock in Figure 6.8.

The transpositions of pitch that we have dealt with up to now are related to frequency, while these transpositions of cyclic rhythms are related to time. In order to avoid confusion, we shall refer to transpositions of cyclic rhythms as ***time-shifts***. We will employ the same notation of T_k, however. The time-shift T_3 shown in Figure 6.8, moves the note onsets of the *cinquillo* rhythm ahead 3 hours. As with transpositions of the chromatic clock, we see that this time-shifting of a cyclic rhythm corresponds to a rotation of the clock marked by the note onsets. In the case of T_3 for the 8-hour clock, the rotation is clockwise by an angle of $3 \cdot 2\pi/8$ radians ($135°$).

On the right of Figure 6.8 we show an alternative way of viewing the time-shift T_3. We can view this time-shift as employing a different starting time for the time-shifted rhythm. The starting time for the *cinquillo* rhythm is hour 0, marked by the arrow pointing to hour 0 in the figure. If we set a new starting time, hour $-3 \equiv 5 \mod 8$, indicated by the other arrow in the figure, then the cyclic rhythm starting at hour -3 is equivalent to the time-shifted rhythm. This alternative view of time-shifting lies at the heart of the *clave* rhythms used in Afro-Latin music.

6.3.4 Afro-Latin Clave Rhythms

A lovely application of rotation of a cyclic rhythm, and its interpretation in terms of a different starting time, is found in the Afro-Latin rhythms called *clave* rhythms. These rhythms can be played with percussion instruments called *claves,* but they are also played with many other instruments.

A clave rhythm combines a cyclic rhythm with a rotated version of that rhythm. For example, in Figure 6.9, the first clock on the left of the figure shows a cyclic rhythm called *son*. Its note onsets

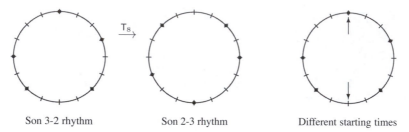

| Son 3-2 rhythm | Son 2-3 rhythm | Different starting times |

Figure 6.9. The son clave rhythm. Left: son 3-2 rhythm rotated by T_8 to give son 2-3 rhythm. Right: son 3-2 rhythm begins at hour 0, while son 2-3 rhythm begins at hour 8.

are at the hours $0, 3, 6, 10, 12$ on a 16-hour clock. It spans two measures. The first measure begins at hour 0, while the second measure begins at hour 8. Because there are 3 note onsets in the first measure, and 2 note onsets in the second measure, this rhythm is called the son 3-2 rhythm. The second clock shown in Figure 6.9 is a rotated version of the son 3-2 rhythm, obtained by applying the time-shift T_8. Applying T_8 to the hours $0, 3, 6, 10, 12$ produces the hours $8, 11, 14, 18, 20$, which in mod 16 arithmetic are equivalent to the hours $8, 11, 14, 2, 4$. Written in standard order, these hours are $2, 4, 8, 11, 14$ and those are the hours of the note onsets shown on the second clock. This rotated rhythm has 2 note onsets in the first measure, and 3 note onsets in the second measure, so it is called a son 2-3 rhythm. In a *son clave* rhythm, the 3-2 rhythm and the 2-3 rhythm are played simultaneously. Since the rotated rhythm can be viewed as starting at hour 8, the son 2-3 rhythm can begin at hour 8 *in the midst of the son 3-2 rhythm.* Typically, the son 3-2 rhythm is played repeatedly by one instrument, and then at a moment when the son 3-2 rhythm has reached hour 8 mod 16, the son 2-3 rhythm is begun by another instrument. The third clock in Figure 6.9 shows the timing for this son clave rhythm.

Musical expressions of cyclic rhythms

The clocks used to describe the patterns of note onsets for cyclic rhythms only specify the timing of the onsets of the notes. They do not uniquely determine the relative note durations. This implies that there can be multiple ways of musically expressing the same cyclic rhythm. In Figure 6.10 we show two different musical expressions for the son 3-2 rhythm. Although one can hear subtle differences when these notes are played, the fact that they are expressing the same rhythm is clearly audible.

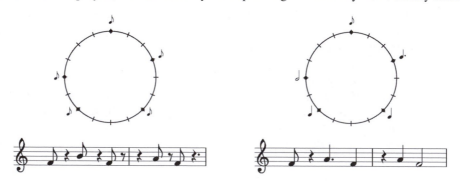

Figure 6.10. Two musical expressions of son rhythm.

A musical example of the son clave

The son clave can be implemented with various instruments. Some instruments will play the 3-2 son, while other instruments will play the 2-3 son. To further enhance the music, additional instruments will play rhythms augmenting and complementing the son rhythms. We show one example of this in Figure 6.11. The instruments employed are an alto steel drum, an electric bass, a conga drum, and a

Figure 6.11. Musical example using son clave. Clocks on left show timing of note onsets for four instruments. Top clock shows electric bass (El. B.) beginning at hour 0 mod 16, and alto steel drum (A. St. Dr.) beginning at hour 8 mod 16. Middle clock shows conga drum (Co.) beginning at hour 0 mod 16, and bottom clock shows quijada (Qui.) beginning at hour 0 mod 16. These latter two instruments are playing rhythms that augment and complement the son rhythms. Score on right is a musical expression of this rhythmic organization.

quijada.[3] The electric bass begins the music with the son 3-2 rhythm. In measure 2, at time 8 mod 16 on the 16-hour clock, the steel drum begins playing the son 2-3 rhythm. The conga drum and quijada, both of which begin at hour 0 mod 16 in the third measure, are playing two different cyclic rhythms that contrast nicely with the son clave. Notice also that the score calls for the 8 measures shown to be played twice. Consequently, at the end of the first 8 measures, there are breaks for two measures in the playing of the steel drums, conga and quijada. The steel drums begin again at hour 8 mod 16, and the other two instruments begin again at hour 0 mod 16. Because of the mathematical precision of the rhythmic organization, these breaks do not affect the coherence of the music's rhythm.

In this example, we used an eclectic mix of instruments. Some of them, like the electric bass, are not traditional ones in the Afro-Latin musical genre. Nevertheless, the Afro-Latin quality of the music is quite evident. The use of these Afro-Latin rhythms is now very widespread in music, especially popular music. The son clave has been described by the rhythm scholar, Godfried Toussaint, as "the rhythm that conquered the world."[4] Howard Goodall's excellent music documentary, *How Music Works — Rhythm*, discusses some of the ideas we have described for the son clave, and shows how

---

[3] A "quijada" is a percussion instrument held in the hand and shaken, with a rattling sound.
[4] G. Toussaint, The Rhythm that Conquered the World: What Makes a "Good" Rhythm Good?, *Percussive Notes*, Nov. 2011, pp. 52–59.

universal the son clave has become in popular music.

Other claves

Besides the son clave, there are many other Afro-Latin claves. In Figure 6.12, we show three other ones, the *rumba*, the *bossa nova,* and the *gahu*. All of these claves are slight variants of the son clave. They each have a 3-2 version and a time-shifted 2-3 version. So they can be used to generate complex rhythmic patterns in the same way as the son clave.

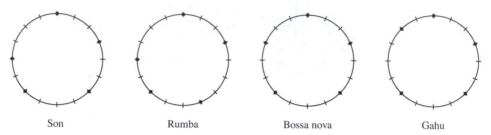

Figure 6.12. Four Afro-Latin clave rhythms.

6.3.5 Phasing

An interesting use of time-shifting of rhythms occurs in the *phasing* technique of Steve Reich. In Figure 6.13 we show the timing diagram for the complex pattern of hand claps that make up Reich's composition, *Clapping Music*. The pattern is created by two sequences of claps. The first sequence,

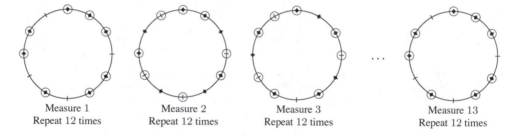

Figure 6.13. Timing diagram for Steve Reich's *Clapping Music*. Solid dots mark onset times for one sequence of claps. Small circles mark onset times for another sequence of claps. Second sequence is repeatedly time-shifted by T_{-1} until it returns to its original form in measure 13. First sequence repeats, unchanged, in every measure.

indicated by solid dots, repeats a total of 156 times: 12 times for each of 13 measures. The second sequence, indicated by small circles, begins in synchrony with the first sequence. In measure 1, both sequences of claps have onsets

$$0 \quad 1 \quad 2 \quad 4 \quad 5 \quad 7 \quad 9 \quad 10$$

for a 12-hour clock.[5] They each repeat 12 times in measure 1. In measure 2, the first sequence is again repeated 12 times. However, the second sequence is time-shifted by T_{-1}. As can be seen from the open circles shown for measure 2 in Figure 6.13, the onsets for the second sequence are

$$0 \quad 1 \quad 3 \quad 4 \quad 6 \quad 8 \quad 9 \quad 11$$

---

[5]Such a clock of onsets would be consistent with a time signature of either $\frac{12}{8}$ or $\frac{6}{4}$. Reich, however, does not indicate a time signature in the score, since he wants performers to avoid putting any rhythmic emphases on the claps. Such rhythmic emphases often being implicit for musicians when time signatures are given.

which is the result of T_{-1} applied to the onsets from measure 1. This time-shifted second sequence is also repeated 12 times, but now it is slightly asynchronous (out of phase) with the first sequence. Again, in measure 3, the second sequence is time-shifted again by T_{-1}, while the first sequence is left the same. Both sequences repeat 12 times in measure 3. This pattern of shifting the second sequence by T_{-1} continues through each subsequent measure, until measure 13. At measure 13 the cumulative shifting by T_{-1} has produced a total shift of T_{-12}, but $T_{-12} = T_0$ because $-12 \equiv 0$ mod 12. Since T_0 is just repetition, the second sequence has returned to its original onsets, in perfect synchrony with the first sequence. This perfect synchrony provides the resolution that ends the piece.

You can listen to a couple of different versions of *Clapping* at the book's web site, including a striking video performance of the piece, by a chorus of hand clappers. In this video you can see the conductor mark off precisely when each new measure begins, and the consequent transition of half of the chorus of clappers to a new cyclic rhythm. *Clapping Music* is just one of Steve Reich's compositions that uses rhythmic phasing. Some additional compositions can be accessed at the book's web site, under the video links for this chapter.

Exercises

6.3.1. In **(a)** through **(d)** below, a shortest duration note is specified, along with a time signature and the number of measures, for a musical passage. In each case, find the number of hours to be used on a clock for describing the rhythm of note onsets. We have done **(a)** for you.

(a) shortest duration: eighth note; time signature $\frac{4}{4}$; measures: 2.   Ans. 16-hour clock.

(b) shortest duration: eighth note; time signature $\frac{3}{4}$; measures: 3.

(c) shortest duration: quarter note; time signature $\frac{4}{4}$; measures: 3.

(d) shortest duration: sixteenth note; time signature $\frac{4}{4}$; measures: 1.

6.3.2. In **(a)** through **(d)** below, a shortest duration note is specified, along with a time signature and the number of measures, for a musical passage. In each case, find the number of hours to be used on a clock for describing the rhythm of note onsets.

(a) shortest duration: eighth note; time signature $\frac{3}{4}$; measures: 4.

(b) shortest duration: sixteenth note; time signature $\frac{4}{4}$; measures: 2.

(c) shortest duration: quarter note; time signature $\frac{5}{4}$; measures: 2.

(d) shortest duration: sixteenth note; time signature $\frac{6}{8}$; measures: 1.

6.3.3. For the following score fragment:

draw a single clock with note onsets for both measures marked by dots.

6.3.4. For the following score fragment:

draw a single clock with note onsets for both measures marked by dots.

6.3.5. For the following score fragment:

draw a single clock with note onsets for both measures marked by dots.

6.3.6. For the following score fragment:

draw a single clock with note onsets for both measures marked by dots.

6.3.7. In the beginning of the piano part of the Adagio section of Mendelssohn's *Sonata No. 2 in D, Op. 58, for cello and piano*, there is an interesting sequence of rhythmic passages. See Figure 6.14. For the first passage, which lies between the first two arrows in the figure, draw a 20-hour clock with note onsets marked by dots. Show that the subsequent three passages, marked off between arrows, are cycling around this clock, but that the second passage jumps ahead by 4 hours when the first passage finishes its cycle.

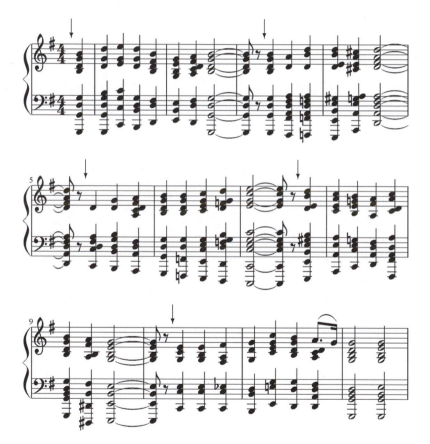

Figure 6.14. Beginning of piano part in Adagio section of Mendelssohn's *Sonata No. 2 in D, Op. 58, for cello and piano*. Arrows indicate beginnings and endings of rhythmic passages. (Note: in this section, Mendelssohn begins the music in key of G-major.)

6.3.8. The previous exercise referred to a rhythmic passage in the beginning of the piano part of Mendelssohn's *Sonata No. 2 in D, Op. 58, for cello and piano*. This passage is shown in Figure 6.14, between the first two arrows in the figure. Find the sequence of chord progressions in this passage. (Note: as pointed out in the caption of the figure, the key for this passage is G-major.)

6.3.9. The following graphs show that note onsets on a 16-hour clock are to be time-shifted by T_3.

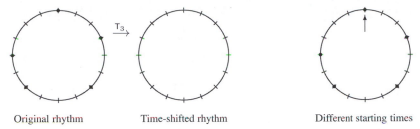

Original rhythm   Time-shifted rhythm   Different starting times

(a) Draw dots on the middle clock indicating hours for note onsets of the time-shifted rhythm. **(b)** The clock on the right shows an arrow pointing to the starting time for the original rhythm. Draw an arrow on this clock pointing to the starting time for the time-shifted rhythm. **(c)** Find a formula, using mod-arithmetic, for time-shifting T_3.

6.3.10. The following graphs show that note onsets on a 16-hour clock are to be time-shifted by T_{-3}.

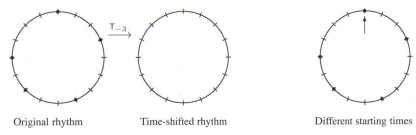

Original rhythm   Time-shifted rhythm   Different starting times

(a) Draw dots on the middle clock indicating hours for the note onsets of the time-shifted rhythm. **(b)** The clock on the right shows an arrow pointing to the starting time for the original rhythm. Draw an arrow on this clock pointing to the starting time for the time-shifted rhythm. **(c)** Find a formula, using mod-arithmetic, for time-shifting T_{-3}.

6.3.11. The following graphs show that note onsets on a 16-hour clock are to be time-shifted by T_4.

Original rhythm   Time-shifted rhythm   Different starting times

(a) Draw dots on the middle clock indicating hours for note onsets of the time-shifted rhythm. **(b)** The clock on the right shows an arrow pointing to the starting time for the original rhythm. Draw an arrow on this clock pointing to the starting time for the time-shifted rhythm. **(c)** Find a formula, using mod-arithmetic, for time-shifting T_4.

6.3.12. The following graphs show that note onsets on a 16-hour clock are to be time-shifted by T_{-2}.

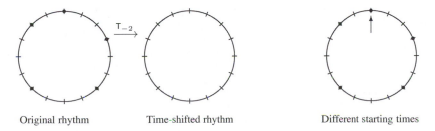

Original rhythm   Time-shifted rhythm   Different starting times

(a) Draw dots on the middle clock indicating hours for note onsets of the time-shifted rhythm. **(b)** The clock on the right shows an arrow pointing to the starting time for the original rhythm. Draw an arrow on this clock pointing to the starting time for the time-shifted rhythm. **(c)** Find a formula, using mod-arithmetic, for time-shifting T_{-2}.

Phasing

The following four exercises deal with the method of rhythmic phasing.

6.3.13. In Figure 6.15, we show the hand clapping composition, *Cinquillo Phasing,* which uses rhythmic phasing based on the *cinquillo rhythm*. Draw the clock diagrams that describe the onsets for the claps in Measures 1, 2, 3, and 8. Describe the method of time-shifting that is being used to create the phasing.

Figure 6.15. Rhythmic phasing composition, *Cinquillo Phasing*. Each measure repeated 8 times.

6.3.14. In Figure 6.16, we show another example of a hand clapping composition, *Cinquillo Phasing 2,* which uses rhythmic phasing. Draw the clock diagrams that describe the onsets for the claps in Measures 1, 2, 3, and 8. Describe the method of time-shifting that is being used to create the phasing.

Figure 6.16. Another rhythmic phasing composition, *Cinquillo Phasing 2*. Each measure repeated 8 times.

6.3.15. For two rhythmic sequences, described by the same time onset clock, we define the ***rhythmic distance*** between them to be the number of time onsets where there is just 1 onset from the two sequences (not 2 onsets together, or 0 onsets). Using this definition of rhythmic distance, complete the following tables:

(a) Rhythmic distances for the 9 measures of the composition, *Cinquillo Phasing,* in Figure 6.15:

Measure:	1	2	3	4	5	6	7	8	9
Rhythmic Distance:	0	6	4						0

(b) Rhythmic distances for the 9 measures of the composition, *Cinquillo Phasing 2,* in Figure 6.16:

Measure:	1	2	3	4	5	6	7	8	9
Rhythmic Distance:	0	2	4						0

(c) Rhythmic distances for the 13 measures of Steve Reich's *Clapping Music.* [Hint: You can use note onset diagrams, as in Figure 6.13.]

Measure:	1	2	3	4	5	6	7	8	9	10	11	12	13
Rhythmic Distance:	0	8	6										0

6.3.16. In Figure 6.17, we show the composition *Phased Son*, which uses rhythmic phasing. **(a)** Describe the method of time-shifting used to create the phasing. **(b)** Complete the following table of rhythmic distances for the 9 measures of this composition:

Measure:	1	2	3	4	5	6	7	8	9
Rhythmic Distance:	0	8	10						0

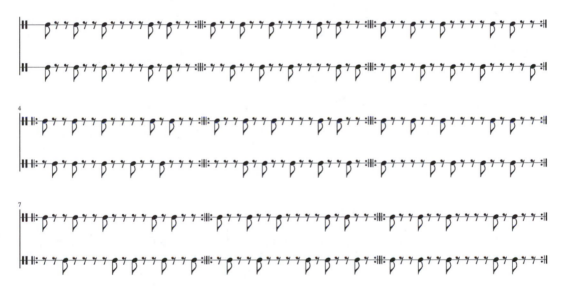

Figure 6.17. Rhythmic phasing composition, *Phased Son*. Each measure repeated 8 times.

Some examples of pitch-rhythmic serialism

Later in the chapter, we shall describe the composing method of *total serialism*. As a warmup to this discussion, in the next four exercises we illustrate how transpositions of pitches and time-shifts of rhythm can be applied to an initial set of notes. Thereby generating more notes in a sequential, or serial, fashion.

6.3.17. For the following score:

obtain notes for the last two measures in the following way. Apply transposition T_5 to pitches, and time-shift T_2 to note onsets. Apply both transformations to the entire set of notes in the first two measures.

6.3.18. For the following score:

obtain notes for the last two measures in the following way. Apply transposition T_7 to pitches, and time-shift T_5 to note onsets. Apply both transformations to the entire set of notes in the first two measures.

6.3.19. For the following score:

obtain notes for the last four measures in the following way. Apply transposition T_{-5} to pitches, and time-shift T_{-3} to note onsets, for the entire set of notes in the first two measures. Then apply transposition T_5 to pitches, and time-shift T_2 to note onsets, for the entire set of notes in the third and fourth measures.

6.3.20. For the following score:

obtain notes for the last four measures in the following way. Apply transposition T_4 to pitches, and time-shift T_2 to note onsets, for the entire set of notes in the first two measures. Then apply transposition T_{-5} to pitches, and time-shift T_{-2} to note onsets, for the entire set of notes in the third and fourth measures.

6.4 Rhythmic Inversion

We saw in the previous section that a cyclic rhythm can be time-shifted, using the same mathematical technique of modular arithmetic that we used for transposition of pitch classes. In this section, we describe how cyclic rhythms can also be inverted, using the same mathematics as for chromatic inversion of pitch classes.

6.4.1 Inversion of Cyclic Rhythms

A cyclic rhythm can be notated by dots at hours of an N-hour clock. Therefore, we can employ the same formulas for inversion of the rhythm as we used for chromatic inversion of pitch classes. For example, the son clave rhythm can be plotted on a 16-hour clock. We then define a transformation $\mathcal{I}_k$ by

$$\mathcal{I}_k : \quad m \to k - m \mod 16$$

for each $k = 0, 1, 2, \ldots, 15$. Each $\mathcal{I}_k$ is called an ***inversion*** of the cyclic rhythm. Although we also use the notation $\mathcal{I}_k$ for chromatic inversions, the context should make it clear whether $\mathcal{I}_k$ refers to a chromatic inversion of pitch classes or to an inversion of cyclic rhythms.

Example 6.4.1. Inversion of the son clave. We show in Figure 6.18 an application of the inversion $\mathcal{I}_3$ to the son clave. The note onsets for the son clave are $0, 3, 6, 10, 12$. They are mapped by $\mathcal{I}_3$ as follows:

$$\begin{array}{cccccc} m\text{:} & 0 & 3 & 6 & 10 & 12 \\ 3 - m \mod 16\text{:} & 3 & 0 & 13 & 9 & 7 \end{array}$$

Writing the note onsets in the bottom row in standard order, we have $0, 3, 7, 9, 13$ for the note onsets of the inverted rhythm. This inverted rhythm is graphed on the right of Figure 6.18. Below the son

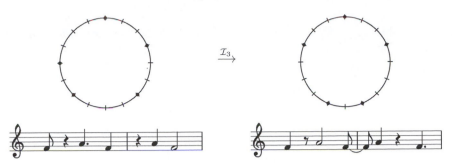

Figure 6.18. Inversion of son clave. Note onsets for son clave are graphed on left, with musical example below. Inversion $\mathcal{I}_3$ is applied to son clave note onsets to obtain graph of note onsets on right, with musical example below.

clave and its inverted rhythm, we have shown musical examples. Listening to these two examples, the reader may find it difficult to discern the connection between them—inverted rhythms are not as easy to discern as inverted sequences of pitches.

Example 6.4.2. Inversion of bossa nova clave. We show in Figure 6.19 an application of the inversion $\mathcal{I}_5$ to the bossa nova clave. The note onsets for the bossa nova clave are $0, 3, 6, 10, 13$ which are mapped by $\mathcal{I}_5$ as follows:

$$
\begin{array}{lccccc}
m\text{:} & 0 & 3 & 6 & 10 & 13 \\
5 - m \;\; \text{mod } 16\text{:} & 5 & 2 & 15 & 11 & 8
\end{array}
$$

Writing the note onsets in the bottom row in standard order, we have $2, 5, 8, 11, 15$ for the note onsets of the inverted rhythm. This inverted rhythm is graphed on the right of Figure 6.19. Below the bossa nova clave and its inverted rhythm, we have shown musical examples.

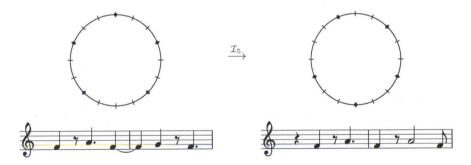

Figure 6.19. Inversion of bossa nova clave. Note onsets for bossa nova clave are graphed on left, with musical example below. Inversion $\mathcal{I}_5$ is applied to bossa nova clave note onsets to obtain graph of note onsets on right, with musical example below.

Inversions of Afro-Latin rhythms are used, sometimes back-to-back, in order to provide an interesting contrast and texture to the entire rhythm. In Section 6.7.3, we describe the composing technique of *total serialism,* which makes use of both inversions and time-shifting of rhythms.

6.4.2 Rhythmic and Pitch Transformation Groups

In total, when an N-hour clock is being used, there are $2N$ possible rhythmic transformations:

$$
\begin{array}{ccccc}
\mathsf{T}_0 & \mathsf{T}_1 & \mathsf{T}_2 & \cdots & \mathsf{T}_{N-1} \\[4pt]
\mathcal{I}_0 & \mathcal{I}_1 & \mathcal{I}_2 & \cdots & \mathcal{I}_{N-1}
\end{array}
\tag{6.5}
$$

We shall refer to this set of $2N$ transformations as the ***rhythmic transformation group*** $\mathfrak{R}_N$. For example, if an 8-hour clock is used, then $\mathfrak{R}_8$ consists of these 16 transformations:

$$
\begin{array}{cccccccc}
\mathsf{T}_0 & \mathsf{T}_1 & \mathsf{T}_2 & \mathsf{T}_3 & \mathsf{T}_4 & \mathsf{T}_5 & \mathsf{T}_6 & \mathsf{T}_7 \\
\mathcal{I}_0 & \mathcal{I}_1 & \mathcal{I}_2 & \mathcal{I}_3 & \mathcal{I}_4 & \mathcal{I}_5 & \mathcal{I}_6 & \mathcal{I}_7
\end{array}
\tag{6.6}
$$

While, if a 16-hour clock is used, then $\mathfrak{R}_{16}$ consists of these 32 transformations:

$$
\begin{array}{ccccc}
\mathsf{T}_0 & \mathsf{T}_1 & \mathsf{T}_2 & \ldots & \mathsf{T}_{15} \\
\mathcal{I}_0 & \mathcal{I}_1 & \mathcal{I}_2 & \ldots & \mathcal{I}_{15}
\end{array}
\tag{6.7}
$$

For pitch classes, there is an analogous collection of transformations. We have considered transpositions and chromatic inversions of pitch classes in this chapter. However, we also discussed the retrograde transformation R in Chapter 3. We used retrograde on specific notes in that chapter, but it can certainly be applied to a set of pitch classes. Any sequence of notes, say C_4, F_4, G_4, corresponds to a sequence of pitch classes, $\mathbf{C}, \mathbf{F}, \mathbf{G}$. Applying the retrograde R to the sequence of notes produces G_4, F_4, C_4, which corresponds to an analogous application of R to the pitch classes: $\mathbf{G}, \mathbf{F}, \mathbf{C}$.

We can combine R with transposition and inversion, producing transformations such as $\mathsf{R}\,\mathsf{T}_5$ or $\mathsf{R}\,\mathcal{I}_4$. The transformation $\mathsf{R}\,\mathsf{T}_5$ consists of applying R first, followed by T_5. For instance, suppose we have the pitch classes $\mathbf{C}, \mathbf{F}, \mathbf{G}$. These pitch classes have the hours $0, 5, 7$. Applying R gives hours $7, 5, 0$, and then applying T_5 gives

$$
\begin{array}{lccc}
m: & 7 & 5 & 0 \\
5 + m \ \ \mathrm{mod}\ 12: & 0 & 10 & 5
\end{array}
$$

So $\mathsf{R}\,\mathsf{T}_5$ maps the pitch classes $\mathbf{C}, \mathbf{F}, \mathbf{G}$ to $\mathbf{C}, \mathbf{A}^\sharp, \mathbf{F}$. Or, if we use $\mathsf{R}\,\mathcal{I}_4$ instead, then R again gives $7, 5, 0$, while $\mathcal{I}_4$ maps those hours as follows:

$$
\begin{array}{lccc}
m: & 7 & 5 & 0 \\
4 - m \ \ \mathrm{mod}\ 12: & 9 & 11 & 4
\end{array}
$$

So $\mathsf{R}\,\mathcal{I}_4$ maps the pitch classes $\mathbf{C}, \mathbf{F}, \mathbf{G}$ to the pitch classes $\mathbf{A}, \mathbf{B}, \mathbf{E}$. In the last section of this chapter, we describe a composing method, *pitch serialism,* which makes use of these pitch class transformations.

For pitch class transformations, we use the 12-hour chromatic clock. When a 12-hour clock is being used, there are 48 possible transformations:

$$
\begin{array}{ccccc}
\mathsf{T}_0 & \mathsf{T}_1 & \mathsf{T}_2 & \ldots & \mathsf{T}_{11} \\
\mathcal{I}_0 & \mathcal{I}_1 & \mathcal{I}_2 & \ldots & \mathcal{I}_{11} \\
\mathsf{R}\,\mathsf{T}_0 & \mathsf{R}\,\mathsf{T}_1 & \mathsf{R}\,\mathsf{T}_2 & \ldots & \mathsf{R}\,\mathsf{T}_{11} \\
\mathsf{R}\,\mathcal{I}_0 & \mathsf{R}\,\mathcal{I}_1 & \mathsf{R}\,\mathcal{I}_2 & \ldots & \mathsf{R}\,\mathcal{I}_{11}.
\end{array}
\tag{6.8}
$$

These transformations constitute the ***pitch class transformation group*** $\mathfrak{C}_{12}$. There are also cyclic rhythms that use a 12-hour clock, and the rhythmic transformation group $\mathfrak{R}_{12}$ acts on those cyclic rhythms. We have left out the retrograde transformation from $\mathfrak{R}_{12}$ because it does not act on note onsets. It reverses the order of pitches, and that is not the same as reversing the order of note onsets. In fact, the inversion $\mathcal{I}_0$ reverses the order of note onsets, so retrograde is not needed for (nor applicable to) note onset reversal.

It may seem that we could have infinitely many different ways of combining the basic transformations in $\mathfrak{C}_{12}$. However, when other transformations are combined, they can be reduced to one of these 48 transformations. For example, suppose we have $\mathsf{T}_3 \mathcal{I}_4$. This combination will map any hour m as follows:

$$ m \xrightarrow{\mathsf{T}_3} 3 + m \xrightarrow{\mathcal{I}_4} 4 - (3 + m) = 1 - m $$

hence it maps m to $1 - m$, the same as $\mathcal{I}_1$ would. Thus, we can write $\mathsf{T}_3 \mathcal{I}_4 = \mathcal{I}_1$, and so $\mathsf{T}_3 \mathcal{I}_4$ belongs to the transformation group $\mathfrak{C}_{12}$. As another example, suppose we have $\mathcal{I}_9 \mathsf{T}_{11}$. This combination will map any hour m as follows:

$$ m \xrightarrow{\mathcal{I}_9} 9 - m \xrightarrow{\mathsf{T}_{11}} 11 + (9 - m) = 20 - m \equiv 8 - m \mod 12. $$

Thus, $\mathcal{I}_9 \mathsf{T}_{11}$ maps m to $8 - m$, the same as $\mathcal{I}_8$ would. We have $\mathcal{I}_9 \mathsf{T}_{11} = \mathcal{I}_8$, and so $\mathcal{I}_9 \mathsf{T}_{11}$ belongs to the transformation group $\mathfrak{C}_{12}$. Other possible combinations, and their equivalences with members of $\mathfrak{C}_{12}$, are considered in the exercises.

Remark 6.4.1. It should be noted that one cannot arbitrarily switch the order of combined transformations. For example, $\mathsf{T}_3 \mathcal{I}_4 \neq \mathcal{I}_4 \mathsf{T}_3$. To see this, we recall that $\mathsf{T}_3 \mathcal{I}_4 = \mathcal{I}_1$ was shown above, and we also have $\mathcal{I}_4 \mathsf{T}_3 = \mathcal{I}_7$. Since it is clear that $\mathcal{I}_1 \neq \mathcal{I}_7$, we conclude that $\mathsf{T}_3 \mathcal{I}_4 \neq \mathcal{I}_4 \mathsf{T}_3$. In general, we cannot switch the order of an inversion $\mathcal{I}_k$ with a transposition T_j. It is also not possible, in general, to switch the order of two inversions. That is because $\mathcal{I}_j \mathcal{I}_k = \mathsf{T}_{k-j}$, while $\mathcal{I}_k \mathcal{I}_j = \mathsf{T}_{j-k}$, and generally $k - j \neq j - k$. One can, however, always switch the retrograde R with either an inversion or a transposition. See Exercise 6.4.2. Moreover, we can always switch the order of two transpositions, since $\mathsf{T}_j \mathsf{T}_k = \mathsf{T}_{j+k} = \mathsf{T}_{k+j} = \mathsf{T}_k \mathsf{T}_j$.

Exercises

6.4.1. Show that the following equations hold for transformations in $\mathfrak{C}_{16}$:

$$ \mathsf{T}_2 \mathcal{I}_3 = \mathcal{I}_1, \qquad \mathcal{I}_8 \mathsf{T}_{10} = \mathcal{I}_2, \qquad \mathcal{I}_{10} \mathcal{I}_6 = \mathsf{T}_{12}, \qquad \mathcal{I}_{11} \mathsf{T}_{10} = \mathcal{I}_5. $$

6.4.2. For the pitch class transformation group $\mathfrak{C}_{12}$, explain why $\mathsf{R}\mathsf{T}_k = \mathsf{T}_k \mathsf{R}$ and $\mathsf{R}\mathcal{I}_k = \mathcal{I}_k \mathsf{R}$ hold for every T_k and every $\mathcal{I}_k$. Also explain why $\mathsf{R}\,\mathsf{R} = \mathsf{T}_0$ (the identity, or simple repetition, transformation).

6.4.3. Show that the following equations hold for transformations in $\mathfrak{C}_{12}$:

$$ \mathsf{R}\,\mathsf{T}_6 \,\mathsf{R}\,\mathcal{I}_8 = \mathcal{I}_2, \qquad \mathsf{R}\,\mathcal{I}_9 \,\mathsf{R}\,\mathsf{T}_6 = \mathcal{I}_3, \qquad \mathsf{R}\,\mathsf{T}_{10} \,\mathsf{R}\,\mathsf{T}_9 = \mathsf{T}_7, \qquad \mathsf{R}\,\mathcal{I}_8 \,\mathsf{R}\,\mathcal{I}_{11} = \mathsf{T}_3. $$

6.4.4. Find the transformations in $\mathfrak{C}_{12}$ that each of the following combination of transformations is equal to:

$$ \mathcal{I}_3 \mathsf{T}_5, \qquad \mathsf{R}\,\mathsf{T}_4 \,\mathcal{I}_8, \qquad \mathsf{R}\,\mathsf{T}_9 \,\mathsf{R}\,\mathsf{T}_{11}, \qquad \mathsf{T}_7 \,\mathsf{R}\,\mathcal{I}_6. $$

6.4.5. Find the transformations in $\mathfrak{C}_{12}$ that each of the following combination of transformations is equal to:

$$ \mathcal{I}_3 \mathsf{T}_4, \qquad \mathsf{R}\,\mathsf{T}_4 \,\mathcal{I}_9, \qquad \mathsf{R}\,\mathsf{T}_{10} \,\mathsf{R}\,\mathsf{T}_5, \qquad \mathsf{T}_7 \,\mathsf{R}\,\mathcal{I}_8. $$

6.4.6. Draw the positions of note onsets for the following transformations in $\mathfrak{R}_{16}$:

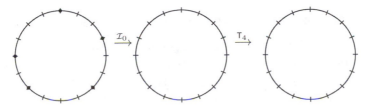

6.4.7. Draw the positions of note onsets for the following transformations in $\mathfrak{R}_{16}$:

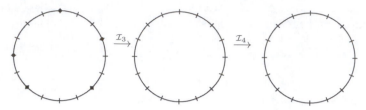

6.4.8. For the 2 measures shown on the left below, apply the indicated transformation to obtain note onsets, and use those note onsets to fill out the score on the right (without changing note pitches):

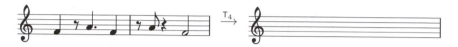

6.4.9. For the 2 measures shown on the left below, apply the indicated transformation to obtain note onsets, and use those note onsets to fill out the score on the right (without changing note pitches):

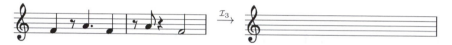

6.4.10. For the score shown on the left below:

fill in the empty score on the right by applying the inversion $\mathcal{I}_2$ to the note onsets and the transformation T_4 to the pitch classes of the notes.

6.4.11. For the score shown on the left below:

fill in the empty score on the right by applying the transformation T_2 to the note onsets and the transformation $\mathsf{R}\,\mathcal{I}_1$ to the pitch classes of the notes.

6.5   Construction of Scales and Cyclic Rhythms

What makes a musical scale sound good? Or a cyclic rhythm sound good? Partial answers to these questions can be found by investigating the requirement that notes on a scale, or note onsets in a cyclic rhythm, are as evenly spaced apart as possible.[6] We will show that this requirement leads, by way of a mathematical procedure known as the ***Euclidean algorithm,*** to the construction of many of the commonly used musical scales and cyclic rhythms.

6.5.1   The Euclidean Algorithm

First, we need to briefly describe the Euclidean algorithm for finding the greatest common divisor of two positive whole numbers. The ***greatest common divisor*** of two positive whole numbers is

---

[6]We will more completely answer this question in the next section, and in the next chapter—by examining the amount of consonance and dissonance in scales, and the number of triadic chords they contain.

the largest positive whole number that divides both of these numbers. For example, the greatest common divisor of 24 and 10 is 2. The greatest common divisor of 12 and 5 is 1. The ancient Greek mathematician, Euclid, found an algorithm (a step-by-step procedure) for determining the greatest common divisor of two positive whole numbers. Here is how it works with 24 and 10. First, subtract 10 successively from 24 until you get a remainder less than 10:

$$24 = 10 + 14$$

$$24 = (2 \cdot 10) + 4. \tag{6.9}$$

Now, repeat this process with 10 and 4. Repeatedly subtract 4 from 10, until you get a remainder less than 4:

$$10 = 4 + 6$$

$$10 = (2 \cdot 4) + 2. \tag{6.10}$$

Then, repeat the process with 4 and 2. This time the process ends with a remainder of 0:

$$4 = 2 + 2$$

$$4 = (2 \cdot 2) + 0.$$

Therefore, 2 divides 4. Equation (6.10) shows that it must also divide 10. Because 2 divides 4 and 10, Equation (6.9) shows that it also divides 24. Thus, the algorithm guarantees that 2 divides both 10 and 24. Moreover, if any positive whole number d divides 24 and 10, then d must also divide 4 because Equation (6.9) tells us that $4 = 24 - 2 \cdot 10$. Furthermore, since d divides 10 and 4, then d must also divide 2 because Equation (6.10) tells us that $2 = 10 - 2 \cdot 4$. Thus, d divides 2 whenever d divides 10 and 24. It follows that 2 is the largest divisor of both 24 and 10.

As another very important example of the Euclidean algorithm, consider the numbers 12 and 5. In this case, we subtract 5 successively from 12 until we get a remainder less than 5:

$$12 = 5 + 7$$

$$12 = (2 \cdot 5) + 2. \tag{6.11}$$

We then subtract 2 successively from 5 until we get a remainder less than 2:

$$5 = 2 + 3$$

$$5 = (2 \cdot 2) + 1. \tag{6.12}$$

We then conclude the algorithm by subtracting 1 repeatedly from 2 until we get a remainder of 0:

$$2 = 1 + 1$$

$$2 = (2 \cdot 1) + 0. \tag{6.13}$$

In this case, we have found that 1 is the greatest common divisor of 12 and 5.

6.5.2 Constructing Musical Scales

We now show how the Euclidean algorithm corresponds to a method for constructing musical scales. To perform this construction, we use the numbers 1 and 0 to indicate whether a pitch class belongs to a scale or not. If a pitch class belongs to the scale, then we use 1 to indicate that the hour for that pitch class is to be marked on the chromatic clock. If a pitch class does not belong to the scale, then we use 0 to indicate that the hour for that pitch class is not marked on the chromatic clock. For example, the C-major scale is represented by the following sequence: 1 0 1 0 1 1 0 1 0 1 0 1. The first 1 indicates the hour 0, and so hour 0 is marked on the chromatic clock, denoting the pitch class C which belongs to the C-major scale. The next number is 0, and that indicates that hour 1 is not marked, since the pitch class C$^\sharp$ does not belong to the C-major scale. The reader should check that the rest of the numbers in the sequence do serve to mark off the hours corresponding to the pitch classes in the C-major scale. In Figure 6.20, we show this sequence and a few others.

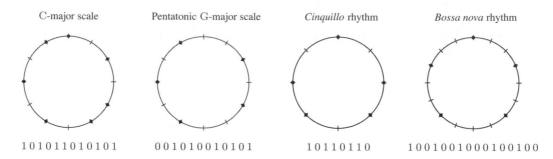

C-major scale	Pentatonic G-major scale	*Cinquillo* rhythm	*Bossa nova* rhythm
1 0 1 0 1 1 0 1 0 1 0 1	0 0 1 0 1 0 0 1 0 1 0 1	1 0 1 1 0 1 1 0	1 0 0 1 0 0 1 0 0 0 1 0 0 1 0 0

Figure 6.20. Sequences of 1's and 0's and their clock markings for scales and rhythms.

As we noted above, the scales and rhythms that sound good are to be constructed using the requirement that notes in a scale, or note onsets in a cyclic rhythm, are as evenly spaced apart as possible. This will correspond to having 1's with equal numbers (or nearly equal as possible) of 0's in between them. Here are a couple of examples.

Example 6.5.1. Pentatonic scale. Suppose we want a pentatonic scale consisting of 5 pitch classes chosen from the 12 pitch classes of the chromatic scale. We will begin our construction by simply writing down five 1's, followed by seven 0's:

$$1\,1\,1\,1\,1,\ 0\,0\,0\,0\,0\,0\,0 \qquad (12 = 5 + 7)$$

Notice that this initial set of 1's and 0's corresponds to the first step in calculating the greatest common divisor of 12 and 5. We now place a 0 from the remainder list of seven 0's in front of each of the 1's:

$$1\,0\ 1\,0\ 1\,0\ 1\,0\ 1\,0,\ 0\,0 \qquad (12 = 2 \cdot 5 + 2)$$

Again, this second step corresponds to the second step in calculating the greatest common divisor of 12 and 5. The remainder 2 for this second step corresponds to the 2 remaining 0's. We now move those 2 remaining 0's in front of the first 2 sub-sequences of the form 1 0, and we have

$$1\,0\,0\ 1\,0\,0,\ 1\,0\ 1\,0\ 1\,0 \qquad (5 = 2 + 3)$$

The remainder 3 in the Euclidean algorithm step corresponds to the remainder of 3 subsequences of form 1 0. We now place 2 of those subsequences in front of the 2 subsequences of form 1 0 0:

$$1\,0\,0\,1\,0\ 1\,0\,0\,1\,0,\ 1\,0 \qquad (5 = 2 \cdot 2 + 1) \qquad (6.14)$$

Continuing with this process, we obtain these sequences:

$$1\,0\,0\,1\,0\,1\,0,\ 1\,0\,0\,1\,0 \qquad (2 = 1 + 1)$$

$$1\,0\,0\,1\,0\,1\,0\,1\,0\,0\,1\,0, \qquad (2 = 2 \cdot 1 + 0). \qquad (6.15)$$

The process ends when we reach the single sequence corresponding to the 0-remainder step in the Euclidean algorithm. The final sequence we obtained corresponds to the hours $0, 3, 5, 7, 10$. This gives pitch classes for a pentatonic scale. In fact, if we apply the transposition T_9 to these hours and use $\mod 12$ arithmetic, we get the hours $9, 0, 2, 4, 7$. These hours give the pentatonic C-major scale. Thus, our process has produced a rotated (transposed) version of the pentatonic C-major scale.

Example 6.5.2. Diatonic scale. Suppose we want 7 distinct pitch classes chosen from the 12 pitch classes of the chromatic scale. Applying our Euclidean algorithm technique, we obtain:

$$
\begin{aligned}
&1\,1\,1\,1\,1\,1\,1,\ 0\,0\,0\,0\,0 && (12 = 7 + 5) \\
&10\ 10\ 10\ 10\ 10,\ 11 && (7 = 5 + 2) \\
&101\ 101,\ 10\ 10\ 10 && (5 = 2 + 3) \\
&10110\ 10110,\ 10 && (5 = 2 \cdot 2 + 1) && (6.16) \\
&1011010,\ 10110 && (2 = 1 + 1) \\
&101101010110, && (2 = 2 \cdot 1 + 0). && (6.17)
\end{aligned}
$$

The final sequence corresponds to the hours $0, 2, 3, 5, 7, 9, 10$. This gives pitch classes for the C-Dorian mode. Moreover, by applying the transposition T_2 to these hours, we obtain $2, 4, 5, 7, 9, 11, 0$, which on the chromatic clock gives the pitch classes for the C-major scale.

Remark 6.5.1. When the greatest common divisor is 1, then it is possible to stop the Euclidean algorithm at the equation where the remainder of 1 is first reached, and just attach the one remainder sequence to the end. In Example 6.5.1, we can stop at Equation (6.14) and get the sequence

$$1\,0\,0\,1\,0\,1\,0\,0\,1\,0\,1\,0.$$

This sequence corresponds to the hours $0, 3, 5, 8, 10$. Applying T_4 gives hours $4, 7, 9, 0, 2$ which are the hours of the C-major pentatonic scale. In Example 6.5.2, we can stop at Equation (6.16) and get

$$1\,0\,1\,1\,0\,1\,0\,1\,1\,0\,1\,0.$$

This sequence corresponds to hours $0, 2, 3, 5, 7, 8, 10$. These are the hours for the natural C-minor scale. It is interesting that this sequence of hours, when viewed as marking 7 note onsets on a 12-hour clock, corresponds to a cyclic rhythm called *Mpre* that is used by the *Ashanti* people of Ghana. Moreover, if we apply the transposition T_9 and use $\mod 12$ arithmetic, we get the hours $9, 11, 0, 2, 4, 5, 7$ which are the hours for the C-major scale. These latter hours, when viewed as note onsets on a 12-hour clock, correspond to the Afro-Latin *Bembé* rhythm. The *Bembé* rhythm is so widely employed in African-derived music that it has been referred to as the *standard* rhythm pattern by several musicologists.[7]

6.5.3 Constructing Cyclic Rhythms

The construction of cyclic rhythms uses the same procedure as we just used for constructing scales. In our constructions, we shall always stop at the first equation that reaches a remainder of 1, as described in Remark 6.5.1. Here are a few examples.

---

[7]These musicologists include: Desmond Tai (in *Studies in African Music*, by A.M. Jones, Oxford, 1959); Willie Anku ("Circles and time: A theory of structural organization of rhythm in African music," *Music Theory Online*, **6**(1), Jan. 2000); and D. Locke (*Drum Gahu: An Introduction to African Rhythm*, White Cliffs Media, 1998).

Example 6.5.3. Suppose we want 5 note onsets on an 8-hour clock. Then we begin with a sequence of five 1's followed by three 0's, and proceed as follows:

$$1\,1\,1\,1\,1,\ 0\,0\,0 \qquad (8 = 5 + 3)$$

$$1\,0\ 1\,0\ 1\,0,\ 1\,1 \qquad (5 = 3 + 2)$$

$$1\,0\,1\ 1\,0\,1,\ 1\,0 \qquad (3 = 2 + 1)$$

Our final sequence is then

$$1\,0\,1\,1\,0\,1\,1\,0.$$

This sequence corresponds to the note onsets $0, 2, 3, 5, 6$ on an 8-hour clock. These are the note onsets for the *cinquillo rhythm* (see Figure 6.20).

Example 6.5.4. Suppose we want 5 note onsets on a 9-hour clock. Then we begin with a sequence of five 1's followed by four 0's, and proceed as follows:

$$1\,1\,1\,1\,1,\ 0\,0\,0\,0 \qquad (9 = 5 + 4)$$

$$1\,0\ 1\,0\ 1\,0\ 1\,0,\ 1 \qquad (5 = 4 + 1)$$

Our final sequence is then

$$1\,0\,1\,0\,1\,0\,1\,0\,1.$$

This sequence corresponds to the note onsets $0, 2, 4, 6, 8$ on a 9-hour clock. These are the note onsets for the popular Arabic rhythm, *Agsag-Samai*. If started on the second onset (time-shifted by T_{-2}):

$$1\,0\,1\,0\,1\,0\,1\,1\,0$$

it is a drumming pattern used by the *Venda* in South Africa, and a Rumanian folk-dance rhythm.

Example 6.5.5. Suppose we would like 5 note onsets on a 16-hour clock. Then we begin with a sequence of five 1's followed by eleven 0's, and proceed as follows:

$$1\,1\,1\,1\,1,\ 0\,0\,0\,0\,0\,0\,0\,0\,0\,0\,0 \qquad (16 = 5 + 11)$$

$$1\,0\ 1\,0\ 1\,0\ 1\,0\ 1\,0,\ 0\,0\,0\,0\,0\,0 \qquad (16 = 2 \cdot 5 + 6)$$

$$1\,0\,0\ 1\,0\,0\ 1\,0\,0\ 1\,0\,0\ 1\,0\,0,\ 0 \qquad (16 = 3 \cdot 5 + 1).$$

Our final sequence is then

$$1\,0\,0\,1\,0\,0\,1\,0\,0\,1\,0\,0\,1\,0\,0\,0.$$

This sequence gives note onsets $0, 3, 6, 9, 12$ on a 16-hour clock. These note onsets are a time-shift of the *bossa nova* rhythm. In fact, if we apply time-shift T_{10} to these note onsets, we obtain (using mod 16 arithmetic): $10, 13, 0, 3, 6$. In standard order, those hours are $0, 3, 6, 10, 13$, the note onsets for the *bossa nova* rhythm (see Figure 6.20).

In this section, we have seen that several important musical scales and cyclic rhythms can be obtained from a process corresponding to the Euclidean algorithm (if we allow for transposition or time-shifting). Godfried Toussaint has catalogued a huge number of cyclic rhythms that can be constructed in this way. For example, we got the cyclic rhythms described above from one of his papers.[8] In the exercises, we will discuss several other examples that Toussaint describes in this paper, as well as some other examples with musical scales.

---

[8]G. Toussaint, The *Euclidean* Algorithm Generates Traditional Musical Rhythms. *Proc. of BRIDGES: Mathematical Connections in Art, Music and Science.* Banff, 2005, pp. 47–56.

Exercises

6.5.1. For each of the following pairs of positive whole numbers, use the Euclidean algorithm to find their greatest common divisor:

(a) $24, 84$

(b) $36, 120$

(c) $200, 440$

6.5.2. For each of the following pairs of positive whole numbers, use the Euclidean algorithm to find their greatest common divisor:

(a) $30, 64$

(b) $80, 144$

(c) $200, 880$

6.5.3. Suppose you want a scale with 8 pitch classes out of the 12 pitch classes of the chromatic scale. Show that the Euclidean algorithm method generates the B-octatonic scale:

$$B \quad C \quad D \quad E^\flat \quad F \quad G^\flat \quad G^\sharp \quad A \quad B$$

provided you take the tonic note to be B.

6.5.4. Suppose you want a scale with 6 pitch classes out of the 12 pitch classes of the chromatic scale. Show that the Euclidean algorithm method generates the whole tone scale:

$$C \quad D \quad E \quad F^\sharp \quad G^\sharp \quad A^\sharp \quad C.$$

6.5.5. Suppose you want a cyclic rhythm with 4 note onsets on a 12-hour clock. Show that the Euclidean algorithm generates the sequence:
$$1\,0\,0\,1\,0\,0\,1\,0\,0\,1\,0\,0.$$
If each 1 represents a loud clap and each 0 represents a soft clap, then this sequence corresponds to the $\frac{12}{8}$ time *Fandango* clapping pattern used in the Flamenco music of southern Spain.

6.5.6. Suppose you want a cyclic rhythm with 3 note onsets on a 7-hour clock. Show that the Euclidean algorithm generates the sequence:
$$1\,0\,1\,0\,1\,0\,0.$$
This sequence corresponds to a *Ruchenitza* folk rhythm from Bulgaria.

6.5.7. Suppose you want a cyclic rhythm with 3 note onsets on an 8-hour clock. Show that the Euclidean algorithm generates the sequence:
$$1\,0\,0\,1\,0\,0\,1\,0.$$
This sequence corresponds to the *tresillo* rhythm from Cuba.

6.5.8. Suppose you want a chord made up of 3 distinct pitch classes from the 12 pitch classes of the chromatic scale. Show that the Euclidean algorithm method yields the chord **C-E-G**$^\sharp$.

6.5.9. The chord **C-E-G**$^\sharp$ contains the following set of pitch classes: $S = \{C, E, G^\sharp\}$. Show that this set S is left unchanged when the transpositions T_0, T_4, and T_8, are applied to each of its pitch classes. (The set S is said to have 3 transpositional symmetries.)

6.5.10. Show that the set S of pitch classes defined in the previous exercise is left unchanged when the pitch class chromatic inversions $\mathcal{I}_0, \mathcal{I}_4$, and $\mathcal{I}_8$, are applied to each of its pitch classes. (The set S is said to have 3 chromatic inversional symmetries.)

6.5.11. Answer these questions:

(a) What chord do you get from **C-E-G**$^\sharp$ if you lower **G**$^\sharp$ by one half step? What chord do you get from **C-E-G**$^\sharp$ if you raise **G**$^\sharp$ by one half step?

(b) What chord do you get from **C-E-G$^\sharp$** if you lower **E** by one half step? What chord do you get from **C-E-G$^\sharp$** if you raise **E** by one half step?

(c) What chord do you get from **C-E-G$^\sharp$** if you lower **C** by one half step? What chord do you get from **C-E-G$^\sharp$** if you raise **C** by one half step?

This exercise shows that there are six major and minor chords that are as close as possible—only one half step difference for just a single note—to the chord **C-E-G$^\sharp$**.

6.5.12. Suppose you want a cyclic rhythm with 4 note onsets on a 9-hour clock. Show that the Euclidean algorithm generates the sequence:

$$1\ 0\ 1\ 0\ 1\ 0\ 1\ 0\ 0.$$

This sequence corresponds to the *Aksak* rhythm from Turkey.

6.5.13. Suppose you want a cyclic rhythm with 7 note onsets on a 16-hour clock. Show that the Euclidean algorithm generates the sequence:

$$1\ 0\ 0\ 1\ 0\ 1\ 0\ 1\ 0\ 0\ 1\ 0\ 1\ 0\ 1\ 0.$$

This sequence is closely related to the *Samba* rhythm from Brazil. If the rhythm is started on the last onset (time-shifting by T_2), then the *Samba* rhythm is obtained.

6.5.14. Suppose you want a cyclic rhythm with 9 note onsets on a 16-hour clock. Show that the Euclidean algorithm generates the sequence:

$$1\ 0\ 1\ 1\ 0\ 1\ 0\ 1\ 0\ 1\ 1\ 0\ 1\ 0\ 1\ 0.$$

This sequence is closely related to several different rhythms. If it is time-shifted by T_2, so that it begins on the last onset, then it is the bell pattern used in the *Ngbaka-Maibo* rhythm of the Central African Republic. If the rhythm is started on the fourth onset (time-shifting by T_{-5}), then it is the cow-bell rhythm used in Brazilian *Samba,* and is also a rhythm played in West and Central Africa.

6.5.15. Sometimes a cyclic rhythm will correspond to a partial application of the Euclidean algorithm. Suppose, for example, that you want a cyclic rhythm with 10 note onsets on a 16-hour clock. Show that the sequence of 1's and 0's obtained at the step corresponding to $4 = 2 + 2$ gives the note onsets for the rhythm played by the *quijada* in Figure 6.11 on p. 181.

6.6 Comparing Musical Scales and Cyclic Rhythms

In this section we look at a method for analyzing a musical scale, based on the relative frequency of harmonic intervals between the pitch classes within the scale. We relate this relative frequency to the amount of dissonance that the scale allows for. This analysis allows us to musically compare different scales. Interval frequencies will also be used to compare different cyclic rhythms, based on the mathematical equivalence of scales and cyclic rhythms.

6.6.1 Interval Frequencies and Musical Scales

The method we shall use for analyzing musical scales is based on finding the relative frequencies of occurrence of intervals between pitch classes of a given scale. Comparing these frequencies of occurrence, and how they relate to the amount of consonance and dissonance between harmonic intervals, will provide us with a means to compare different scales. We will explain this method by considering two examples, a pentatonic major scale and a diatonic major scale.

Example 6.6.1. Pentatonic C-major. The distinct pitch classes for the pentatonic C-major scale are

$$\text{C} \quad \text{D} \quad \text{E} \quad \text{G} \quad \text{A}$$

corresponding to hours $0, 2, 4, 7, 9$ on the chromatic clock. To calculate the frequencies of occurrence of the intervals in this scale, we have to calculate intervals between each possible pair of pitch classes. To begin, calculate all of the intervals between 0 and the remaining hours $(2, 4, 7, 9)$:

$$\text{From 0 clockwise:} \quad 2 \quad 4 \quad 7 \quad 9.$$

Now, throw out 0 and calculate the intervals between 2 and the remaining hours $(4, 7, 9)$:

From 2 clockwise: 2   5   7.

Then, throw out 2 and calculate the intervals between 4 and the remaining hours $(7, 9)$:

From 4 clockwise: 3   5.

Finally, throw out 4 and calculate the interval between 7 and the last remaining hour (9):

From 7 clockwise: 2.

These calculations can be summarized by the following diagram (where the top row is the initial set of hours for the pitch classes):

$$
\begin{array}{ccccc}
(0 & 2 & 4 & 7 & 9) \\
& 2 & 4 & 7 & 9 \\
& & 2 & 5 & 7 \\
& & & 3 & 5 \\
& & & & 2
\end{array}
\tag{6.18}
$$

The intervals for the second row are obtained from differences between successive numbers in the first row and the number 0 that starts the first row. Likewise, the intervals for the third row are obtained from differences between successive numbers in the second row and the number 2 that starts the second row. This pattern continues until we reach just the single number in the last row. Because of the elegant logical ordering of these calculations, we shall use this method of calculating intervals from now on.

There is one final step. Since we are dealing with pitch classes, we replace the interval numbers in (6.18) by chromatic distances. This gives

$$
\begin{array}{cccc}
2 & 4 & 5 & 3 \\
& 2 & 5 & 5 \\
& & 3 & 5 \\
& & & 2
\end{array}
\tag{6.19}
$$

from which we count the following number of occurrences of chromatic distances:

1: 0   2: 3   3: 2   4: 1   5: 4   6: 0.

The total number of these counts is $0 + 3 + 2 + 1 + 4 + 0 = 10$. Therefore, the following fractions describe the relative proportions of chromatic distances that can occur for this scale:

$$
1: \frac{0}{10} \quad 2: \frac{3}{10} \quad 3: \frac{2}{10} \quad 4: \frac{1}{10} \quad 5: \frac{4}{10} \quad 6: \frac{0}{10}.
$$

These fractions, which are all non-negative and add up to 1, can be interpreted as probabilities. They are probabilities for the occurrence of these chromatic distances, or intervals, using this scale. (Note: we mean just the scale here, not a key—see Remark 6.6.1 below.) A graph of these probabilities is shown on the left of Figure 6.21. The caption for this figure refers to a *dissonance score*. We will discuss dissonance scores after our next example.

Remark 6.6.1. The probabilities for intervals discussed in this last example, and in our next example, are probabilities for a *scale* only. In other words, they are probabilities of occurrences of intervals when notes are chosen randomly from the scale. Probably the best way to view our discussion of scales in this section, is as a further attempt to answer the question posed previously: Why do certain scales sound good? We shall have more to say about that question in the next chapter as well.

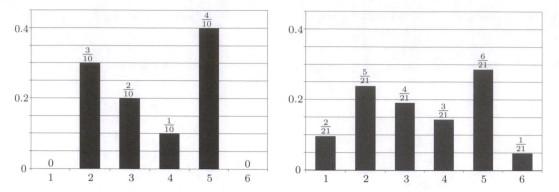

Figure 6.21. Left: Interval probabilities for pentatonic major scale. Dissonance score 2.2. Right: Interval probabilities for diatonic scale. Dissonance score 2.95. (Dissonance scores are explained in Section 6.6.2.)

Example 6.6.2. Diatonic scale. As the most basic example of a diatonic scale, we shall look at the C-major scale. This scale has these distinct pitch classes:

$$C \quad D \quad E \quad F \quad G \quad A \quad B$$

corresponding to hours $0, 2, 4, 5, 7, 9, 11$ on the chromatic clock. Our calculation of the possible intervals between pairs of distinct notes, goes as follows:

$$
\begin{array}{ccccccc}
(0 & 2 & 4 & 5 & 7 & 9 & 11) \\
& 2 & 4 & 5 & 7 & 9 & 11 \\
& & 2 & 3 & 5 & 7 & 9 \\
& & & 1 & 3 & 5 & 7 \\
& & & & 2 & 4 & 6 \\
& & & & & 2 & 4 \\
& & & & & & 2
\end{array}
\tag{6.20}
$$

Therefore, the possible chromatic distances between pairs of pitch classes for this scale are

$$
\begin{array}{cccccc}
2 & 4 & 5 & 5 & 3 & 1 \\
& 2 & 3 & 5 & 5 & 3 \\
& & 1 & 3 & 5 & 5 \\
& & & 2 & 4 & 6 \\
& & & & 2 & 4 \\
& & & & & 2
\end{array}
\tag{6.21}
$$

From this last array of numbers, we count the following number of occurrences of chromatic distances:

$$1: 2 \qquad 2: 5 \qquad 3: 4 \qquad 4: 3 \qquad 5: 6 \qquad 6: 1.$$

The total number of these counts is $2 + 5 + 4 + 3 + 6 + 1 = 21$. Therefore, the following fractions provide the probabilities for the occurrence of chromatic distances (intervals) for this diatonic scale:

$$1: \frac{2}{21} \qquad 2: \frac{5}{21} \qquad 3: \frac{4}{21} \qquad 4: \frac{3}{21} \qquad 5: \frac{6}{21} \qquad 6: \frac{1}{21}.$$

A graph of these probabilities is shown on the right of Figure 6.21.

Comparing the graphs of probabilities for the pentatonic and diatonic scales in Figure 6.21, we can see that the diatonic scale's graph is more spread out. It includes some very dissonant intervals, such as the minor 2nd and the tritone, which are not included within the pentatonic major scale. We now describe one way of quantifying the amount of dissonance within a scale.

6.6.2 Measuring Dissonance for a Scale

Using the interval probabilities that we have calculated for a pentatonic major scale and a diatonic scale, we can quantify the amount of dissonance (harmonic dissonance) contained within these scales. We will view a minor 2nd interval and a tritone as very dissonant harmonic intervals, and a major 2nd, minor 3rd, and major 3rd intervals as slightly dissonant harmonic intervals.[9] The perfect fourth and fifth intervals, however, are assumed to be acoustically consonant (even though a fourth is regarded as dissonant in some contexts in music theory). Based on these ideas, we define an *average dissonance*, or *dissonance score*, using the following definition.

Definition 6.6.1. *We assign these* **dissonance weights** *to each harmonic interval:*

$$1: \ 8 \qquad 2: \ 4 \qquad 3: \ 2 \qquad 4: \ 2 \qquad 5: \ 1 \qquad 6: \ 6. \tag{6.22}$$

These weights are used as multipliers times the probabilities for each harmonic interval value, and a weighted sum is calculated. If the intervals in a given scale have probabilities $p_1, p_2, p_3, p_4, p_5, p_6$, then the **dissonance score** *for that scale is computed as follows:*

$$\text{(dissonance)} = 8p_1 + 4p_2 + 2p_3 + 2p_4 + 1p_5 + 6p_6. \tag{6.23}$$

We shall describe further how the dissonance weights in (6.22) were obtained in a moment. First, we consider a couple of examples of calculating dissonance scores.

Example 6.6.3. For the pentatonic major scale, we obtain

$$\begin{aligned}
\text{(dissonance, pentatonic)} &= 8 \cdot 0 + 4 \cdot \frac{3}{10} + 2 \cdot \frac{2}{10} + 2 \cdot \frac{1}{10} + 1 \cdot \frac{4}{10} + 6 \cdot 0 \\
&= \frac{12 + 4 + 2 + 4}{10} \tag{6.24} \\
&= 2.2.
\end{aligned}$$

Example 6.6.4. For a diatonic scale like the C-major scale (or the natural A-minor scale), we obtain

$$\begin{aligned}
\text{(dissonance, diatonic)} &= 8 \cdot \frac{2}{21} + 4 \cdot \frac{5}{21} + 2 \cdot \frac{4}{21} + 2 \cdot \frac{3}{21} + 1 \cdot \frac{6}{21} + 6 \cdot \frac{1}{21} \\
&= \frac{16 + 20 + 8 + 6 + 6 + 6}{21} \tag{6.25} \\
&= 2.95.
\end{aligned}$$

These are satisfying results. The dissonance score for a diatonic major (or natural minor) scale is equal to 2.95, which is significantly larger than the dissonance score of 2.2 for a pentatonic major scale. This accords well with our intuitive notion that there is more dissonance contained within the diatonic scale.

---

[9]Minor and major 3rd intervals are regarded as consonant in music theory, although their consonances are said to be imperfect. Here we are thinking of the slight beating occurring in overtone harmonics for a minor or major 3rd, a slight amount of acoustic dissonance.

We will now describe in more detail how the dissonance weights in (6.22) were determined. These dissonance weights were largely based on the relative magnitudes of the tonal dissonance values, as defined by Plomp and Levelt.[10] We averaged the tonal dissonance values for intervals that correspond to the pitch class intervals 1 through 6, with greater weight being given to more commonly occurring intervals (e.g., the interval of 1 on the chromatic scale occurs more frequently than the interval of 11, although both have chromatic distance 1). We also raised the value for the tritone to make it closer to the minor 2nd value, since it is generally perceived as musically quite dissonant (perhaps due to its lying right between the chromatic intervals of 5 and 7 which are much more consonant). We will not go into more specific details here. We certainly could have chosen other values for the dissonance weights. However, as long as we give more weight to the minor 2nd and tritone than to the major 2nd and minor and major 3rd, along with little or no weight to the perfect fourth and fifth, then we will get similar results for the dissonance scores of the pentatonic and diatonic scales. The results we have found do not seem inconsistent with musical practice, so we shall use the dissonance weights given in Definition 6.6.1.

6.6.3 Interval Frequencies for Cyclic Rhythms

Because of the mathematical equivalence of cyclic rhythms with pitch classes, we can repeat these calculations for cyclic rhythms. First, we will show a basic example. After that, we provide some reasons why these computations are musically relevant.

Example 6.6.5. Interval frequencies for the *cinquillo* rhythm. On the left of Figure 6.22 we show the note onsets for the *cinquillo* rhythm. They are plotted on an 8-hour clock, with onsets at hours $0, 2, 3, 5, 6$. Following the same type of calculation as we did for pitch classes, we obtain the following array of all possible time intervals between distinct note onsets (where the top row is the initial set of hours for the note onsets):

$$
\begin{matrix}
(0 & 2 & 3 & 5 & 6) \\
 & 2 & 3 & 5 & 6 \\
 & & 1 & 3 & 4 \\
 & & & 2 & 3 \\
 & & & & 1
\end{matrix}
\qquad (6.26)
$$

Since we are dealing with time intervals for a cyclic rhythm, which continually repeats, we replace these time interval numbers by their equivalent smallest possible values on an 8-hour clock. This amounts to replacing any number n that is greater than 4 by $8 - n$. So we have

$$
\begin{matrix}
2 & 3 & 3 & 2 \\
 & 1 & 3 & 4 \\
 & & 2 & 3 \\
 & & & 1
\end{matrix}
\qquad (6.27)
$$

from which we can count the following number of occurrences of time intervals:

$$1\colon 2 \quad 2\colon 3 \quad 3\colon 4 \quad 4\colon 1.$$

[10]R. Plomp and W.J.M. Levelt, Tonal consonance and critical bandwidth. *J. Acoust. Soc. Am.,* **38** (1965), 548–560. Dissonance for instrumental tones are then computed by adding these dissonance quantities for the intervals between the harmonics of their pure tone components. We worked from a dissonance curve calculated by Sethares, shown in Figure 3 of the following link:

Relating Tuning and Timbre

at the book's web site (see Links). This reference provides a concise summary of the paper of Plomp and Levelt and applies it to musical theory. It contains some of the highlights from W. Sethares, *Tuning, Timbre, Spectrum, Scale, Second Edition,* Springer, London, 2005.

The total number of these counts is $2 + 3 + 4 + 1 = 10$. Therefore, the fractions that describe the frequency of occurrence of these time intervals as the cyclic rhythm is repeatedly played are

$$1: \frac{2}{10} \qquad 2: \frac{3}{10} \qquad 3: \frac{4}{10} \qquad 4: \frac{1}{10}.$$

Rather than being interpreted as probabilities, it is important to emphasize that for a cyclic rhythm, a set of fractions like these describes the frequency of occurrence of the time intervals separating note onsets *as the cyclic rhythm is repeatedly played*. A graph of these frequencies is shown on the right of Figure 6.22.

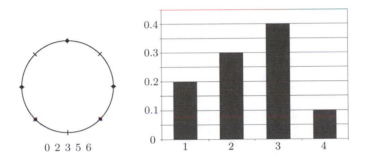

Figure 6.22. Left: Onsets for *cinquillo* rhythm. Right: Frequencies of occurrence of all intervals between onsets.

In this last example, we computed the frequencies of occurrence of all time intervals between any two note onsets for the *cinquillo* rhythm. What do these frequencies of occurrence tell us about the rhythm? One thing to remember is that different emphases can be put on different notes within the rhythm, by playing those notes with louder volume, or different durations, or different timbres (including different instruments). This emphasizes different intervals within the rhythm. In his book on the cyclic *Gahu* rhythm, David Locke describes how these factors come into play:

> The whole and its parts can be heard in a variety of ways depending on conceptual factors such as duration, placement and internal subdivision of beats, the relative prominence given to a particular part, or the moment a phrase is thought to begin. Creative performance depends on a player's ever-renewing interpretation of his part within this kaleidoscopic musical context of shifting aural illusions. Repetition need not be monotonous.[11]

Examples of Gahu drumming can be found at the book's web site. You can hear the drummers putting differing emphases on varying note onsets within the overall cyclic rhythm. We look at the interval frequencies for the *Gahu* rhythm in our next example.

Example 6.6.6. Comparing different cyclic rhythms. Collections of interval frequencies can be used for comparing different rhythms. In Figure 6.23, we show graphs of the interval frequencies for three different cyclic rhythms, the *Son, Rumba,* and *Gahu* rhythms. The rhythm analyst, Godfried Toussaint, has pointed out that *Gahu* is the only one of these African rhythms that has a collection of interval frequencies with these two properties: **(1)** a small maximum (just 0.2); **(2)** no silent intervals (interval frequencies of 0) between non-zero frequency intervals.[12] He remarks that it may be these two features that give the *Gahu* rhythm some of its distinctive qualities, such as its "rhythmic potency," and "tricky quality," as well as its creation of a "spiralling effect," with "ambiguity of phrasing" leading to "aural illusions." Toussaint references the book on *Gahu* rhythm by David Locke, cited above, which illustrates with sound recordings these distinctive features of the *Gahu* rhythm. Toussaint also

---

[11] David Locke, *Drum Gahu: An Introduction to African Rhythm,* White Cliffs Media (1998), p. 7.

[12] G. Toussaint, Computational Geometric Aspects of Rhythm, Melody, and Voice Leading. *Computational Geometry* (2010), **43**, pp. 2–22.

points out that a rhythm has "drive" when it does not contain relatively long silent intervals (interval frequencies of 0 for larger interval numbers). Consequently, the *Son* and *Rumba* rhythms have less drive than the *Gahu* rhythm.

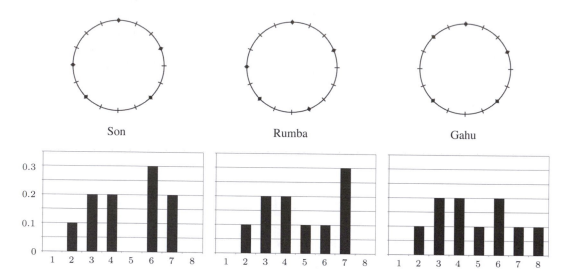

Figure 6.23. Note onsets and interval frequencies for *Son*, *Rumba*, and *Gahu* rhythms.

Remark 6.6.2. We would not want to give the impression that these properties of the *Gahu* rhythm make it better than other cyclic rhythms. For instance, although the *son* rhythm does not seem to have either of the two complexity properties (**1**) and (**2**), in its *clave* expression it has become an extremely popular rhythm in world music. Our discussion of rhythms in this chapter only scratches the surface of an extremely rich and varied component of music.

Exercises

6.6.1. Calculate and graph the probabilities of intervals between the distinct pitch classes of the chromatic scale. Also find its dissonance score.

6.6.2. Calculate and graph the probabilities of intervals between the distinct pitch classes of the following harmonic minor scale:

$$\text{C \quad D \quad E}^\flat \text{ \quad F \quad G \quad A}^\flat \text{ \quad B.}$$

Also find its dissonance score.

6.6.3. Calculate and graph the probabilities of intervals between the distinct pitch classes of the following whole tone scale:

$$\text{C \quad D \quad E \quad F}^\sharp \text{ \quad G}^\sharp \text{ \quad A}^\sharp.$$

Also find its dissonance score.

6.6.4. Calculate and graph the probabilities of intervals between the distinct pitch classes of the following melodic minor scale:

$$\text{C \quad D \quad E}^\flat \text{ \quad F \quad G \quad A \quad B.}$$

Also find its dissonance score.

6.6.5. Calculate the probabilities of intervals between the distinct pitch classes of the following harmonic major scale:

$$\text{C \quad D \quad E \quad F \quad G \quad A}^\flat \text{ \quad B.}$$

Also find its dissonance score.

6.6.6. Calculate the probabilities of intervals between the distinct pitch classes of the following octatonic scale:

$$\text{C} \quad \text{D}^{\flat} \quad \text{E}^{\flat} \quad \text{E} \quad \text{F}^{\sharp} \quad \text{G} \quad \text{A} \quad \text{B}^{\flat}.$$

Also find its dissonance score.

6.6.7. During the Renaissance, it was common to use a C-major scale, but with the seventh note allowed to be either B or $\text{B}^{\flat}$ in order to avoid tritones. For example, F would be played with $\text{B}^{\flat}$, rather than B. To compute the probabilities of intervals for distinct pitch classes for this scale, do the following: **(a)** Calculate the numbers of occurrences of intervals between distinct pitch classes for the C-major scale:

$$\text{C} \quad \text{D} \quad \text{E} \quad \text{F} \quad \text{G} \quad \text{A} \quad \text{B}$$

and calculate the numbers of occurrences of intervals between distinct pitch classes for this flatted-seventh scale:

$$\text{C} \quad \text{D} \quad \text{E} \quad \text{F} \quad \text{G} \quad \text{A} \quad \text{B}^{\flat}.$$

(b) Combine the two counts, but do not count tritone intervals. **(c)** From this combined count, form fractions that describe the probabilities for all possible intervals, and find the dissonance score.

6.6.8. Verify that the graphs of interval frequencies for the rhythms in Figure 6.23 are correct.

6.6.9. Calculate and graph the frequency of intervals for the *Bossa nova* rhythm, in Figure 6.12 on p. 182.

6.6.10. The *Shiko* rhythm has note onsets of 0, 4, 6, 10, 12 on a 16-hour clock. Calculate and graph the frequency of intervals for this rhythm.

6.6.11. The *Soukous* rhythm has note onsets of 0, 3, 6, 10, 11 on a 16-hour clock. Calculate and graph the frequency of intervals for this rhythm.

6.6.12. Consider the two sets of hours, $\{0, 1, 4, 6\}$ and $\{0, 1, 3, 7\}$, on a 12-hour clock. These sets could indicate either pitch classes or note onsets. Show, for both of these two sets, that the numbers of intervals between distinct hours is:

$$1 : 1 \quad 2 : 1 \quad 3 : 1 \quad 4 : 1 \quad 5 : 1 \quad 6 : 1.$$

Remark 6.6.3. In music theory, two sets of pitch classes are called ***Z-related*** if the numbers of intervals between distinct pitch classes in both sets are exactly the same. (A rationale for using the letter Z is given in Remark 6.6.4.) Based on the previous exercise, we can say that these two sets of pitch classes

$$\{\text{C}, \text{C}^{\sharp}, \text{E}, \text{F}^{\sharp}\} \quad \text{and} \quad \{\text{C}, \text{C}^{\sharp}, \text{D}^{\sharp}, \text{G}\}$$

are Z-related.

6.6.13. **(a)** Explain why the sets of pitch classes for any two major or minor chords are Z-related. **(b)** Explain why any two sets of pitch classes which are transpositions or chromatic inversions of each other are Z-related. **(c)** Explain why the Z-related sets, $\{\text{C}, \text{C}^{\sharp}, \text{E}, \text{F}^{\sharp}\}$ and $\{\text{C}, \text{C}^{\sharp}, \text{D}^{\sharp}, \text{G}\}$, are *not* transpositions or chromatic inversions of each other.

6.6.14. Show that the two sets of pitch classes in Exercise 6.6.7 are Z-related. Find a transposition that maps the first scale (C-major) to the second scale.

6.6.15. Show that the sets

$$\{\text{E}, \text{F}, \text{G}^{\sharp}, \text{A}, \text{B}\} \quad \text{and} \quad \{\text{D}^{\sharp}, \text{F}^{\sharp}, \text{A}, \text{B}^{\flat}, \text{B}\}$$

are Z-related, but are *not* transpositions or chromatic inversions of each other.

6.6.16. Show that the sets

$$\{\text{C}, \text{C}^{\sharp}, \text{D}^{\sharp}, \text{G}, \text{A}^{\sharp}, \text{B}\} \quad \text{and} \quad \{\text{E}, \text{F}, \text{G}, \text{G}^{\sharp}, \text{A}, \text{B}\}$$

are Z-related, but are *not* transpositions or chromatic inversions of each other.

Remark 6.6.4. The Z-related sets in Remark 6.6.3 and in Exercise 6.6.16 are from Webern's *Three Short Pieces, Op. 11, Piece 1*. The Z-related sets in Exercise 6.6.15 are from Bartok's *Suite Op. 14*. For more details, see the book by Alan Forte, *The Structure of Atonal Music*, Yale Univ. Press (1973), pp. 21–24. Alan Forte introduced the notion of Z-related sets. He said there was no particular reason for using the letter Z. However, it is worth noting that Z is often used by mathematicians to denote the set of whole numbers (integers); Z being the first letter of the German noun, *Zählen*, the plural form of Number.

6.7 Serialism

The transformation groups discussed in Section 6.4, especially the pitch class transformation group $\mathfrak{C}_{12}$, provide the mathematical foundation for a variety of composing techniques developed during the 20th century. These techniques share some important similarities that has led them to be grouped in a category called *serialism*. We will begin with the most widely used method in serialism, which we shall call *pitch serialism*. In a given composition, pitch serialism applies members of the pitch class transformation group $\mathfrak{C}_{12}$ to collections of pitch classes in a sequential, or serial, fashion to create new collections of pitch classes. As extensions of pitch serialism, there are other serialist techniques that also apply transformation groups to rhythm, dynamics, and articulation. We shall discuss these latter techniques, called *total serialism*, at the end of this section.

6.7.1 Pitch Serialism

The method of pitch serialism was developed by the composers Arnold Schoenberg, Anton Webern, and Alban Berg, in the first half of the 20th century. Here we will sketch a basic outline of pitch serialism. It is most commonly practiced by specifying an initial set of 12 notes, chosen from each of the 12 pitch classes of the chromatic scale. This set of 12 notes is called the *initial tone row* for the composition. In the first three measures of the score in Figure 6.24, there is an example of an initial tone row. The pitch classes for this tone row, and their hours on the chromatic clock, are

$$
\begin{array}{cccccccccccc}
\mathbf{C} & \mathbf{E} & \mathbf{G} & \mathbf{F} & \mathbf{A} & \mathbf{B} & \mathbf{G}^{\sharp} & \mathbf{F}^{\sharp} & \mathbf{A}^{\sharp} & \mathbf{D} & \mathbf{C}^{\sharp} & \mathbf{D}^{\sharp} \\
0 & 4 & 7 & 5 & 9 & 11 & 8 & 6 & 10 & 2 & 1 & 3.
\end{array}
\qquad (6.28)
$$

All of the pitch classes for the chromatic scale are listed in this initial tone row, each appearing precisely once.

After the initial tone row is specified, then pitch classes for subsequent rows are created by serially applying transformations from the pitch class transformation group $\mathfrak{C}_{12}$. For example, if we apply the pitch class chromatic inversion $\mathcal{I}_7$ to the hours of the initial tone row in (6.28), then we have

$$
\begin{array}{rcccccccccccc}
m\colon & 0 & 4 & 7 & 5 & 9 & 11 & 8 & 6 & 10 & 2 & 1 & 3 \\
7 - m \ \mathrm{mod}\ 12\colon & 7 & 3 & 0 & 2 & 10 & 8 & 11 & 1 & 9 & 5 & 6 & 4
\end{array}
\qquad (6.29)
$$

The bottom row of hours tells us the new pitch classes, as follows:

$$
\begin{array}{cccccccccccc}
7 & 3 & 0 & 2 & 10 & 8 & 11 & 1 & 9 & 5 & 6 & 4 \\
\mathbf{G} & \mathbf{D}^{\sharp} & \mathbf{C} & \mathbf{D} & \mathbf{A}^{\sharp} & \mathbf{G}^{\sharp} & \mathbf{B} & \mathbf{C}^{\sharp} & \mathbf{A} & \mathbf{F} & \mathbf{F}^{\sharp} & \mathbf{E}
\end{array}
$$

From each of these pitch classes, we can select any pitch *that lies within the range of the instrument to be played*. For example, pitch class **G** contains the pitches G_1 through G_7. For a piano, any of those pitches could be selected for the first note. For a vibraphone, however, only G_3, G_4, or G_5, could be selected. Similar freedom is allowed for choosing pitches from the remainder of the pitch classes. In Figure 6.24, we show one possible selection of pitches in measures 4 to 6.

To get the pitch classes for the next row, we apply another transformation from $\mathfrak{C}_{12}$. This transformation is applied to the previous row, so we are acting in a serial fashion. We used the retrograde-inversion $\mathsf{R}\,\mathcal{I}_5$. Applying the retrograde R to the previous sequence of hours we get

$$4 \quad 6 \quad 5 \quad 9 \quad 1 \quad 11 \quad 8 \quad 10 \quad 2 \quad 0 \quad 3 \quad 7.$$

We then apply the pitch class chromatic inversion $\mathcal{I}_5$ to these hours, obtaining

$$
\begin{array}{llllllllllll}
m: & 4 & 6 & 5 & 9 & 1 & 11 & 8 & 10 & 2 & 0 & 3 & 7 \\
5-m \bmod 12: & 1 & 11 & 0 & 8 & 4 & 6 & 9 & 7 & 3 & 5 & 2 & 10
\end{array}
$$

(6.30)

The bottom row of hours tells us the new pitch classes, as follows:

$$
\begin{array}{cccccccccccc}
1 & 11 & 0 & 8 & 4 & 6 & 9 & 7 & 3 & 5 & 2 & 10 \\
\mathbf{C}^\sharp & \mathbf{B} & \mathbf{C} & \mathbf{G}^\sharp & \mathbf{E} & \mathbf{F}^\sharp & \mathbf{A} & \mathbf{G} & \mathbf{D}^\sharp & \mathbf{F} & \mathbf{D} & \mathbf{A}^\sharp
\end{array}
$$

In Figure 6.24, we show the notes that we chose from these pitch classes in measures 7 to 9.

Figure 6.24. Example of serialist composition, written for vibraphone. Measures 1 to 3 contain notes for initial tone row. Notes in measures 4 to 6 were chosen from pitch classes obtained by applying chromatic inversion $\mathcal{I}_7$ to pitch classes of initial tone row. Notes in measures 7 to 9 were chosen from pitch classes obtained by applying retrograde-inversion $\mathsf{R}\,\mathcal{I}_5$ to pitch classes for measures 4 to 6.

This example was meant to give the essential idea behind pitch serialism. In the hands of great composers it involves considerably more elaboration and effort. For some individual instruments, there can be harmony from several notes played simultaneously. There can also be harmony involving multiple instruments playing together, with different tone rows and transformations used for each instrument. Generally speaking, however, the harmonies involved are not based on the chord progressions that are typical for diatonic music. In fact, by using an initial tone row consisting of single notes from each of the 12 distinct pitch classes of the chromatic scale, and then applying transformations from $\mathfrak{C}_{12}$ to it, one can avoid establishing a tonal center or tonic note. Since all notes from the chromatic scale are used in equal ratios, there will be no perceivable tonic if the composer also avoids diatonic features such as major and minor triadic chords. Historically, the method of pitch serialism was developed from ideas of Schoenberg, who felt that new formal techniques of composition had progressed beyond the control of traditional functions of harmony and tonality.[13] Pitch serialism

---

[13] For example, Schoenberg wrote the following: "The method of composing with twelve tones purports reinstatement of the effects formerly furnished by the structural functions of harmony.... In the works of Strauss, Mahler, and, even more, Debussy, one can already observe reasons for the advance of new formal techniques. Here it is already doubtful ... whether there is a tonic in power which has control over all these centrifugal tendencies of the harmonies." (Source: *Music in the Western World: A History in Documents*, Piero Weiss and Richard Taruskin (Eds.), Schirmer Books, 1984, pp. 435–436.)

was created as an alternative approach to creating melodies and harmony, outside of the procedures employed in diatonic music.

A couple of examples of pitch serialism

The composers who use pitch serialism have created a large body of work. Here we will give references to just a couple of the most famous works in this genre. First, you can listen to Berg's *Violin Concerto, Movement I*, at the book's web site. This piece is often cited as an example of Berg's lyrical use of the tone row technique. At the web site we provide some additional discussion of this, as well as a spectrogram of the opening of the concerto.

Second, you can also listen to Schoenberg's *Piano Concerto, Movement I*, at the book's web site. This piece has no relation to diatonic music. In particular, there are no diatonic chord progressions. That makes the piece difficult for many listeners to follow. Nevertheless, there is a great beauty to the piano tones, and significant emotion provided to their playing by the pianist, Mitsuko Uchida.

6.7.2   Musical Matrices

The method of pitch serialism employs an initial tone row, and then successively applies transformations from $\mathfrak{C}_{12}$ to generate new tone rows. Once the initial tone row is specified, there is a concise and striking description of all the various possibilities available to a composer. This description is called a ***musical matrix***. We shall first give an example of a musical matrix and how to interpret it. We then describe the mathematics underlying its construction.

In order for the labelling of rows and columns of our musical matrices to match the traditional labelling used by composers, we need to use the chromatic inversions IT_k discussed in Section 6.2. To make our notation concise, we shall write I_j for IT_j. So, for instance, $\mathsf{I}_3 = \mathsf{IT}_3$, and $\mathsf{I}_7 = \mathsf{IT}_7$. One special case of this notation is for I_0. Since T_0 leaves all hours unchanged, we have $\mathsf{I}_0 = \mathsf{IT}_0 = \mathsf{I}$, and therefore $\mathsf{I}_0 = \mathsf{I}$.

We will find it convenient to use an alternative expression for I_0, in terms of the pitch class inversions $\{\mathcal{I}_k\}$. Since I_0 will be acting on pitch classes, it must equal $\mathcal{I}_k$ for some k. This basic chromatic inversion $\mathsf{I} = \mathsf{I}_0$ leaves some note unchanged. Suppose that note has hour m on the chromatic clock. Then I_0 maps hour m to hour m. Therefore, we have

$$m \xrightarrow{\mathsf{I}_0} k - m = m.$$

Consequently, $k = 2m$, so we have

$$\mathsf{I}_0 = \mathcal{I}_{2m}. \tag{6.31}$$

We will need this formula when we describe how musical matrices are constructed.

We are now prepared to describe musical matrices. As an example, suppose we begin with the following initial tone row of pitch classes:

$$
\begin{array}{cccccccccccc}
\mathbf{D} & \mathbf{F} & \mathbf{A} & \mathbf{G} & \mathbf{B} & \mathbf{C} & \mathbf{A^\sharp} & \mathbf{G^\sharp} & \mathbf{F^\sharp} & \mathbf{E} & \mathbf{C^\sharp} & \mathbf{D^\sharp} \\
2 & 5 & 9 & 7 & 11 & 0 & 10 & 8 & 6 & 4 & 1 & 3.
\end{array} \tag{6.32}
$$

The musical matrix that describes the effect of every transformation from $\mathfrak{C}_{12}$ on the hours in this initial tone row is shown in Figure 6.25. This musical matrix is interpreted in the following way. The initial tone row is displayed in the first row of hours, and labeled by T_0 (since T_0 acting on this row just reproduces it). Reading down the left column are all the possible transpositions in $\mathfrak{C}_{12}$, and each row shows how they map the initial tone row to a new tone row. For example, the second transposition listed is T_9 and it maps the hours in the initial tone row as follows:

$$
\begin{array}{lcccccccccccc}
m\text{:} & 2 & 5 & 9 & 7 & 11 & 0 & 10 & 8 & 6 & 4 & 1 & 3 \\
9 + m \;\bmod 12\text{:} & 11 & 2 & 6 & 4 & 8 & 9 & 7 & 5 & 3 & 1 & 10 & 0
\end{array}
$$

	I_0	I_3	I_7	I_5	I_9	I_{10}	I_8	I_6	I_4	I_2	I_{11}	I_1	
T_0	2	5	9	7	11	0	10	8	6	4	1	3	RT_0
T_9	11	2	6	4	8	9	7	5	3	1	10	0	RT_9
T_5	7	10	2	0	4	5	3	1	11	9	6	8	RT_5
T_7	9	0	4	2	6	7	5	3	1	11	8	10	RT_7
T_3	5	8	0	10	2	3	1	11	9	7	4	6	RT_3
T_2	4	7	11	9	1	2	0	10	8	6	3	5	RT_2
T_4	6	9	1	11	3	4	2	0	10	8	5	7	RT_4
T_6	8	11	3	1	5	6	4	2	0	10	7	9	RT_6
T_8	10	1	5	3	7	8	6	4	2	0	9	11	RT_8
T_{10}	0	3	7	5	9	10	8	6	4	2	11	1	RT_{10}
T_1	3	6	10	8	0	1	11	9	7	5	2	4	RT_1
T_{11}	1	4	8	6	10	11	9	7	5	3	0	2	RT_{11}
	RI_0	RI_3	RI_7	RI_5	RI_9	RI_{10}	RI_8	RI_6	RI_4	RI_2	RI_{11}	RI_1	

Figure 6.25. Musical matrix for initial tone row in (6.32). Hours for initial tone row are in first row, labeled by T_0 on left side. Hours for all mappings of initial tone row by transformations in $\mathfrak{C}_{12}$ are listed in this matrix, according to labeling scheme described in the text.

The hours in the last line above are precisely the hours listed in the second row of hours of the musical matrix, labeled on the left by T_9. The pitch classes for this row are

$$\textbf{B} \quad \textbf{D} \quad \textbf{F}^\sharp \quad \textbf{E} \quad \textbf{G}^\sharp \quad \textbf{A} \quad \textbf{G} \quad \textbf{F} \quad \textbf{D}^\sharp \quad \textbf{C}^\sharp \quad \textbf{A}^\sharp \quad \textbf{C}.$$

If a row is read in reverse, from right to left, then that gives a retrograde-transposition of the initial tone row. For example, reading the second row in reverse gives the tone row obtained by applying RT_9 to the initial tone row. Consequently the label RT_9 is written on the right side of the second row of the matrix.

The columns in the musical matrix give all the pitch class chromatic inversions of the initial tone row. For example, the first column of hours is labeled at the top by I_0. We found above that, if $I_0 = I$ maps hour m to hour m, then $I_0 = \mathcal{I}_{2m}$. Since I_0 must map 2 to 2, we conclude that $I_0 = \mathcal{I}_4$. If $\mathcal{I}_4$ is applied to the hours in the initial tone row, we have the following mapping:

$$
\begin{array}{lcccccccccccc}
m\text{:} & 2 & 5 & 9 & 7 & 11 & 0 & 10 & 8 & 6 & 4 & 1 & 3 \\
4-m \ \bmod 12\text{:} & 2 & 11 & 7 & 9 & 5 & 4 & 6 & 8 & 10 & 0 & 3 & 1
\end{array}
\tag{6.33}
$$

The hours in the last line above are precisely the hours listed in the first column of hours of the musical matrix in Figure 6.25. The second column is labeled at the top by I_3, because when T_3 is applied to the first column, we have the following mappings:

$$
\begin{array}{lcccccccccccc}
m\text{:} & 2 & 11 & 7 & 9 & 5 & 4 & 6 & 8 & 10 & 0 & 3 & 1 \\
3+m \ \bmod 12\text{:} & 5 & 2 & 10 & 0 & 8 & 7 & 9 & 11 & 1 & 3 & 6 & 4
\end{array}
$$

The hours in the last line above are precisely the hours listed in the second column of hours of the musical matrix in Figure 6.25. If a column is read in reverse, from bottom to top, then that gives a retrograde-chromatic inversion of the initial tone row. For example, reading the second column in reverse gives the tone row obtained by applying $R\,I_3$ to the initial tone row. Consequently the label $R\,I_3$ is written at the bottom of the second column of the matrix.

It is important to note that all of the 12 possible pitch class chromatic inversions are used for the 12 columns of the matrix. This follows from the fact that each $I_j = I\,T_j$ is a chromatic inversion, and the 12 chromatic inversions listed at the top are all distinct. (As they must be, since the first row of the musical matrix has 12 distinct hours.)

A method for constructing musical matrices

A method for constructing musical matrices can be deduced from a careful examination of the musical matrix given in Figure 6.25. The method, which we call the MUSICAL MATRIX ALGORITHM, is the following 4-step procedure.

MUSICAL MATRIX ALGORITHM

Step 1. Write down an initial tone row. The hours for this tone row make up the first row of the musical matrix, labeled as T_0 on the left and $R\,T_0$ on the right.

Step 2. Perform a chromatic inversion I_0 of the hours for the first tone row. Denoting the first hour of this tone row by m, this inversion is $I_0 = \mathcal{I}_{2m}$. The resulting hours form the first column of the musical matrix, labeled at the top by I_0 and at the bottom by $R\,I_0$.

Step 3. For each hour k in the first column of the musical matrix, the transposition T_{k-m} maps the first hour m to hour k. Perform this transposition T_{k-m} on all of the hours of the initial tone row, to get the hours for the row that begins with hour k. Label the row that begins with hour k by T_{k-m} on the left, and $R\,T_{k-m}$ on the right.

Step 4. Complete the labeling of the columns of the musical matrix. For each hour j in the first row of the musical matrix, the transposition T_{j-m} maps the first hour m to hour j. Label the column that begins with hour j by I_{j-m} at the top, and $R\,I_{j-m}$ at the bottom.

This step-by-step method, or algorithm, is perfect for implementation by a computer. An excellent program for computing musical matrices is available at

$$\text{Musical Matrix Calculator} \tag{6.34}$$

on the book's web site (see Software). When you use this program, be sure and check the option TRADITIONAL METHOD in order to label your musical matrix using the notation we have described. The program uses letters A and B to stand for the numbers 10 and 11, but allows you to enter 10 and 11 in their place. It also uses the letter P rather than T to label rows. For example, P_3 would be used instead of T_3. The letter P is meant to indicate "prime series," which is another expression that may be used instead of "tone row." Also, it writes R_k rather than $R\,T_k$. These slight notational differences from our labeling are of no significance.

Additional examples related to musical matrices

We have described a step-by-step method for computing musical matrices, and given a reference for an excellent computer program that does these computations. For some readers, however, it might be more satisfying if we showed some further examples of computing musical matrices. Unfortunately, a musical matrix consists of 144 hours, which makes it rather tedious to describe its computation in a step-by-step manner for specific cases. Fortunately, there are matrices with much smaller numbers of entries that can be constructed using the MUSICAL MATRIX ALGORITHM.

We shall work with tone rows chosen from the set of four pitch class hours, $S = \{0, 3, 6, 9\}$, and the subset of $\mathfrak{C}_{12}$ consisting of only these transformations:

$$
\begin{array}{cccc}
\mathsf{T}_0 & \mathsf{T}_3 & \mathsf{T}_6 & \mathsf{T}_9 \\[4pt]
\mathsf{I}_0 & \mathsf{I}_3 & \mathsf{I}_6 & \mathsf{I}_9 \\[4pt]
\mathsf{R}\,\mathsf{T}_0 & \mathsf{R}\,\mathsf{T}_3 & \mathsf{R}\,\mathsf{T}_6 & \mathsf{R}\,\mathsf{T}_9 \\[4pt]
\mathsf{R}\,\mathsf{I}_0 & \mathsf{R}\,\mathsf{I}_3 & \mathsf{R}\,\mathsf{I}_6 & \mathsf{R}\,\mathsf{I}_9.
\end{array}
\tag{6.35}
$$

Since all of the indices used in this subset of $\mathfrak{C}_{12}$ are evenly divisible by 3, these transformations will always produce tone rows whose hours belong to the set S. The situation is just like the 12 hours on the chromatic clock and the transformations $\mathfrak{C}_{12}$. The only difference is that, instead of the pitch classes being spaced in single hour increments around the chromatic clock, they are spaced at three-hour increments. We shall denote the transformations shown in (6.35) as $\mathfrak{C}_S$.

Following *Step 1* of the MUSICAL MATRIX ALGORITHM, we specify an initial tone row from the set S:

$$
\mathsf{T}_0 : \quad 3 \quad 0 \quad 9 \quad 6. \tag{6.36}
$$

This row will be the first row of the musical matrix, labeled by T_0 on the left and $\mathsf{R}\,\mathsf{T}_0$ on the right.

Following *Step 2*, the first column of the musical matrix is created by applying the pitch class chromatic inversion, $\mathsf{I}_0 = \mathcal{I}_6$, to this initial tone row:

$$
\begin{array}{llcccc}
m\colon & & 3 & 0 & 9 & 6 \\[4pt]
6 - m \ \bmod 12\colon & & 3 & 6 & 9 & 0
\end{array}
$$

We now have the first row and column of the musical matrix:

$$
\begin{array}{ccccccc}
 & & \mathsf{I}_0 & & & & \\
\mathsf{T}_0 & 3 & 0 & 9 & 6 & \mathsf{R}\,\mathsf{T}_0 \\
 & 6 & & & & \\
 & 9 & & & & \\
 & 0 & & & & \\
 & \mathsf{R}\,\mathsf{I}_0 & & & &
\end{array}
$$

Following *Step 3*, we fill in the rest of the hours that make up the matrix. We know that the second row begins with 6, and the initial tone row begins with 3. Since $6 - 3 = 3$, we apply T_3 to the initial tone row:

$$
\begin{array}{llcccc}
m\colon & & 3 & 0 & 9 & 6 \\[4pt]
3 + m \ \bmod 12\colon & & 6 & 3 & 0 & 9
\end{array}
$$

We now have the first two rows, and the first column, of the musical matrix:

$$
\begin{array}{ccccccc}
 & & \mathsf{I}_0 & & & & \\
\mathsf{T}_0 & 3 & 0 & 9 & 6 & \mathsf{R}\,\mathsf{T}_0 \\
\mathsf{T}_3 & 6 & 3 & 0 & 9 & \mathsf{R}\,\mathsf{T}_3 \\
 & 9 & & & & \\
 & 0 & & & & \\
 & \mathsf{R}\,\mathsf{I}_0 & & & &
\end{array}
$$

The remaining rows are figured out in the same way. The third row begins with 9, and $9 - 3 = 6$. So the third row is obtained by applying T_6 to the initial tone row:

$$m: \quad 3 \quad 0 \quad 9 \quad 6$$

$$6 + m \mod 12: \quad 9 \quad 6 \quad 3 \quad 0$$

For the fourth row, we have $0 - 3 = -3 \equiv 9 \mod 12$. So the fourth row is obtained by applying T_9 to the initial tone row:

$$m: \quad 3 \quad 0 \quad 9 \quad 6$$

$$9 + m \mod 12: \quad 0 \quad 9 \quad 6 \quad 3$$

These calculations give us all of the hour values for the matrix:

$$
\begin{array}{ccccc}
 & \mathsf{I}_0 & & & \\
T_0 & 3 & 0 & 9 & 6 & \mathsf{R}T_0 \\
T_3 & 6 & 3 & 0 & 9 & \mathsf{R}T_3 \\
T_6 & 9 & 6 & 3 & 0 & \mathsf{R}T_6 \\
T_9 & 0 & 9 & 6 & 3 & \mathsf{R}T_9 \\
 & \mathsf{R}\,\mathsf{I}_0 & & &
\end{array}
\qquad (6.37)
$$

Following **Step 4**, we complete the labelling of the matrix. The transposition T_9 maps 3 to 0, so we shall label the second column by I_9 and $\mathsf{R}\,\mathsf{I}_9$ on the bottom. Likewise, T_6 maps 3 to 9, so we label the third column by I_6 on the top and $\mathsf{R}\,\mathsf{I}_6$ on the bottom. Finally, T_3 maps 3 to 6, so we label the fourth column by I_3 on the top and $\mathsf{R}\,\mathsf{I}_3$ on the bottom. The completely labeled matrix is the following:

$$
\begin{array}{ccccc}
 & \mathsf{I}_0 \quad \mathsf{I}_9 \quad \mathsf{I}_6 \quad \mathsf{I}_3 & & & \\
T_0 & 3 \quad 0 \quad 9 \quad 6 & \mathsf{R}T_0 \\
T_3 & 6 \quad 3 \quad 0 \quad 9 & \mathsf{R}T_3 \\
T_6 & 9 \quad 6 \quad 3 \quad 0 & \mathsf{R}T_6 \\
T_9 & 0 \quad 9 \quad 6 \quad 3 & \mathsf{R}T_9 \\
 & \mathsf{R}\,\mathsf{I}_0 \quad \mathsf{R}\,\mathsf{I}_9 \quad \mathsf{R}\,\mathsf{I}_6 \quad \mathsf{R}\,\mathsf{I}_3 & & &
\end{array}
\qquad (6.38)
$$

The reader may wish to use the computer program listed in (6.34) to check our work here. If you enter the numbers 3 0 9 6 leaving all other entries blank, and then choose the option TRADITIONAL METHOD, you will obtain the same matrix as above.

6.7.3 Total Serialism

We conclude this section with a brief discussion of a composing method called *total serialism*. Total serialism applies transformation groups to all of the discrete aspects of a musical composition, such as its pitches, its rhythm, its dynamics, and its articulations. We will not go into great detail about this method as it is quite difficult to apply, and is rather controversial as to its musical value.

We have seen that pitch classes can be mapped to other pitch classes using various transformations. So, we now turn to how transformations can be applied to other discrete aspects of musical composition. First, consider the rhythm for a composition. Suppose a 12-hour clock is used for describing the initial rhythm. Members of the rhythmic transformation group $\mathfrak{R}_{12}$ can then be used for serially transforming this initial rhythm to create subsequent rhythms for the composition. These transformations of rhythm are used in combination with the pitch transformations from $\mathfrak{C}_{12}$.

Second, suppose we allow for 12 different levels of dynamics, starting from **pppp** at the lowest level up to **ffff** at the highest level. We can assign these levels to the hours of a 12-hour clock as follows:

Hour 0: **pppp**	Hour 1: **ppp**	Hour 2: **pp**	Hour 3: **p**
Hour 4: *quasi* **mp**	Hour 5: **mp**	Hour 6: **mf**	Hour 7: *quasi* **f**
Hour 8: **f**	Hour 9: **ff**	Hour 10: **fff**	Hour 11: **ffff**

The transformation group $\mathfrak{R}_{12}$ can then be used to make serial transformations of dynamics. An initial specification of dynamics is made, over a measure or set of measures, and then transformations from $\mathfrak{R}_{12}$ are successively applied in order to create new dynamics for subsequent measures.

Third, like dynamics, articulation indicators like staccato and *sforzando,* are also finite in number. A list of 12 articulation indicators indexed by the hours 0 through 11 can be made on a 12-hour clock. This allows the transformation group $\mathfrak{R}_{12}$ to be applied to these articulation indicators. Such a list can be found in an article by Jonathan Cross.[14] The article describes total serialism in more detail, including examples by composers who used it, such as Pierre Boulez.

As we stated above, the method of total serialism is difficult to employ effectively and is rather controversial as to its musical value. Perhaps the main reason for these problems is that we do not perceive rhythmic and dynamic relationships in the same way that we perceive pitch relationships. For example, an inversion on a rhythm is very difficult to relate to the original rhythm. As another example, the relationship of a minor 3rd (a 3-hour shift on the chromatic clock) is more clearly audible to a musician than a 3-hour shift between **pppp** and **p**.

Exercises

6.7.1. Complete the following musical matrix, using operations from $\mathfrak{C}_S$.

$$
\begin{array}{c}
\quad\;\; I_0 \\
T_0 \quad 6 \quad 3 \quad 9 \quad 0 \quad R\,T_0 \\
9 \\
3 \\
0 \\
R\,I_0
\end{array}
$$

6.7.2. Complete the following musical matrix, using operations from $\mathfrak{C}_S$.

$$
\begin{array}{c}
\quad\;\; I_0 \\
T_0 \quad 9 \quad 3 \quad 6 \quad 0 \quad R\,T_0 \\
3 \\
0 \\
6 \\
R\,I_0
\end{array}
$$

---

[14]J. Cross, Composing with numbers: sets, rows and magic squares. Chap. 8 in *Music and Mathematics: From Pythagoras to Fractals* (Eds. J. Fauvel, R. Flood, R. Wilson), Oxford, 2003. (See especially Figure 4.)

6.7.3. From the following initial tone row:

$$T_0 \quad 9 \quad 0 \quad 3 \quad 6 \quad R\,T_0$$

construct the musical matrix, using transformations from $\mathfrak{C}_S$.

6.7.4. From the following initial tone row:

$$T_0 \quad 6 \quad 9 \quad 3 \quad 0 \quad R\,T_0$$

construct the musical matrix, using transformations from $\mathfrak{C}_S$.

6.7.5. Someone decides to use the computer program given in (6.34) to compute a musical matrix for the following sequence of four pitch classes:

$$7 \quad 4 \quad 1 \quad 10.$$

The program produces this musical matrix:

	I_0	I_9	I_6	I_3	
T_0	7	4	1	10	$R\,T_0$
T_3	10	7	4	1	$R\,T_3$
T_6	1	10	7	4	$R\,T_6$
T_9	4	1	10	7	$R\,T_9$
	$R\,I_0$	$R\,I_9$	$R\,I_6$	$R\,I_3$	

This matrix uses only pitch classes from the set $\{1, 4, 7, 10\}$, so it is giving a valid musical matrix for that set. However, it is applying transformations from the group $\mathfrak{C}_S$ given in (6.35). Explain what is going on here.

6.7.6. Can the group $\mathfrak{C}_S$ be used to create a musical matrix for the set $\{1, 3, 6, 8\}$? What about the set $\{2, 5, 8, 11\}$? Describe the property that characterizes those sets of pitch classes for which $\mathfrak{C}_S$ can be used to construct musical matrices. List all such sets.

Musical matrices for the whole tone scale

The following four exercises deal with musical matrices constructed from the pitch classes of the whole tone scale: $\{0, 2, 4, 6, 10\}$. The transformations used for these pitch classes are restricted to the following ones:

$$
\begin{array}{cccccc}
T_0 & T_2 & T_4 & T_6 & T_8 & T_{10} \\[4pt]
I_0 & I_2 & I_4 & I_6 & I_8 & I_{10} \\[4pt]
R\,T_0 & R\,T_2 & R\,T_4 & R\,T_6 & R\,T_8 & R\,T_{10} \\[4pt]
R\,I_0 & R\,I_2 & R\,I_4 & R\,I_6 & R\,I_8 & R\,I_{10}.
\end{array}
\tag{6.39}
$$

We shall denote the group of transformations in (6.39) by $\mathfrak{C}_w$.

6.7.7. Complete the following musical matrix, using operations from $\mathfrak{C}_w$.

	I_0						
T_0	2	0	8	4	10	6	$R\,T_0$
	4						
	8						
	0						
	6						
	10						
	$R\,I_0$						

6.7.8. Complete the following musical matrix, using transformations from $\mathfrak{C}_w$.

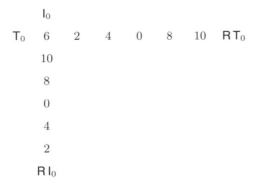

6.7.9. From the following initial tone row:

$$T_0 \quad 8 \quad 4 \quad 2 \quad 0 \quad 6 \quad 10 \quad R\,T_0$$

construct the musical matrix, using transformations from $\mathfrak{C}_w$.

6.7.10. From the following initial tone row:

$$T_0 \quad 10 \quad 2 \quad 4 \quad 6 \quad 8 \quad 0 \quad R\,T_0$$

construct the musical matrix, using transformations from $\mathfrak{C}_w$.

6.7.11. From the following initial tone row:

$$T_0 \quad 10 \quad 8 \quad 4 \quad 2 \quad 3 \quad 5 \quad 7 \quad 6 \quad 9 \quad 11 \quad 0 \quad 1 \quad R\,T_0$$

construct the musical matrix, using transformations from $\mathfrak{C}_{12}$.

6.7.12. From the following initial tone row:

$$T_0 \quad 7 \quad 9 \quad 6 \quad 4 \quad 8 \quad 10 \quad 5 \quad 3 \quad 2 \quad 0 \quad 1 \quad 11 \quad R\,T_0$$

construct the musical matrix, using transformations from $\mathfrak{C}_{12}$.

6.7.13. For all of the musical matrices that we have calculated, the diagonal values from the top left to the bottom right are always some constant value. For example, in the musical matrix in Figure 6.25 the constant value is 2, while in the musical matrix in (6.38) the constant value is 3. Explain why this phenomenon always occurs.

6.7.14. How many possible musical matrices are there if one uses the 6 hours, $\{0, 2, 4, 6, 8, 10\}$, of the whole tone scale?

6.7.15. How many possible musical matrices are there if one uses the 12 hours, $\{0, 1, 2, 3, \ldots, 11\}$, of the chromatic scale?

Chapter 7

A Geometry of Harmony

Of the three elements—harmony, rhythm, and melody—which have such a profound influence on tonal music, harmony is perhaps the most important, because it is the most potent. You can play the same chord with millions of different rhythms. It can cope with them all without needing to change. A melody is uninteresting if it does not move harmonically. That implies that the impact of harmony is much greater than that of rhythm and melody.—And it exists in every tonal work. There are thousands of distinctions between Bach, Wagner, Tchaikovsky and Debussy, but they have one thing in common: harmonic impact. This implies that a chord exerts a kind of vertical pressure on the horizontal movement of the music. When a chord moves, the horizontal flow of the music is changed. This has nothing to do with Bach or Chopin or anyone else; in my opinion it is a law of nature.

—Daniel Barenboim

Previously we used some basic geometric ideas—circles and triangles, rotations and reflections—to examine pitch organization in music. In this chapter we explore more advanced applications of geometric ideas relating to chord progressions, consonance, dissonance, and the sound qualities of different keys and modes. Our focus will be the network of chords and tones, called the *Tonnetz*, devised by the mathematician Leonhard Euler in the eighteenth century. The *Tonnetz* was later used, in the context of just intonation, by the music theorists Arthur von Oettingen and Hugo Riemann in the 19th century. It was then applied, in the context of equal temperament, by a number of music theorists at the end of the 20th century. We will focus exclusively on the *Tonnetz* in this latter context of equal temperament. We shall see that this version of the *Tonnetz* provides a geometric picture of pitch relationships (such as consonance and dissonance), as well as some of the more commonly used progressions of triadic chords.

7.1 Riemann's Chromatic Inversions

We begin our discussion of the *Tonnetz* by showing how to construct it as a network of triadic chords, using three special chromatic inversions. In Chapter 6 we described the geometric interpretation of chromatic inversions of major and minor triadic chords based on mirror reflections of the chromatic clock. We also showed that any line through opposite hours, or opposite half-hours, of the chromatic clock could be used as a mirror. The three chromatic inversions used for constructing the *Tonnetz* make use of three special mirrors. Each of these mirrors bisects one of the three sides of the triangle corresponding to a major or minor chord. (See Figure 7.1.)

These chromatic inversions are denoted by $\mathcal{P}$, $\mathcal{R}$, and $\mathcal{L}$. We will explain the reason for this notation in a moment. We shall refer to $\mathcal{P}$, $\mathcal{R}$, and $\mathcal{L}$ as **neo-Riemannian chordal inversions**. They were originally studied in the 19th century by the music theorists, Arthur von Oettingen and Hugo Riemann, in the context of just intonation. Later, toward the end of the 20th century, they were studied again by a number of music theorists, in the context of twelve-tone equal-tempered tuning. In this latter context, they have a simpler mathematical structure that applies more widely to Western music. Therefore, we shall examine the neo-Riemannian chordal inversions $\mathcal{P}$, $\mathcal{R}$, and $\mathcal{L}$ within the system of

217

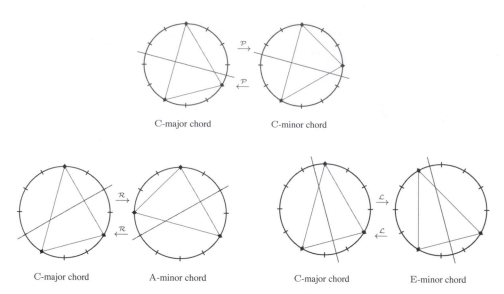

Figure 7.1. The neo-Riemannian chromatic inversions $\mathcal{P}, \mathcal{R}, \mathcal{L}$. For $\mathcal{P}$, the mirror through the origin bisects the line segment of chromatic distance 5. For $\mathcal{R}$, it bisects the line segment of chromatic distance 4. For $\mathcal{L}$, it bisects the line segment of chromatic distance 3.

twelve-tone equal-tempered tuning. This allows us to make use of the mod 12 arithmetic (chromatic clock) description of chords given in Chapters 3 and 6.

We now examine these neo-Riemannian chordal inversions more closely. Our examination will reveal why they are denoted by the letters $\mathcal{P}$, $\mathcal{R}$, and $\mathcal{L}$, and give some indication of their musical significance. In this initial examination, we will discuss major and minor chords that are in root position. We do this only to provide the most straightforward explanation for the notations $\mathcal{P}$, $\mathcal{R}$, and $\mathcal{L}$. It should be kept in mind, however, that chordal triangles, such as those shown in Figure 7.1, can be interpreted as triples of pitch classes. That will allow us to apply the chromatic inversions $\mathcal{P}$, $\mathcal{R}$, and $\mathcal{L}$ in a wide variety of musical settings. We shall apply them not only to chords in root position, but also to inverted chords, arpeggiated chords, and chords used as descriptors of harmonic frameworks for musical passages.

The neo-Riemannian chordal inversion $\mathcal{P}$

First, we examine the inversion $\mathcal{P}$. As shown in Figure 7.1, $\mathcal{P}$ maps the C-major chord to the C-minor chord: C $\xrightarrow{\mathcal{P}}$ c. It also reverses itself by mapping the C-minor chord to the C-major chord: c $\xrightarrow{\mathcal{P}}$ C. In music theory, because the C-major chord and C-minor chords have the same root C, they are called *parallel* major and minor chords. The neo-Riemannian chordal inversion $\mathcal{P}$ associates these parallel major and minor chords via the mappings C $\xrightarrow{\mathcal{P}}$ c and c $\xrightarrow{\mathcal{P}}$ C. It lowers the middle tone of the C-major chord by one half step, thereby producing its parallel minor chord c. It also raises the middle tone of the minor chord c by one half step, thereby producing its parallel major chord C. The raising or lowering of the middle tone is produced by a mirror through the side of chromatic distance 5 in the chordal triangle of a triadic chord. Therefore, the neo-Riemannian chordal inversion $\mathcal{P}$ will always pair two parallel major and minor chords. Consequently, the letter $\mathcal{P}$ is used to denote this inversion.

Here are a few examples of the use of $\mathcal{P}$ in music. One use is in the chromatic embellishment of the widely used chord progression $\mathbf{V} \rightarrow \mathbf{I} \rightarrow \mathbf{V}$. An example of this progression, in the key of C-major, is shown in the first measure of Figure 7.2. The specific chords are G, followed by C/G, followed by G. Applying $\mathcal{P}$ to C gives the minor chord c, and in the second measure the chord c/G is placed between the chords C/G and G. This noticeably alters the sound of the chord progressions, the flatted note serving as a passing tone from the highest pitch of the chord C/G to the highest pitch of the

Figure 7.2. Chromatic embellishment illustrating use of neo-Riemannian chordal inversion $\mathcal{P}$. Third chord in second measure is obtained by applying $\mathcal{P}$ to chord preceding it.

chord G. Another example — this time using $\mathcal{P}$ in the harmonic background for a musical passage — is shown in Figure 7.3. The extremely low pitch note, $A_1^\flat$, is held as a constant tone throughout this whole passage. According to the eminent musicologist, Deryck Cooke, in performance it can actually be heard sounding as an $A^\flat$-major chord.[1] Its first five harmonics are the pitches:

$$A_1^\flat, \ A_2^\flat, \ E_3^\flat, \ A_3^\flat, \ C_4$$

which create the sound of an $A^\flat$-major chord. In contrast to this, the melody line for Wotan is tracing out many notes from the $A^\flat$-minor chord ($A^\flat$-$C^\flat$-$E^\flat$). Therefore, the harmony of the passage's sonority relates to the pairing of the chords $A^\flat$ and $a^\flat$ via the mapping $\mathcal{P}$.

Figure 7.3. Passage from Wagner's *Die Walküre (The Valkyrie)*, WWV 86B, Movement II (when Wotan sings *"Als junger Liebe ..."*). Subscript 8 below bass clef in lower staff indicates pitches on that staff are an octave lower, hence $A^\flat$ note shown there is $A_1^\flat$.

As a third example, we describe how the transformation $\mathcal{P}$ relates to *mode mixtures*.[2] One type of mode mixture is the use of the major chord parallel to the tonic chord of a minor key. For example, the chorus of Bach's contata *"Gottlob! nun geht das Jahr zu Ende"* (BWV 28), ends with an A-major chord. Given that the key for this chorus is A-minor, this can be viewed as an example of the mapping $a \xrightarrow{\mathcal{P}} A$. The use of the raised third note of the natural A-minor scale, $C^\sharp$, within the A-major chord is called a **Picardy third**. Kostka and Payne point out that most compositions in minor from the early 1500s until around 1750 were ended by this procedure of using the major chord parallel to the minor's tonic chord. In every case, we can view it as an instance of a mapping of the form $i \xrightarrow{\mathcal{P}} I$. The mapping $\mathcal{P}$ can also be viewed as describing a large scale mode mixture such as the famous transition from the C-minor tonic key to the parallel C-major key in Beethoven's Fifth Symphony. Here we can write $c \xrightarrow{\mathcal{P}} C$ as a succinct description of this transition.

---

[1] D. Cooke, *The Language of Music*, Oxford Univ. Press (1959), p. 58.
[2] Our description is based on Chapter 21 of S. Kostka and D. Payne, *Tonal Harmony, Sixth Edition*.

The neo-Riemannian chordal inversion $\mathcal{R}$

Second, we examine the inversion $\mathcal{R}$. Figure 7.1 shows that $\mathcal{R}$ maps the C-major chord to the A-minor chord: C $\xrightarrow{\mathcal{R}}$ a. It also reverses itself by mapping the A-minor chord to the C-major chord: a $\xrightarrow{\mathcal{R}}$ C. The root C of the C-major chord is the tonic for the key of C-major. The root A of the A-minor chord is the tonic for the key of A-minor. This A-minor key is the relative minor for the C-major key. Likewise, the C-major key is the relative major for the A-minor key. The neo-Riemannian chordal inversion $\mathcal{R}$ pairs the tonic chords for these two related keys, via the mappings C $\xrightarrow{\mathcal{R}}$ a and a $\xrightarrow{\mathcal{R}}$ C. Every major key can be obtained from a transposition of the C-major key, and the tonic of every relative minor key is obtained by subtracting 3 hours on the chromatic clock from the tonic of its major key. It follows that the neo-Riemannian chordal inversion $\mathcal{R}$ always pairs the tonic chord of a major key with the tonic chord of its relative minor key. Consequently, the letter $\mathcal{R}$ is used to denote this inversion.

Remark 7.1.1. There is an equivalent way to see that $\mathcal{R}$ always pairs the tonic chord of a major key with the tonic chord of its relative minor key. We simply observe that the diagram for $\mathcal{R}$ in Figure 7.1 can be labeled with hours that begin on the top of the clock with the hour for the root of the major chord. Then $\mathcal{R}$ will map that major chord to its relative minor (and vice versa), just as it maps the chord C to the chord a (and vice versa).

Another example of applying $\mathcal{R}$ is

$$F \xrightarrow{\mathcal{R}} d. \tag{7.1}$$

This mapping holds because the hour for the root F of the F-major chord is 5, and subtracting 3 from the hour 5 for F yields hour 2, the hour for the root D of the d-minor chord. This result will be used in Example 7.1.1, when we describe how Beethoven's Ninth Symphony includes several applications of the neo-Riemannian chordal inversions, $\mathcal{R}$ and $\mathcal{L}$.

The neo-Riemannian chordal inversion $\mathcal{L}$

Third, we examine the neo-Riemannian chordal inversion $\mathcal{L}$. Figure 7.1 shows that $\mathcal{L}$ maps the C-major chord to the E-minor chord: C $\xrightarrow{\mathcal{L}}$ e. It also reverses itself by mapping the E-minor chord to the C-major chord: e $\xrightarrow{\mathcal{L}}$ C. As shown in the figure, the geometric effect of the mapping C $\xrightarrow{\mathcal{L}}$ e is to lower the root C of the C-major chord by one half step, while mapping the notes E and G to each other (so they remain unchanged in the resulting minor chord). Recall that the tone B is the *leading tone* for the tonic C in the C-major key. The resulting E-minor chord has the leading tone B for the root C of the C-major chord as its highest pitch note, while retaining the notes E and G from the C-major chord. When $\mathcal{L}$ is applied in reverse to the E-minor chord, it preserves the notes E and G, while mapping the highest pitch tone B to the note C that it leads into in the C-major key. This relation holds for all other major and minor chords that are paired via $\mathcal{L}$. In each case, $\mathcal{L}$ will map the root of a major chord to the leading tone of the major key for which that root is tonic, by lowering that root by one half step (and also raise the highest pitch note of a minor chord by one half step while preserving the other two notes). This connection with leading tones provides the rationale for using the letter $\mathcal{L}$ to denote this inversion.

Here is a musical example of the use of $\mathcal{L}$. In the passage from Wagner shown in Figure 7.3, when Wotan sings *"Mut: von jäher Wün,"* the score indicates the notes:

$$A_2^\flat, C_3^\flat, F_3^\flat, F_3^\flat, F_3^\flat.$$

These are notes in the pitch classes $\mathbf{F}^\flat$, $\mathbf{A}^\flat$, $\mathbf{C}^\flat$ for the $F^\flat$-major chord. When $\mathcal{L}$ is applied to the $F^\flat$-major chord, it lowers its root $F^\flat$ by one half step to get the note $E^\flat$, while preserving the notes

$A^\flat$ and $C^\flat$. The resulting chord is $A^\flat$-minor ($A^\flat$-$C^\flat$-$E^\flat$). Consequently, $\mathcal{L}$ maps the $F^\flat$-major chord to the $A^\flat$-minor chord: $F^\flat \xrightarrow{\mathcal{L}} a^\flat$. In the score, following the last $F^\flat$, the melody line for Wotan from *"sche"* to *"die"* is largely tracing out notes from the $A^\flat$-minor chord. Therefore, this part of the melody line for Wotan is an instance of the mapping $F^\flat \xrightarrow{\mathcal{L}} a^\flat$. Moreover, the beginning melody line for Wotan, from *"Als junger Liebe"* to *"Macht mein,"* is tracing out notes from $a^\flat$ as well (recall that a note of $E^\flat$ is implicit due to the 3rd harmonic of $A_1^\flat$ held throughout). Hence the transition from the first two measures to the beginning of the third measure is described by the mapping $a^\flat \xrightarrow{\mathcal{L}} F^\flat$. The entire passage then corresponds to the self-reversing property of $\mathcal{L}$: $a^\flat \xrightarrow{\mathcal{L}} F^\flat \xrightarrow{\mathcal{L}} a^\flat$.

As another example of $\mathcal{L}$ we have

$$F \xrightarrow{\mathcal{L}} a.$$

This mapping holds because when the root F is lowered by a half step, we obtain E. The minor chord, in root position, with E as its highest pitch note, is the chord A-C-E. In other words, the chord a. Since $\mathcal{L}$ is self-reversing, we also have

$$a \xrightarrow{\mathcal{L}} F. \tag{7.2}$$

This mapping will be needed for the following musical example.

Example 7.1.1. The mappings $\mathcal{R}$ and $\mathcal{L}$ in Beethoven's Ninth Symphony. A musical example that uses both $\mathcal{R}$ and $\mathcal{L}$ is shown in Figure 7.4. In this example from Beethoven, the first measure contains three C-major chords. The chord in the second measure is A-minor. Hence, the chord progression is $C \xrightarrow{\mathcal{R}} a$. The chords in the third measure are all F-major chords. Hence, from the second to the third measure, the chord progression is $a \to F$. Since $\mathcal{L}$ maps a to F, we can express this chord progression as $a \xrightarrow{\mathcal{L}} F$. Finally, the chord in the fourth measure is D-minor. So, in going from the third measure to the fourth measure, we have the chord progression $F \xrightarrow{\mathcal{R}} d$. This sequence of chord progressions:

$$C \xrightarrow{\mathcal{R}} a \xrightarrow{\mathcal{L}} F \xrightarrow{\mathcal{R}} d$$

only tells part of the story. In Example 7.2.2 we will elaborate on this theme.

Figure 7.4. Score for string section in Beethoven's Ninth Symphony (movement II, measures 143 to 146). The sequence of chord progressions, from each measure to the next, is $C \xrightarrow{\mathcal{R}} a \xrightarrow{\mathcal{L}} F \xrightarrow{\mathcal{R}} d$.

Remark 7.1.2. We have seen in the examples above that there is a connection between chord progressions and neo-Riemannian chordal inversions. We shall pursue this point in more detail as we proceed.

Exercises

7.1.1. Complete the following mappings. Part **(a)** has been done as an example.

(a) F $\xrightarrow{P}$ f   **(b)** f $\xrightarrow{\mathcal{R}}$ **(c)** B $\xrightarrow{\mathcal{L}}$ **(d)** a $\xrightarrow{P}$ **(e)** d$^\flat$ $\xrightarrow{\mathcal{R}}$ **(f)** A$^\flat$ $\xrightarrow{\mathcal{L}}$

7.1.2. Complete the following mappings.

(a) E $\xrightarrow{P}$ **(b)** d $\xrightarrow{\mathcal{R}}$ **(c)** E $\xrightarrow{\mathcal{L}}$ **(d)** g $\xrightarrow{P}$ **(e)** G $\xrightarrow{\mathcal{R}}$ **(f)** E$^\flat$ $\xrightarrow{\mathcal{L}}$

7.1.3. Complete the following mappings.

(a) e$^\flat$ $\xrightarrow{P}$ **(b)** a $\xrightarrow{\mathcal{R}}$ **(c)** b $\xrightarrow{\mathcal{L}}$ **(d)** B $\xrightarrow{P}$ **(e)** c $\xrightarrow{\mathcal{R}}$ **(f)** D $\xrightarrow{\mathcal{L}}$

7.1.4. Complete the following mappings.

(a) g$^\flat$ $\xrightarrow{P}$ **(b)** G$^\flat$ $\xrightarrow{\mathcal{R}}$ **(c)** a $\xrightarrow{\mathcal{L}}$ **(d)** D $\xrightarrow{P}$ **(e)** e $\xrightarrow{\mathcal{R}}$ **(f)** f $\xrightarrow{\mathcal{L}}$

7.1.5. In the spectrogram shown in Figure 7.5, the first arrow (the higher up one) points to a 2nd harmonic of a $C_3^\flat$ tone. Point out where this tone occurs as a note in the score shown in Figure 7.3, and discuss how the harmonics shown in the spectrogram relate to the mapping A$^\flat$ $\xrightarrow{P}$ a$^\flat$.

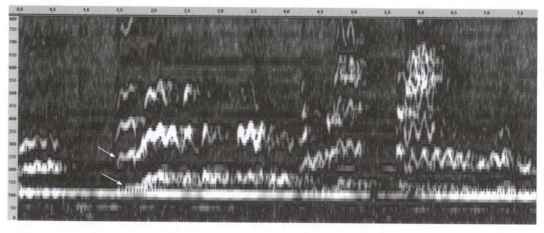

Figure 7.5. Spectrogram of recording of measures 3 and 4 of the passage from Wagner's *Die Walküre (The Valkyrie)* in Figure 7.3. Two arrows point to 1st and 2nd harmonics of $C_3^\flat$ tone of Wotan singing *"von."*

7.1.6. In the spectrogram shown in Figure 7.5, the second arrow (the lower down one) points to the 1st harmonic of a $C_3^\flat$ tone (enharmonic with B_2). Explain what is causing the interference pattern of alternating thin white and dark vertical segments that the arrow is pointing to.

7.1.7. Show that $T_1 \mathcal{R} = \mathcal{R} T_1$. Hint: Show that $T_1 \mathcal{R}$ and $\mathcal{R} T_1$ map the C-major chord to the same chord, and map the C-minor chord to the same chord. Why is that sufficient?

7.1.8. Show that $T_1 \mathcal{P} = \mathcal{P} T_1$. Hint: Show that $T_1 \mathcal{P}$ and $\mathcal{P} T_1$ map the C-major chord to the same chord, and map the C-minor chord to the same chord.

7.1.9. Show that $T_1 \mathcal{L} = \mathcal{L} T_1$.

7.1.10. Using the fact that $T_2 = T_1 T_1$, show that $T_2 \mathcal{R} = \mathcal{R} T_2$. Use similar reasoning to show that $T_k \mathcal{R} = \mathcal{R} T_k$ for each integer $k = 3, 4, \ldots, 11$. Then show that we also have $T_k \mathcal{R} = \mathcal{R} T_k$ for all integers k.

7.1.11. Show that $T_k \mathcal{P} = \mathcal{P} T_k$ for all integers k. Also show that $T_k \mathcal{L} = \mathcal{L} T_k$ for all integers k.

7.2 A Network of Triadic Chords

We now discuss how the three neo-Riemannian chordal inversions are used to construct a network of all the triadic chords. We will follow this discussion with several musical examples.

Suppose we start with the C-major chord, C. If we apply $\mathcal{P}$, then we have $C \xrightarrow{\mathcal{P}} c$. Since c is a minor chord with root C, its relative major chord will have its root 3 half steps above C, which is $D^\sharp$ or $E^\flat$. We will write $E^\flat$. Thus, $\mathcal{R}$ maps c to $E^\flat$. If we now apply $\mathcal{L}$ to the chord $E^\flat$, then its root $E^\flat$ will move down one half step to D. The resulting minor chord is G-$B^\flat$-D, the chord g.

Summarizing our calculations, we have found the following sequence of mappings:

$$C \xrightarrow{\mathcal{P}} c \xrightarrow{\mathcal{R}} E^\flat \xrightarrow{\mathcal{L}} g. \tag{7.3}$$

If we now apply $\mathcal{P}$, $\mathcal{R}$, and $\mathcal{L}$, in sequence again, we obtain an interesting result. First, observe that the parallel major chord to g is G. So we have $g \xrightarrow{\mathcal{P}} G$, and (7.3) extends to

$$C \xrightarrow{\mathcal{P}} c \xrightarrow{\mathcal{R}} E^\flat \xrightarrow{\mathcal{L}} g \xrightarrow{\mathcal{P}} G. \tag{7.4}$$

Second, applying $\mathcal{R}$ and $\mathcal{L}$ in succession gives:

$$G \xrightarrow{\mathcal{R}} e \xrightarrow{\mathcal{L}} C. \tag{7.5}$$

To see that (7.5) holds we note that lowering the root G by 3 half steps gives E, so $\mathcal{R}$ must map G to e. Furthermore, the map $e \xrightarrow{\mathcal{L}} C$ is shown in Figure 7.1.

Combining (7.4) and (7.5), we obtain our interesting result:

$$C \xrightarrow{\mathcal{P}} c \xrightarrow{\mathcal{R}} E^\flat \xrightarrow{\mathcal{L}} g \xrightarrow{\mathcal{P}} G \xrightarrow{\mathcal{R}} e \xrightarrow{\mathcal{L}} C. \tag{7.6}$$

The succession of mappings in (7.6) is interesting because, starting with chord C, and applying $\mathcal{P}$, $\mathcal{R}$, $\mathcal{L}$ in sequence two times, we have returned to the chord C. In mathematics, the succession of mappings in (7.6) is called **cyclic**. We can arrange the cyclic pattern of mappings in (7.6) in a geometric diagram that has the shape of a regular hexagon. See the regular hexagon on the left side of Figure 7.6. On this hexagon, we have drawn each mapping as a line segment without any arrowheads. That is because each mapping is self-reversing. For example, the line segment connecting the minor chord c with the major chord C is labeled by $\mathcal{P}$, and should be thought of as indicating both $c \xrightarrow{\mathcal{P}} C$ and $C \xrightarrow{\mathcal{P}} c$. The mapping $\mathcal{P}$ is like a two-lane highway connecting these chords. Likewise, there is a two-lane highway $\mathcal{R}$ connecting chords c and $E^\flat$, and a two-lane highway $\mathcal{L}$ connecting $E^\flat$ and g.

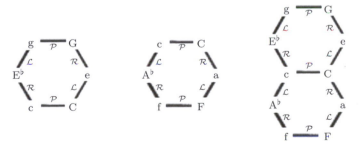

Figure 7.6. Left: Regular hexagon of triadic chords obtained by applying $\mathcal{P}, \mathcal{R}, \mathcal{L}$ twice, beginning from triadic chord C. Middle: Regular hexagon of triadic chords obtained by applying $\mathcal{P}, \mathcal{R}, \mathcal{L}$ twice, beginning from triadic chord F. Right: Merging these two hexagons along their common edge.

If we start from a different major chord, say F, and again apply $\mathcal{P}, \mathcal{R}, \mathcal{L}$ twice, then we get another cyclic pattern of mappings. We leave it as an exercise for the reader to check that it is

$$\text{F} \xrightarrow{\mathcal{P}} \text{f} \xrightarrow{\mathcal{R}} \text{A}^\flat \xrightarrow{\mathcal{L}} \text{c} \xrightarrow{\mathcal{P}} \text{C} \xrightarrow{\mathcal{R}} \text{a} \xrightarrow{\mathcal{L}} \text{F}. \tag{7.7}$$

The regular hexagon for this cyclic pattern of mappings is shown in the middle of Figure 7.6. Because we used the same method to generate both of these hexagons, we draw them as congruent figures.

The two congruent hexagons shown in Figure 7.6 can be connected together, because they share a common edge. This edge is the two-lane highway $\mathcal{P}$ connecting the chords c and C. On the right of Figure 7.7 we show the connected hexagons. The mappings $\mathcal{P}, \mathcal{R},$ and $\mathcal{L}$ are used several times for these two hexagons. There is an interesting feature of their orientation that is worth noting. The neo-Riemannian chordal inversion $\mathcal{P}$ always connects chords horizontally (slope 0 line segments). The neo-Riemannian chordal inversion $\mathcal{L}$ always connects chords by a positively sloped line segment. Lastly, the neo-Riemannian chordal inversion $\mathcal{R}$ always connects chords by a negatively sloped line segment. This will always be the case because we shall construct all of the hexagons in our chordal network in the same way. We always start from a major chord and apply $\mathcal{P}, \mathcal{R}, \mathcal{L}$ twice to return to this same major chord.

Since there are 12 major chords, we will get 12 distinct regular hexagons, all congruent to each other. The complete collection, all linked together by their common edges, is shown in Figure 7.7. This chordal network is called the *Tonnetz*. An interesting subtlety of the *Tonnetz* is that the chords eventually repeat themselves, as indicated by placing these repeated chords in parentheses. This repetition is an artifact of our displaying the network of hexagons on a flat surface (and our use of equal temperament). It turns out that to have a one-to-one correspondence between the corners of these hexagons and the 24 major and minor chords, the 12 distinct hexagons would have to be placed on a donut-shaped surface (a torus). We will not go into the details of this construction, because even if it is clearly understood, using a donut-shaped surface for displaying the *Tonnetz* would make matters quite difficult to visualize. Therefore, we shall work with the form of the *Tonnetz* shown in Figure 7.7. We just have to accustom ourselves to the fact that chords can be listed multiple times.

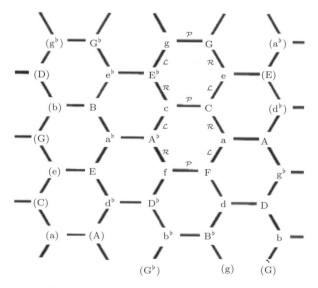

Figure 7.7. *Tonnetz* connecting all 24 major and minor chords. If a chord has parentheses around it, then it is a repetition of a chord elsewhere in the *Tonnetz*. The positioning of $\mathcal{P}, \mathcal{R},$ and $\mathcal{L}$ follows the pattern indicated in two of the hexagons ($\mathcal{P}$ for slope-0 edges, $\mathcal{R}$ for negatively sloped edges, $\mathcal{L}$ for positively sloped edges).

7.2.1 Musical Examples

We now discuss several musical examples that use the *Tonnetz* shown in Figure 7.7.

Example 7.2.1. Our first example is from a work of Haydn, his *Piano Sonata in E-minor* (Hob XVI:34). (See Figure 7.8.) For the five measures shown, the chord progressions from each measure to the next are the following:

$$\text{b} \xrightarrow{\mathcal{L}} \text{G} \xrightarrow{\mathcal{R}} \text{e} \xrightarrow{\mathcal{L}} \text{C} \xrightarrow{\mathcal{R}} \text{a}.$$

This illustrates an alternating sequence of the neo-Riemannian chordal inversions $\mathcal{L}$ and $\mathcal{R}$.

Figure 7.8. Passage from Haydn's *Piano Sonata in E-minor*, Hob XVI:34, measures 72 to 76. Chords progress from one measure to next by alternately applying $\mathcal{L}$ and $\mathcal{R}$.

Example 7.2.2. More on the mappings $\mathcal{R}$ and $\mathcal{L}$ in Beethoven's Ninth. As a second example, recall the discussion of Movement II of Beethoven's Ninth Symphony in Example 7.1.1. In that example, we showed that the following alternating sequencing of $\mathcal{R}$ and $\mathcal{L}$:

$$\text{C} \xrightarrow{\mathcal{R}} \text{a} \xrightarrow{\mathcal{L}} \text{F} \xrightarrow{\mathcal{R}} \text{d}$$

describes the chord progressions from one measure to the next, beginning with measure 143. In fact, this alternation of $\mathcal{R}$ and $\mathcal{L}$ continues all the way to measure 176, giving this sequencing of 18 chord progressions:[3]

$$\text{C} \xrightarrow{\mathcal{R}} \text{a} \xrightarrow{\mathcal{L}} \text{F} \xrightarrow{\mathcal{R}} \text{d} \xrightarrow{\mathcal{L}} \text{B}^\flat \xrightarrow{\mathcal{R}} \text{g} \xrightarrow{\mathcal{L}} \text{E}^\flat$$

$$\xrightarrow{\mathcal{R}} \text{c} \xrightarrow{\mathcal{L}} \text{A}^\flat \xrightarrow{\mathcal{R}} \text{f} \xrightarrow{\mathcal{L}} \text{D}^\flat \xrightarrow{\mathcal{R}} \text{b}^\flat \xrightarrow{\mathcal{L}} \text{G}^\flat \qquad (7.8)$$

$$\xrightarrow{\mathcal{R}} \text{e}^\flat \xrightarrow{\mathcal{L}} \text{B} \xrightarrow{\mathcal{R}} \text{a}^\flat \xrightarrow{\mathcal{L}} \text{E} \xrightarrow{\mathcal{R}} \text{d}^\flat \xrightarrow{\mathcal{L}} \text{A}.$$

It is quite interesting that this alternating sequence runs through more than three-fourths of the 24 major and minor chords in the *Tonnetz*. If one more row of three alternating pairs $\mathcal{R}$, $\mathcal{L}$ were added to the rows in (7.8), the entire set of 24 chords in the *Tonnetz* would be obtained.

Example 7.2.3. Chord progressions and the *Tonnetz*. The *Tonnetz* network of chords captures a large percentage of the typical chord progressions used in diatonic music. For example, in a major key, the chord progressions

$$\textbf{vi} \rightarrow \textbf{IV} \qquad \text{and} \qquad \textbf{IV} \rightarrow \textbf{ii}$$

are typical ones, as shown in Figure 3.10 on p. 61. In the key of C-major, these progressions are

$$\text{a} \xrightarrow{\mathcal{L}} \text{F} \qquad \text{and} \qquad \text{F} \xrightarrow{\mathcal{R}} \text{d}.$$

---

[3] A.S. Crans, T.M. Fiore, and R. Satyendra, *Musical Actions of Dihedral Groups*, *American Mathematical Monthly*, **116**, June-July 2009, p. 489.

Since all major scales are simply transpositions of the C-major scale, we can use the correspondences:

$$\mathbf{I} \leftrightarrow \text{C}, \ \mathbf{ii} \leftrightarrow \text{d}, \ \mathbf{iii} \leftrightarrow \text{e}, \ \mathbf{IV} \leftrightarrow \text{F}, \ \mathbf{V} \leftrightarrow \text{G}, \ \mathbf{vi} \leftrightarrow \text{a}$$

to compare relations between major and minor chords, shown on the *Tonnetz*, with the Roman numeral labelling of chords used for the major key chord progressions in Figure 3.10. Hence, we can write

$$\mathbf{vi} \xrightarrow{\mathcal{L}} \mathbf{IV} \qquad \text{and} \qquad \mathbf{IV} \xrightarrow{\mathcal{R}} \mathbf{ii}.$$

However, not all typical chord progressions are obtained by single applications of a neo-Riemannian chordal inversion. For example, the chord progression $\mathbf{iii} \to \mathbf{vi}$ is a typical one. In the key of C-major, this progression is $\text{e} \to \text{a}$. On the *Tonnetz*, these two chords are not connected by a single edge. There is a connection by two edges, so that we can write $\text{e} \xrightarrow{\mathcal{LR}} \text{a}$. Therefore, in any major key, we have

$$\mathbf{iii} \xrightarrow{\mathcal{LR}} \mathbf{vi}.$$

An even more striking example, is the chord progression $\mathbf{ii} \to \mathbf{V}$. In the key of C-major, this progression is $\text{d} \to \text{G}$. As shown in Figure 7.9, the chords d and G are connected by three edges on the *Tonnetz* (see the lower rectangle in that figure). From those three edges, we infer that $\text{d} \xrightarrow{\mathcal{PRL}} \text{G}$, which translates to any major key as $\mathbf{ii} \xrightarrow{\mathcal{PRL}} \mathbf{V}$.

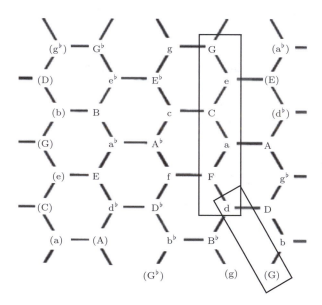

Figure 7.9. *Tonnetz* with edges connecting major and minor chords for key of C-major. Chords enclosed within the rectangles are related to typical chord progressions between major and minor chords in this key via edges in *Tonnetz*. In this diagram, a horizontal edge stands for mapping $\mathcal{P}$, a positively sloped edge stands for mapping $\mathcal{L}$, a negatively sloped edge stands for mapping $\mathcal{R}$.

This connection between the *Tonnetz* and chord progressions raises a number of questions. Such as why some progressions that are possible on the *Tonnetz* are not the typical ones in Figure 3.10 (and also in Figure 3.13), and whether there are better models than the ones described in those figures. In the next section, we continue our discussion of the connection between the *Tonnetz* and typical chord progressions, and attempt to answer these questions.

Exercises

7.2.1. The ending of the Gesualdo madrigal, "Luci serene e chiare," is shown in Figure 7.10. Find the chord progressions for this passage and how they relate to the neo-Riemannian chordal inversions $\mathcal{P}$, $\mathcal{R}$, and $\mathcal{L}$.

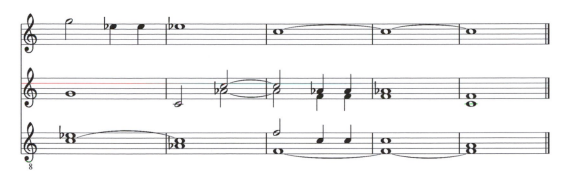

Figure 7.10. Ending of Gesualdo madrigal, "Luci serene e chiare," in modern notation.

7.2.2. Show that each of the neo-Riemannian chordal inversions, $\mathcal{P}$, $\mathcal{L}$, $\mathcal{R}$, for a given hexagon, can be realized by reflections through a mirror passing through opposite sides of the hexagon.

7.2.3. An excerpt from Wagner's *Die Walküre (The Valkyrie)* is shown in Figure 7.11. Find the chord progressions for this passage and how they relate to the neo-Riemannian chordal inversions $\mathcal{P}$, $\mathcal{R}$, and $\mathcal{L}$.

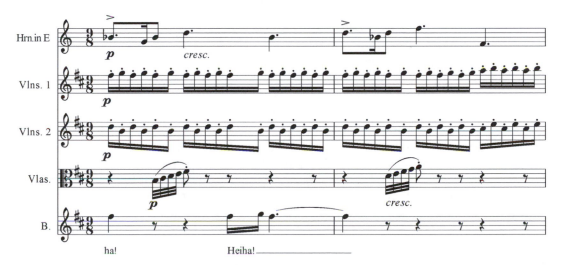

Figure 7.11. Excerpt from Wagner's *Die Walküre (The Valkyrie)*, WWV 86B, Movement II. Notes for Horn in E must be lowered by eight half steps to get their sounding pitches. For example, the first note played by the Horn in E actually sounds as a D-note, since subtracting 8 from hour 10 for $B^\flat$ yields 2 mod 12. Likewise, the second note actually sounds as a B-note, since subtracting 8 from hour 7 yields $-1 \equiv 11$ mod 12.

7.2.4. Verify the statement made at the end of Example 7.2.2 by finding the row of three alternating pairs of mappings, $\mathcal{R}$ and $\mathcal{L}$, that extends the sequence in (7.8) so that it ends with the chord C.

7.2.5. (a) Starting from chord C in the *Tonnetz*, show that alternating $\mathcal{P}$ and $\mathcal{R}$ several times eventually returns to chord C. However, only 8 of the 24 chords in the *Tonnetz* are linked in this way. **(b)** Starting from the chord C in the *Tonnetz*, show that alternating $\mathcal{P}$ and $\mathcal{L}$ several times eventually returns to chord C. However, only 6 of the 24 chords in the *Tonnetz* are linked in this way.

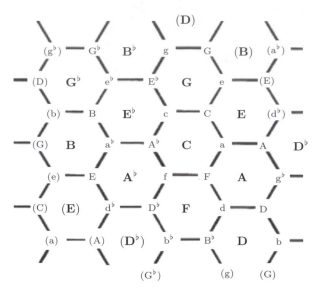

Figure 7.12. Pitch class placement within *Tonnetz*. Interior region of each hexagon is denoted by the single pitch class common to all six chords at vertices of the hexagon. If a pitch class has parentheses around it, then it is a repetition of same pitch class elsewhere in the figure.

7.3 Embedding Pitch Classes within the *Tonnetz*

In this section, we conclude our discussion of the *Tonnetz* by showing how it relates to consonance and dissonance between musical tones, and how it provides a geometric framework for understanding typical chord progressions in diatonic music. The first step in this process is to show how the 12 pitch classes of the chromatic scale provide "name tags" for the 12 hexagons of the *Tonnetz*. A more mathematical description is that these 12 pitch classes precisely correspond, in a way that makes musical sense, to the regions inside the 12 hexagons of the *Tonnetz*.

To see how this correspondence works, recall the hexagons of chords shown in Figure 7.6 on p. 223. The hexagon on the left of that figure was obtained from the sequence of chord progressions in expression (7.6). That sequence links the chords C, c, $E^\flat$, g, G, and e. The unique pitch class shared by all of those chords is **G**. Consequently we shall refer to the region interior to this hexagon as **G**. Likewise, the hexagon in the middle of Figure 7.6 was obtained from the sequence of chord progressions in expression (7.7), linking the chords F, f, $A^\flat$, c, C, and a. The unique pitch class shared by all of those chords is **C**. Consequently, we shall refer to the region interior to this latter hexagon as **C**. In Figure 7.12, we show the *Tonnetz* with each of the interior regions of its hexagons labeled by the unique pitch class common to all of its chords.

Figure 7.12 contains a wealth of information. It is worth studying carefully.[4] First, we look at how the placement of hexagons for distinct tones on the *Tonnetz* relates to their acoustic consonance. This will help us to understand why some scales sound more consonant than others. Second, we exploit the interior regions of the *Tonnetz*'s hexagons to motivate an additional set of chordal transformations beyond the set of three that we have dealt with so far. We shall see that this extended set gives a more parsimonious geometric description of the typical chord progressions in diatonic music. By typical chord progressions, we mean those described either by the models in Figures 3.10 or 3.13, or those described by the two alternative models given in the next section.

---

[4]It may help to print the PDF file for Figure 7.12 for future reference. It is located at the book's web site.

7.3.1 Geometry of Acoustic Consonance and Dissonance

The *Tonnetz* consists of a network of 12 hexagons, the interiors of which correspond to the 12 pitch classes of the chromatic scale. The relative closeness of two hexagons in the *Tonnetz* is an indicator of the acoustic consonance of the intervals separating notes from those two pitch classes. We illustrate this with two diagrams shown in Figure 7.13.

Figure 7.13. Left: Acoustically consonant intervals on *Tonnetz*. Right: Acoustically dissonant intervals.

On the left diagram, we have shown the three ways that two hexagons can lie adjacent to each other on the *Tonnetz*. These alignments correspond to the acoustically consonant intervals. If two hexagons lie vertically adjacent to each other, indicated by the vertical double-arrow, then the interval between the pitch classes for those two hexagons is either a perfect fifth (up arrow) or perfect fourth (down arrow), modulo octave equivalence. For example, in Figure 7.12, the hexagons with interiors **C** and **G** are aligned vertically, and pairs of notes from these two pitch classes are separated by either perfect fifths or perfect fourths. Some examples are shown in Figure 7.14. The other two ways that

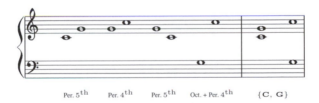

Figure 7.14. Note pairs activating regions **C** and **G** on *Tonnetz*. First measure: four melodic intervals activating **C** then **G**, or **G** then **C**. Second measure: two harmonic intervals activating **C** and **G** simultaneously.

hexagons can lie adjacent to each other are the diagonally oriented adjacent positions shown on the left of Figure 7.13. The negatively sloped double arrow indicates a pairing of two hexagonal regions whose pitch classes are separated by a minor third (or major sixth), modulo octave equivalence. For example, the regions with interiors **C** and **A**, or the regions **G**$^\flat$ and **E**$^\flat$. The positively sloped double arrow indicates a pairing of two hexagonal regions whose pitch classes are separated by a major third (or minor sixth), modulo octave equivalence. For example, the regions with interiors **C** and **E**, or the regions **F** and **A**.

Since acoustically consonant intervals activate adjacent hexagonal regions, it follows that acoustically dissonant intervals activate hexagonal regions that are not adjacent. On the right diagram in Figure 7.13, we have shown two ways that acoustically dissonant intervals activate regions. A minor second interval, modulo octave equivalence, will activate regions that are separated by an intermediate edge, as illustrated by the horizontal double arrow. For example, look at the positioning of the hexagons for **G**$^\flat$ and **G** in Figure 7.12. Moreover, the hexagons for **E** and **F** are also separated by an intermediate edge, and they differ by a minor second. A tritone will activate two hexagonal regions that lie along a negatively sloped diagonal and are separated by an intermediate hexagon, as indicated by the second double arrow shown on the right of Figure 7.13. For example, look at the positions of the hexagons in Figure 7.12 for $\mathbf{G}^\flat$ and $\mathbf{C}$, or $\mathbf{B}$ and $\mathbf{F}$.

Because the *Tonnetz* is displayed on a flat plane in Figure 7.12, with regions listed multiple times, some care must be exercised in using the geometric description of acoustically consonant and dissonant intervals. For instance, the hexagonal region ($\mathbf{E}$) on the lower left of the *Tonnetz* and the hexagonal region $\mathbf{C}$ are lying along a diagonal and are separated by an intermediate hexagon, even though they are not acoustically dissonant. However, those two hexagons are not the closest hexagons for these two pitch classes. The hexagonal regions $\mathbf{C}$ and $\mathbf{E}$ on the right side of the *Tonnetz* are adjacent hexagons, hence $\mathbf{C}$ and $\mathbf{E}$ are shown geometrically to be acoustically consonant. If we require that we always use the closest possible hexagonal regions on the *Tonnetz*, then this ambiguity is resolved.[5] This ambiguity only occurs for separated hexagons lying along a *positively* sloped diagonal; separated hexagons lying along negatively sloped diagonals definitely correspond to dissonant intervals.

7.3.2 Analyzing Scales

The regions of activation for pitch classes, along with the network of chords in the *Tonnetz*, provides a concise geometric method for analyzing different scales. We will look at three examples: a pentatonic major scale, a diatonic major scale, and a 6-note scale from a piano piece by John Cage.

Example 7.3.1. Pentatonic major scale. These five successive black keys on the piano:

$$\mathbf{G}^\flat \quad \mathbf{A}^\flat \quad \mathbf{B}^\flat \quad \mathbf{D}^\flat \quad \mathbf{E}^\flat \tag{7.9}$$

give the pentatonic $\mathbf{G}^\flat$-major scale. On the left of Figure 7.15, we show the regions of activation on the *Tonnetz* of the pitch classes for this scale. We can see clearly, from the largely vertical orientation of this set of regions, that the pentatonic $\mathbf{G}^\flat$-major scale is very consonant as a whole. This is consistent with the interval probabilities and dissonance score that we calculated for a pentatonic major scale in Section 6.6. (See Figure 6.21 on page 200.) Notice also that there are only two triadic chords, $e^\flat$ and $\mathbf{G}^\flat$, that lie within the polygon that marks off these regions. These are the only two triadic chords that can be generated with notes from the pentatonic $\mathbf{G}^\flat$-major scale. This example illustrates how the *Tonnetz* provides a concise visual display of these two features of pentatonic major scales: lots of consonance, and few chords.

Example 7.3.2. Diatonic major scale. On the right of Figure 7.15, we show the regions of activation for the archetypical diatonic major scale, the C-major scale. In contrast to the pentatonic major scale, we can see that there is a lot more horizontal width, in addition to vertical height, for the total region. This allows for considerably more dissonance in this scale, and that is consistent with the interval probabilities and dissonance score that we calculated for this scale in Section 6.6. Notice also how many more major and minor chords lie within the trapezoid that marks the regions of activation for this scale. The trapezoid encloses these chords in a snake-like progression:

$$\mathrm{G} \xrightarrow{\mathcal{R}} \mathrm{e} \xrightarrow{\mathcal{L}} \mathrm{C} \xrightarrow{\mathcal{R}} \mathrm{a} \xrightarrow{\mathcal{L}} \mathrm{F} \xrightarrow{\mathcal{R}} \mathrm{d}$$

on the *Tonnetz*.

We shall refer to a connected series of chords like this, as a ***chordal spine*** for a given collection of regions on the *Tonnetz*. The notion of a chordal spine will prove useful for understanding the differences between various scales or keys, and various modes. Already we have found that the chordal spine for the diatonic C-major scale is three times longer than the minuscule chordal spine for a pentatonic major scale.

This example illustrates how the *Tonnetz* provides a concise, visual display of two important features of diatonic major scales: (1) a balance of consonance and dissonance, and (2) a balanced

---

[5]This ambiguity is again the result of plotting the hexagons for the *Tonnetz* on a flat surface. Requiring that we look at the closest hexagonal regions corresponds to the unique specification of these regions on a donut-shaped surface. On that surface there is only one way to choose a pair of regions for two pitch classes and they are, by default, the closest possible regions.

variety of both major and minor chords. This provides a geometric logic underlying the common practices in diatonic music of alternating consonance with dissonance and using chords extensively.

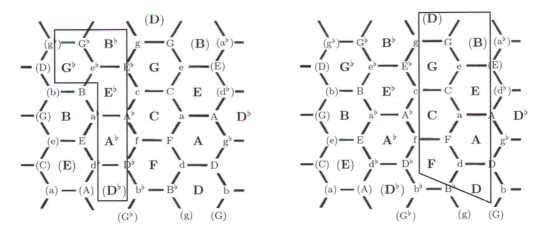

Figure 7.15. Left: Regions of activation for $G^\flat$-major pentatonic scale, lying within a polygon. Right: Regions of activation for C-major scale, lying within a trapezoid. Note: These are the same regions of activation for the natural A-minor scale (the relative minor of C-major).

Example 7.3.3. John Cage's *Dream*. In his piano composition, *Dream*, John Cage uses only notes taken from the following six pitch classes:

$$\text{C} \quad \text{D} \quad \text{E}^\flat \quad \text{G} \quad \text{A}^\flat \quad \text{B}^\flat .$$

We shall refer to this as Cage's *Dream* scale. See the top of Figure 7.16. As shown there, Cage's *Dream* scale has a mirror symmetry about the line drawn through the pitch classes **B** and **F**. It can be construed as the union of the pitch classes contained in the two chords, g and $A^\flat$. On the bottom left of Figure 7.16 we show the regions of activation on the *Tonnetz* for Cage's *Dream* scale. These regions have a chordal spine consisting of edges connecting the four chords, g, $E^\flat$, c, and $A^\flat$. The shape of the regions of activation on the *Tonnetz* for Cage's *Dream* scale is similar to the regions of activation for a natural minor scale, although more truncated vertically. To see this, compare the regions of activation for the natural A-minor scale shown on the right of Figure 7.15 with the regions of activation for Cage's *Dream* scale. Consequently, we would expect Cage's *Dream* scale to have similar interval probabilities as a natural minor scale, but with a bit more dissonance. This is confirmed by comparing its interval probabilities, shown on the bottom right of Figure 7.16, with the interval probabilities for a natural minor scale, shown on the right of Figure 6.21. Furthermore, the dissonance score for Cage's *Dream* scale is:

$$
\begin{aligned}
(\textit{Dream scale dissonance}) &= 8 \cdot \frac{2}{15} + 4 \cdot \frac{3}{15} + 2 \cdot \frac{2}{15} + 2 \cdot \frac{3}{15} + 1 \cdot \frac{4}{15} + 6 \cdot \frac{1}{15} \\
&= \frac{16 + 12 + 4 + 6 + 4 + 6}{15} \\
&= 3.2
\end{aligned}
\tag{7.10}
$$

which is slightly greater than the dissonance score of 2.95 for a natural minor scale alone. The dissonance score for a minor key, with its variable sixth and seventh notes within the natural minor scale, will also be slightly larger than 2.95.

Cage's *Dream* scale is closely related to the natural C-minor scale:

$$\text{C} \quad \text{D} \quad \text{E}^\flat \quad \text{F} \quad \text{G} \quad \text{A}^\flat \quad \text{B}^\flat .$$

except for the missing pitch class **F**. In fact, *Dream* does have a sound quality quite reminiscent of a minor key. It is interesting that the mirror for Cage's *Dream* scale passes through this missing pitch class **F** on the chromatic clock.

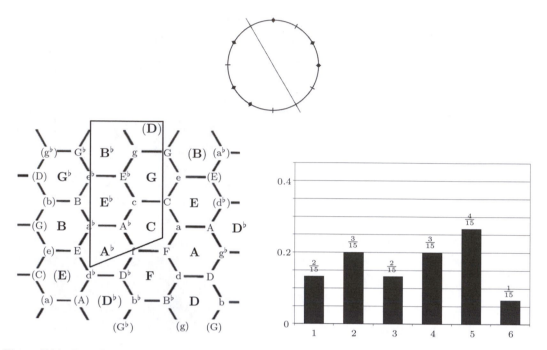

Figure 7.16. Top: Cage's *Dream* scale on chromatic clock, mirror-symmetrical about line through **B** and **F**. Bottom Left: Regions of activation of Cage's *Dream* scale. Bottom Right: Probabilities for its intervals.

Remark 7.3.1. The triadic chords for Cage's *Dream* scale have an amusing significance. If we write them in the following order:

$$c \quad A^\flat \quad g \quad E^\flat$$

then the letters spell Cage! Is this more than a coincidence? It is difficult to say for sure. It is a fact that the first three distinct chords in the piece are c, $A^\flat$, g. Moreover, Cage was famous for his sense of humor.

Cage would not have been the first to embed his name in some fashion in a musical composition. It is known that in some of his compositions, Bach deliberately used a motif made from the notes $B^\flat$, A, C, $B^\natural$. In the German musical notation that he used, $B^\flat$ is denoted as B, while $B^\natural$ is denoted by H. Hence, Bach's motif spells BACH.

Exercises

7.3.1. On the *Tonnetz*, find the chordal spine for the G-major scale.

7.3.2. On the *Tonnetz*, find the chordal spine for the D-major scale.

7.3.3. On the *Tonnetz*, find the chordal spine for the B-major scale.

7.3.4. On the *Tonnetz*, find the chordal spine for the F-major scale.

7.3.5. On the *Tonnetz*, find the chordal spine for the $A^\flat$-major scale.

7.3.6. On the *Tonnetz*, find the chordal spine for the following octatonic scale:

$$C \quad D^\flat \quad E^\flat \quad E \quad G^\flat \quad G \quad A \quad B^\flat \quad C.$$

7.3.7. For the whole tone scale:

$$C \quad D \quad E \quad G^\flat \quad A^\flat \quad B^\flat.$$

find **(a)** its regions of activation on the *Tonnetz*, and **(b)** its chordal spine; **(c)** describe geometrically how this whole tone scale differs from all of the other scales we have considered.

7.3.8. Here is Bach's motif:

Describe the symmetry of this motif.

7.4 Other Chordal Transformations

In this section, we conclude our discussion of neo-Riemannian chordal inversions by examining some additional chordal transformations that apply to a wider range of music. These transformations are defined by movements from one chord to another on the *Tonnetz*.

7.4.1 Using the *Tonnetz* to Define More Transformations

The *Tonnetz* can be thought of as a collection of edges that connect points corresponding to chords. In Figure 7.12 we have shown that these edges mark off hexagonal regions corresponding to the 12 pitch classes of the chromatic scale. If we think of these regions as providing solid ground over which we can move, then we can imagine moving from any chord on a given hexagon to any other chord on that same hexagon. Each of those moves can be used to describe a chord progression. When the move is across a hexagonal region, then one pitch class from each chord is the same. When the move is along the edge of a hexagon (which is also the edge of an adjoining hexagon), then two pitch classes from each chord are the same. For example, the chord progression f $\rightarrow$ C corresponds to a move across the hexagon **C**, and the pitch class **C** belongs to both chords. As another example, the chord progression C $\rightarrow$ E$^\flat$ corresponds to a move across the hexagon **G**, and the pitch class **G** belongs to both chords. On the other hand, C $\rightarrow$ c is a move along the edge of hexagon **C** and also hexagon **G**, and the pitch classes **C** and **G** belong to both of the chords.

Rather than using distinct names for all of these moves, which would be difficult to remember in any case, we shall use the following definition to define them more generically.

Definition 7.4.1. *A **Tonnetz transformation** is a chord progression that corresponds to a move across a single hexagonal region in the* Tonnetz, *or along one hexagonal edge in the* Tonnetz. *Each* Tonnetz *transformation will be denoted generically by* $\mathcal{T}$.

For example, the chord progressions mentioned above will be denoted by f $\xrightarrow{\mathcal{T}}$ C and C $\xrightarrow{\mathcal{T}}$ E$^\flat$. Notice also that the neo-Riemannian chordal inversions are included in this definition. For example, d $\xrightarrow{\mathcal{R}}$ F can be written as d $\xrightarrow{\mathcal{T}}$ F because d and F are connected by an edge of hexagon **F** (also by an edge of hexagon **A**). Not all chord progressions, however, are instances of *Tonnetz* transformations. The chord progression e $\rightarrow$ F, for instance, is not a *Tonnetz* transformation because one cannot go from chord e to chord F by a move across a single hexagon or hexagonal edge in the *Tonnetz*.

We now show how these *Tonnetz* transformations apply to music. As the following examples will show, these transformations provide significant insights into the chordal structure of music.

Example 7.4.1. In Figure 7.17, we have indicated the chordal analysis for the first measure of J.S. Bach's *B-flat Major Prelude (BWV 866)*. The chord progression sequence is

$$\mathbf{I} \to \mathbf{V} \to \mathbf{I} \to \mathbf{V} \to \mathbf{vi} \to \mathbf{iii} \to \mathbf{vi} \to \mathbf{iii}.$$

To more easily follow this sequence on the *Tonnetz* diagram in Figure 7.12, we express this sequence of chord progressions as

$$\mathrm{B}^\flat \to \mathrm{F} \to \mathrm{B}^\flat \to \mathrm{F} \to \mathrm{g} \to \mathrm{d} \to \mathrm{g} \to \mathrm{d}.$$

On the *Tonnetz* diagram in Figure 7.12, we can trace all but one of these chord progressions as moves over the two hexagonal regions **F** and **D** located on the lower right. We obtain the following sequence of *Tonnetz* transformations:

$$\mathrm{B}^\flat \xrightarrow{\mathcal{T}} \mathrm{F} \xrightarrow{\mathcal{T}} \mathrm{B}^\flat \xrightarrow{\mathcal{T}} \mathrm{F} \xrightarrow{\mathcal{TT}} \mathrm{g} \xrightarrow{\mathcal{T}} \mathrm{d} \xrightarrow{\mathcal{T}} \mathrm{g} \xrightarrow{\mathcal{T}} \mathrm{d}.$$

Almost the entire sequence corresponds to single hexagonal moves on the *Tonnetz*. In Roman numeral notation, we have

$$\mathbf{I} \xrightarrow{\mathcal{T}} \mathbf{V} \xrightarrow{\mathcal{T}} \mathbf{I} \xrightarrow{\mathcal{T}} \mathbf{V} \xrightarrow{\mathcal{TT}} \mathbf{vi} \xrightarrow{\mathcal{T}} \mathbf{iii} \xrightarrow{\mathcal{T}} \mathbf{vi} \xrightarrow{\mathcal{T}} \mathbf{iii}.$$

Figure 7.17. Chordal analysis of first measure of J.S. Bach's *B-flat Major Prelude (BWV 866)*. The key is $\mathrm{B}^\flat$-major, and chords are shown in Roman numeral notation based on this key.

Example 7.4.2. In Figure 7.18 we show the chordal analysis of measures 5 and 6 of Chopin's *E Major Prelude, Op. 28 No. 9*. In terms of *Tonnetz* transformations, the chord progressions are

$$\mathrm{E} \xrightarrow{\mathcal{T}} \mathrm{B} \xrightarrow{\mathcal{T}} \mathrm{G} \xrightarrow{\mathcal{T}} \mathrm{C} \xrightarrow{\mathcal{T}} \mathrm{g} \xrightarrow{\mathcal{T}} \mathrm{C} \tag{7.11}$$

where the last progression from chord g to chord C is actually mediated by first passing through the chord C^7. As explained below, the chords in (7.11) do not belong to a single key, so we have not used Roman numeral notation. Nevertheless, these chord progressions all correspond to moves across the hexagonal regions **B** and **G** in the *Tonnetz* shown in Figure 7.12. These moves on the *Tonnetz* provide a geometric logic underlying the harmonic logic of the chord progressions shown in the score.

Remark 7.4.1. The *Tonnetz* transformations in this last example all involve moves across a hexagon, preserving just one pitch class from each of the chords involved. For example, the transformation $\mathrm{E} \xrightarrow{\mathcal{T}} \mathrm{B}$ is the transposition T_7 applied to E to obtain B, and is also a move across hexagon **B**. In contrast, the transformation $\mathrm{B} \xrightarrow{\mathcal{T}} \mathrm{G}$ is an example of a chromatic mediant relation between the two chords B and G. As described in the music theory book by Kostka and Payne,[6] two triadic chords are ***chromatic mediants*** if they satisfy the following three conditions:

(1) They are either both major chords or both minor chords;

---

[6]S. Kostka and D. Payne, *Tonal Harmony, with an introduction to twentieth century music, Sixth Edition*, McGraw-Hill, 2009, p. 327.

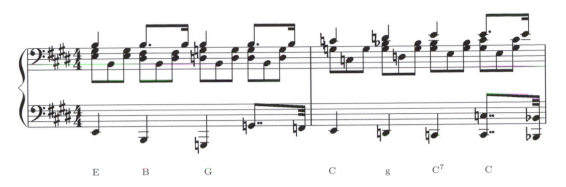

Figure 7.18. Chordal analysis of measures 5 and 6 of Chopin's *E Major Prelude, Op. 28 No. 9.*

(2) Their roots change by ± 3 or ± 4 hours on the chromatic clock (typically intervals of either minor or major thirds);

(3) They share one pitch class.

For B $\xrightarrow{\mathcal{T}}$ G, the chords are both major, the chord root **B** moves to chord root **G** by -4 hours, and the chords share the pitch class **B** (the hexagon over which the *Tonnetz* transformation B $\xrightarrow{\mathcal{T}}$ G moves). We shall see in Example 7.4.4 that B $\xrightarrow{\mathcal{T}}$ G is a progression corresponding to a change of key in the Chopin passage. The reader may wish to check that there are eight versions of chromatic mediant chord changes (depending on whether root motion is up or down by 3 or 4 hours and the chords are both major or both minor), and these are all *Tonnetz* transformations across some hexagon in the *Tonnetz* (the hexagon determined by their one shared pitch class).

In the last two examples above, the progressions for both the Bach and Chopin passages were almost all single *Tonnetz* transformations. This will typically be the case for a large percentage of chord progressions in diatonic music, as we now show.

7.4.2 Modeling Chord Progressions in Diatonic Music

In Example 7.2.3 we described how the *Tonnetz* network of chords captures a large percentage of the typical chord progressions used in diatonic music. For example, we found that

$$a \xrightarrow{\mathcal{L}} F \qquad \text{and} \qquad e \xrightarrow{\mathcal{LR}} a.$$

Now, however, we look again at Figure 7.12 with the idea that we can use *Tonnetz* transformations. We then see that

$$a \xrightarrow{\mathcal{T}} F \qquad \text{and} \qquad e \xrightarrow{\mathcal{T}} a.$$

Since all major scales are simply transpositions of the C-major scale, we can rewrite these two progressions as

$$\mathbf{vi} \xrightarrow{\mathcal{T}} \mathbf{IV} \qquad \text{and} \qquad \mathbf{iii} \xrightarrow{\mathcal{T}} \mathbf{vi}.$$

Figure 7.9 now shows us that most of the typical chord progressions in a major key are connected by single *Tonnetz* transformations, with just a few cases of double applications of $\mathcal{T}$. An interesting case is the chord progression $\mathbf{ii} \rightarrow \mathbf{V}$, which in the key of C-major is d $\rightarrow$ G. In Example 7.2.3, we described this by d $\xrightarrow{\mathcal{PRL}}$ G, since we were restricted to using only edges of hexagons. Now, however, we have d $\xrightarrow{\mathcal{T}}$ G because we can move across the hexagon **D** at the lower right corner of the *Tonnetz* in Figure 7.12.

Another model for chord progressions in both major and minor keys is due to Tymoczko.[7] In this model as well, many of the chord progressions correspond to *Tonnetz* transformations. The typical chord progressions for a major key, according to Tymoczko's model, are shown in Figure 7.19.

$$\mathbf{I} \to \mathbf{vi} \text{ or } \mathbf{IV} \text{ or } \mathbf{ii} \text{ or } \mathbf{vii}^\circ \text{ or } \mathbf{V}$$
$$\mathbf{ii} \to \mathbf{vii}^\circ \text{ or } \mathbf{V}$$
$$\mathbf{IV} \to \mathbf{ii} \text{ or } \mathbf{vii}^\circ \text{ or } \mathbf{V} \text{ or } \mathbf{I}$$
$$\mathbf{V} \to \mathbf{I} \text{ or } \mathbf{vi} \text{ or } \mathbf{IV}$$
$$\mathbf{vi} \to \mathbf{IV} \text{ or } \mathbf{ii} \text{ or } \mathbf{vii}^\circ \text{ or } \mathbf{V} \text{ or } \mathbf{I}$$
$$\mathbf{vii}^\circ \to \mathbf{V} \text{ or } \mathbf{I}$$

Figure 7.19. Typical chord progressions in major key, according to Tymoczko's model. The ordering of chords mapped to is mostly based on descending thirds and fifths for their roots. Chord **iii** is not included. It is viewed as typically an intervening chord within other progressions, such as $\mathbf{V} \to \mathbf{iii} \to \mathbf{I}$.

For example, in a major key, we could have

$$\mathbf{I} \to \mathbf{vi}, \text{ or } \mathbf{I} \to \mathbf{IV}, \text{ or } \mathbf{vi} \to \mathbf{ii}.$$

If we translate those chord progressions to C-major, we have

$$\mathrm{C} \to \mathrm{a}, \quad \mathrm{C} \to \mathrm{F}, \quad \mathrm{a} \to \mathrm{d}.$$

In terms of the *Tonnetz* shown in Figure 7.12, we can express these as *Tonnetz* transformations:

$$\mathrm{C} \xrightarrow{T} \mathrm{a}, \quad \mathrm{C} \xrightarrow{T} \mathrm{F}, \quad \mathrm{a} \xrightarrow{T} \mathrm{d}.$$

Therefore, in any major key, we will have these *Tonnetz* transformations as typical chord progressions:

$$\mathbf{I} \xrightarrow{T} \mathbf{vi}, \quad \mathbf{I} \xrightarrow{T} \mathbf{IV}, \quad \mathbf{vi} \xrightarrow{T} \mathbf{ii}.$$

The typical chord progressions for a minor key, according to Tymoczko's model, are shown in Figure 7.20. Just as with a major key, most of these minor key chord progressions are examples of *Tonnetz* transformations.

$$\mathbf{i} \to \mathbf{VI} \text{ or } \mathbf{iv} \text{ or } \mathbf{ii}^\circ \text{ or } \mathbf{vii}^\circ \text{ or } \mathbf{V}$$
$$\mathbf{ii}^\circ \to \mathbf{vii}^\circ \text{ or } \mathbf{V}$$
$$\mathbf{iv} \to \mathbf{ii}^\circ \text{ or } \mathbf{vii}^\circ \text{ or } \mathbf{V} \text{ or } \mathbf{i}$$
$$\mathbf{V} \to \mathbf{i} \text{ or } \mathbf{VI} \text{ or } \mathbf{iv}$$
$$\mathbf{VI} \to \mathbf{iv} \text{ or } \mathbf{ii}^\circ \text{ or } \mathbf{vii}^\circ \text{ or } \mathbf{V} \text{ or } \mathbf{i}$$
$$\mathbf{vii}^\circ \to \mathbf{V} \text{ or } \mathbf{i}$$

Figure 7.20. Typical chord progressions in a minor key, according to Tymoczko's model.

Not all chord progressions, however, in a major or minor key consist of single *Tonnetz* transformations. For example, in a major key, the chord progression $\mathbf{IV} \to \mathbf{V}$ is described by $\mathbf{IV} \xrightarrow{TT} \mathbf{V}$. Nevertheless, we do find that a great many of the chord progressions described by the models in Figures 7.19 and 7.20, and by the other models in Figures 3.10 and 3.13, are described by single *Tonnetz* transformations.

---

[7]D. Tymoczko, *A Geometry of Music,* Oxford Univ. Press (2011), pp. 227–229.

Remark 7.4.2. Although we have shown that various models for chord progressions largely consist of *Tonnetz* transformations, the models show that not all *Tonnetz* transformations are equally likely. Many *Tonnetz* transformations are not typical chord progressions in any of the models. What we have shown is that ***these models highlight particular Tonnetz transformations as the typical ones used in diatonic music***.

7.4.3 Explaining the Qualitative Difference in Modes

For a given diatonic scale, say the C-major scale, a *mode* will depend on which note is used as the tonic. The names of these modes, and the notes used for their tonics on the C-major scale, are shown in Table 7.1.

Table 7.1 MODES BASED ON C-MAJOR SCALE

Tonic note	Mode
C	Ionian (major)
D	Dorian
E	Phrygian
F	Lydian
G	Mixolydian
A	Aeolian (natural minor)
B	Locrian

Since all of these modes use the same notes in a scale, what makes them sound different? It is the arrangement of chords used, either as chords themselves, or as chordal structure for melodies. The *Tonnetz* provides a concise geometric picture that shows how the chords are arranged in different families depending on the tonic used for each mode. As we pointed out previously, the hexagonal regions for the C-major scale lie adjacent to each other on the *Tonnetz* in a such a way that they create a chordal spine. This chordal spine

$$G - e - C - a - F - d$$

is shown within the trapezoidal region on the right of Figure 7.15. The position of a tonic within this chordal spine will determine the family of chords that are directly related, by just a single application of $\mathcal{T}$, to that tonic. For example, if the tonic is C, then the chords that are directly related to the chord C on the *Tonnetz* are (if we only list chords containing notes from the C-major scale, and include C itself):

$$\{C \quad G \quad e \quad a \quad F\}.$$

On the other hand, if the tonic is D, then the chords directly related to the chord d on the *Tonnetz* are (if we only list chords with notes from the C-major scale, and include d itself):

$$\{d \quad a \quad F \quad G\}.$$

Comparing these two sets, we can see that they differ for the two modes. The difference would be even wider if we had listed chords containing notes that do not belong to the C-major scale.

In Table 7.2 we show a further analysis of modes using *Tonnetz* transformations. This table compares the chords related, via *Tonnetz* transformations, to the tonic chord in a given mode. The table clearly shows the differences in the chord types for the three modes: Major (Ionian), Dorian, and Mixolydian. Adding the Phrygian, Lydian, and Aeolian modes to this table is given as an exercise.

Table 7.2   CHORDS CONNECTED TO TONIC IN THREE MODES OF C-MAJOR SCALE

Mode	Tonic chord	$\mathcal{T}$ changes type	$\mathcal{T}$ preserves type	Use $\mathcal{T}\mathcal{T}$
Major (Ionian)	C	a, e	F, G	d
Dorian	d	F, G	a	C, e
Mixolydian	G	e, d	C	F, a

The Locrian mode is exceptional in that its tonic, B, is not a root for a major or minor chord that lies on the chordal spine for the C-major scale. It is a root for the diminished chord B°, which is not a chord on the *Tonnetz*. The hexagonal regions, **B-D-F**, for the chord B° do not lie in a triangular configuration like those for major and minor chords. In fact, the two pitch classes **B** and **F** form a tritone. The fact that the tonic chord in Locrian mode contains a tritone is one reason, perhaps the main reason, that this mode has only rarely been used.

Example 7.4.3. The Beatles' *Eleanor Rigby*. A famous song that uses a modal melody is the Beatles' *Eleanor Rigby*. The score for this song has a G-major key signature, which typically indicates a key of either G-major or E-minor. The first chord for the song is C-major, but this is followed quickly by multiple repetitions of the chord E-minor.[8] Normally, this would clearly indicate that the key for the song is E-minor. However, the melody for the crucial lyrics that begin with the name "Eleanor Rigby" has a C♯ note within a descending sequence of pitches. This is not standard practice in E-minor, since a minor key would use the natural minor scale for descending pitches. The natural E-minor scale is

$$\text{E F}\sharp\text{ G A B C D E.}$$

This scale does not have the note C♯. The note C♯, however, does belong to the scale for E-Dorian:

$$\text{E F}\sharp\text{ G A B C}\sharp\text{ D E.}$$

Actually, although this song is often cited as being in E-Dorian mode, the issue is more subtle than that. For whenever the chorus enters — at the beginning, and at other times in the song — they begin with the chord of C-major which is not a chord in E-Dorian (due to its root being C-natural). This chord is immediately followed by the string accompaniment playing multiple repetitions of the E-minor chord. This progression, C → e, signals a return to the key of E-minor. So, in fact, *Eleanor Rigby* oscillates between E-minor and E-Dorian. It is a ***multi-modal*** composition.

Example 7.4.4. Changing keys. The *Tonnetz* diagram shown in Figure 7.12 provides a geometric description of how key changes often occur within musical compositions. A key change within a musical composition is called ***modulation***. Recalling that the scale for a given key corresponds to a chordal spine on the *Tonnetz*, we can say that the process of modulation will often involve a *Tonnetz* transformation that moves from a chord on one spine over to a chord on another spine. For example, in the chord progression for Chopin's *E Major Prelude* shown in Figure 7.18, the first two chords, E and B, lie on the chordal spine for the key of E-major. When the *Tonnetz* transformation $\mathcal{T}$ is applied to B to get the chord G, this corresponds to a move across the hexagonal region **B** to an adjacent chordal spine (the leftmost vertical spine on the *Tonnetz* shown in Figure 7.12). By comparing this spine with the one for the C-major key, it appears that Chopin has modulated to G-major. The accidentals in the score for the G-major chord in Figure 7.18 might seem to indicate a modulation to D-major (two sharps in the key signature, after cancellation by the naturals for the G-major chord). However, the next *Tonnetz* move: G $\xrightarrow{\mathcal{T}}$ C, would be impossible on the chordal spine for D-major, but is possible

---

[8]Notice that this chord progression is C $\xrightarrow{\mathcal{L}}$ e.

on the chordal spine for G-major. Hence, the *Tonnetz* provides us with a geometric description of Chopin's modulation from E-major to G-major in this passage. Another example of modulation is the sequence of chord progressions from Beethoven's Ninth Symphony shown in (7.8). By viewing these progressions as moves along edges of the *Tonnetz*, we can see that the chords are sliding along the chordal spines for the keys of C-major, then F-major, then B$^\flat$-major, and so on. The music is smoothly and rapidly modulating through keys moving counter-clockwise around the circle of fifths.

Exercises

7.4.1. Show that each *Tonnetz* transformation for a given hexagon can be realized by a reflection through a mirror line passing through two sides, or two corners, of the hexagon. (Note: this result shows that a *Tonnetz* transformation is self-inversive.)

7.4.2. Draw a graph, using arrows to indicate *Tonnetz* transformations, of the moves from one chord to the next for the Bach passage discussed in Example 7.4.1.

7.4.3. Draw a graph, using arrows to indicate *Tonnetz* transformations, of the moves from one chord to the next for the Chopin passage discussed in Example 7.4.2.

7.4.4. In the J.C. Bach passage shown in Figure 3.30 on page 80, find the four chords that describe the music in the first three measures. Describe, either through words or a diagram, how this sequence of chords is realized by *Tonnetz* transformations.

7.4.5. In the passage from *Für Elise* shown in Figure 3.15 on page 65, find the two chords that describe the chord progression across the fourth measure line. Describe, either through words or a diagram, how this chord progression is realized by a *Tonnetz* transformation.

7.4.6. In the passage from Beethoven's *Moonlight Sonata* given in Exercise 3.3.8 on page 75, find the four chords that describe the music. Discuss how this sequence of chords is realized by *Tonnetz* transformations. (Note: The chords for measures 1 and 2 are c$^\sharp$ and c$^{\sharp 7}$. These are enharmonic with d$^\flat$ and d$^{\flat 7}$. Treat both of these as d$^\flat$ chords, but say that the progression from measures 1 and 2 to measure 3 is mediated by the d$^{\flat 7}$ chord, as we did with the Chopin passage in Example 7.4.2.)

7.4.7. In Figure 7.21 we show another passage from Chopin's *E Major Prelude*. Name the 8 chords that characterize the music within these two measures. Describe, either through words or a diagram, how this sequence of chords is realized by *Tonnetz* transformations.

Figure 7.21. Measures 10 and 11 of Chopin's *E Major Prelude, Op. 28 No. 9*.

7.4.8. The sequence of chords in the introduction to the Beatles' *I am the Walrus* are (omitting repetitions of chords with identical roots, such as some 7th chords, which we will treat as purely transitional):

B   A   G   F   E   D   A   C   D   E   A   C   D   A.

Describe, either through words or a diagram, how this sequence of chords is realized by *Tonnetz* transformations (most by double transformations, some by single transformations). Also, explain why no definite key is implied by this sequence of chords.

7.4.9. Find all the typical chord progressions in the model of Tymoczko for a major key (see Figure 7.19) that are generated by one *Tonnetz* transformation, and find how many require two *Tonnetz* transformations. [Note: since the diminished chord **vii**° is not a chord on the *Tonnetz*, chord progressions involving it are not counted in this exercise.]

7.4.10. **(a)** Find all the typical chord progressions in the model of Tymoczko for a minor key (see Figure 7.20) that are generated by one *Tonnetz* transformation, and find how many require two *Tonnetz* transformations. **(b)** Find all the typical chord progressions for a minor key, using the model given in Figure 3.13, that are generated by one *Tonnetz* transformation. Also, find how many require two *Tonnetz* transformations. [Note: since diminished chords, such as **vii**° or **ii**°, are not chords on the *Tonnetz*, chord progressions involving them are not counted in this exercise.]

7.4.11. Find all the typical chord progressions for a major key, using the model given in Figure 3.10, that are generated by one *Tonnetz* transformation. Also, find how many require two *Tonnetz* transformations. [Note: since the diminished chord **vii**° is not a chord on the *Tonnetz*, chord progressions involving it are not counted in this exercise.]

7.4.12. Describe how the chord progression **IV** → **V** for a given key, say C-major, can be realized by a single reflection through a mirror line on the *Tonnetz*.

7.4.13. Add the cases of Phrygian, Lydian, and Aeolian modes to Table 7.2.

7.4.14. In John Coltrane's improvised solo in *Giant Steps*, there is the following sequence of 26 chord progressions:

$$C^{\sharp} \to E^7 \to A \to C^7 \to F \to b^7 \to E^7 \to$$
$$A \to C^7 \to F \to G^{\sharp 7} \to C^{\sharp} \to g^7 \to C^7 \to$$
$$F \to b \to E^7 \to A \to e^{b 7} \to A^{b 7} \to$$
$$D^{b} \to g^7 \to C^7 \to F \to e^{b} \to A^{b 7} \to C^{\sharp}.$$

(7.12)

Treating each seventh chord as an embellishment of an underlying triadic chord (e.g., treating E^7 as E on the *Tonnetz*, and b^7 as b on the *Tonnetz*), list the chord progressions that are obtained from one *Tonnetz* transformation. List those obtained from two *Tonnetz* transformations. The chord progressions in (7.12) describe modulations through three different keys, which keys are they?

7.4.15. In a transcription of Barney Bigard's solo in the Duke Ellington composition, *Blue Light*,[9] there is the following sequence of 16 chord progressions:

$$G \to E^7 \to A^7 \to D^7 \to G \to G^7 \to C^7 \to E^7 \to$$
$$A^7 \to D^7 \to G \to G^7 \to D^7 \to E^7 \to A^7 \to D^7 \to G.$$

(7.13)

Treating each seventh chord as an embellishment of an underlying triadic chord (e.g., treating E^7 as E on the *Tonnetz*, and A^7 as A on the *Tonnetz*), list the chord progressions that are obtained from one *Tonnetz* transformation. List those obtained from two *Tonnetz* transformations. Draw a diagram of the movement on the *Tonnetz* corresponding to the chord progressions in (7.13). These chord progressions describe modulations back and forth between two different major keys, which keys are they?

[9]See G. Schuller. *The Swing Era: The Development of Jazz. 1930–1945*. Oxford, 1989, Ex. 41b, p. 109.

Chapter 8

Audio Synthesis in Music

...graphically oriented sound manipulation represents a change in paradigm from conventional methods of creating and modifying sounds ...The high level abstractions of the musical score include the pitch of notes and the regular time unit defined by the measure. The high level abstractions in spectrogram "scores" include the frequency content of a signal as displayed in the image and time-grids that can be arbitrarily specified.

—William Sethares

In this chapter we describe how music is created by audio synthesis. We will concentrate on how spectrograms can be used for synthesizing music, by the method of *granular synthesis*. Granular synthesis underlies the AUTOTUNER musical synthesizer. AUTOTUNER's granular synthesis is also known as *phase vocoding*. A phase vocoder was used by Imogen Heap to create her song, *Hide and Seek*. We will discuss this song in Section 8.2. Granular synthesis is also used for time-stretching and time-shrinking of musical sound ***without changing pitch***. We will provide some musical examples of time-stretching and time-shrinking in Section 8.3. The chapter concludes with an entirely different synthesis method: *MIDI synthesis*. MIDI synthesis is widely employed for computer scoring programs. Although it does not typically have the flexibility of granular synthesis, it does have the advantage of more clearly relating to standard musical composition.

The material in this chapter is more technically advanced than the other material in the book. We have included it, however, because this technical wizardry is widely employed for creating music and is probably of considerable interest to many readers. Even if the technical material is not completely understandable on a first reading, we hope the musical examples will be interesting and enjoyable.

8.1   Creating New Music from Spectrograms

In this section we describe granular synthesis of music. The composer creates a new sound recording by "painting a spectrogram." In order to describe this process, we first briefly recall the construction of a spectrogram described in Chapter 4.

We have a sound signal $\{f(t_m)\}$, where $\{t_m\}$ are N discrete values of time, separated by an equal amount of time $\Delta t = t_{m+1} - t_m$ throughout. Time $t_1 = 0$ is usually taken as the starting time. We then have the following N time values:

$$t_1 = 0,\ t_2 = \Delta t,\ t_3 = 2\Delta t, \ldots,\ t_N = (N-1)\Delta t.$$

We multiply the sound signal f by a window function w, beginning at some time τ_j and extending over a short time interval, and then compute from this windowed signal the cosine and sine coefficient estimators $\{\mathcal{C}_w(k\nu_o)\}$ and $\{\mathcal{S}_w(k\nu_o)\}$. In this chapter, we will drop the subscript w on these estimators. Consequently, $\mathcal{C}(k\nu_o)$ stands for the windowed estimator $\mathcal{C}_w(k\nu_o)$, and $\mathcal{S}(k\nu_o)$ stands for the windowed estimator $\mathcal{S}_w(k\nu_o)$. The collection of starting times τ_j, for $j = 1, 2, \ldots J$, are assumed to be spaced evenly apart in time, and the time intervals to overlap significantly (about 95% overlap).

Since the calculated cosine and sine coefficients will vary from one windowing to the next, we attach a subscript j, indicating the windowing beginning at τ_j. Thus, for each $j = 1, 2, \ldots, J$, we have the mapping:

$$\text{(windowing of } f \text{ at } \tau_j) \longrightarrow \{\mathcal{C}_j(k\nu_o)\}, \{\mathcal{S}_j(k\nu_o)\}.$$

We also computed amplitude estimators $\{\mathcal{A}_j(k\nu_o)\}$ from these cosine and sine estimators, using the formula $\mathcal{A}_j(k\nu_o) = \sqrt{[\mathcal{C}_j(k\nu_o)]^2 + [\mathcal{S}_j(k\nu_o)]^2}$ for each frequency $k\nu_o$. Therefore, for each $j = 1, 2, \ldots, J$, we have the mapping:

$$\text{(windowing of } f \text{ at } \tau_j) \longrightarrow \{\mathcal{A}_j(k\nu_o)\}.$$

To synthesize sound we reverse this procedure. This is done in two steps.

Step 1. This step creates individual sound segments, called *sound grains*. The second step combines these sound grains to produce the complete waveform for the sound. To create sound grains, we specify at each τ_j a finite collection of cosine coefficients, $\{\mathcal{C}_j(k\nu_o)\}_{k=1}^M$, and sine coefficients, $\{\mathcal{S}_j(k\nu_o)\}_{k=1}^M$. (This can be done in a variety of ways, some of which we will describe in the sections that follow.) The synthesis of each sound grain is performed by producing for each time interval, starting at τ_j, a windowed sum of cosines and sines. At each time value t_m in the time interval starting at τ_j, we define the value of the sound grain g_j at t_m by the following formula:

$$g_j(t_m) = \mathsf{w}_j(t_m) \sum_{k=1}^L \left[\mathcal{C}_j(k\nu_o) \cos(2\pi \cdot k\nu_o t_m) + \mathcal{S}_j(k\nu_o) \sin(2\pi \cdot k\nu_o t_m) \right]. \tag{8.1}$$

Each w_j is a window function that is 0 at the beginning and end of the time interval, analogous to the Blackman window function used in Chapter 4. Multiplying by this window function prevents the sound grain g_j from having either an abrupt onset or an abrupt end. (See Figure 8.1.) As pointed out in the caption to this figure, an abrupt onset or abrupt end would produce annoying clicking sounds, and when these abrupt transitions are present the sound is often machine-like rather than tonal. By introducing a gentle onset and gentle end to the signal, via multiplication by the window function, the sound quality of the signal is rendered more tonal, more musical. In Exercise 8.1.2 we give more examples of windowing and sound perception, including a windowing that approximates a piano tone.

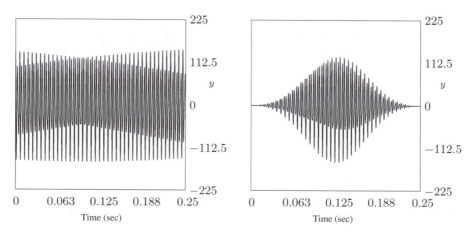

Figure 8.1. Left: Synthesized signal containing two harmonics, no windowing. Sound has clearly audible clicks at beginning and end, and a machine-like electronic timbre. Right: Same signal, multiplied by window function. Sound is like a short breath on a flute, or similar instrument, with no clicking. Gentle increase then decrease of sound volume creates breath-like timbre, more like a traditional instrument.

In practice, it is usually easier to specify amplitudes, rather than cosine and sine coefficients. Recalling the phase angle ϕ_k used in Formula (4.11) on page 112, we have:

$$\alpha_k = A_k \cos(\phi_k), \quad \beta_k = A_k \sin(\phi_k) \tag{8.2}$$

for amplitude A_k. These equations came from the expression $A_k \cos(2\pi k\nu_o t - \phi_k)$ by invoking the cosine subtraction identity. In these equations, α_k corresponds to the cosine coefficient $C_j(k\nu_o)$, and β_k corresponds to the sine coefficient $S_j(k\nu_o)$. We will indicate the index j in the phase angle as well, and write $\phi_{j,k}$ instead of just ϕ_k. In Equation (8.2), we replace α_k by $C_j(k\nu_o)$, and β_k by $S_j(k\nu_o)$, and the amplitude A_k by $A_j(k\nu_o)$. This gives us the following equations:

$$C_j(k\nu_o) = A_j(k\nu_o) \cos(\phi_{j,k}) \quad \text{and} \quad S_j(k\nu_o) = A_j(k\nu_o) \sin(\phi_{j,k}). \tag{8.3}$$

The equations in (8.3) must hold for a phase angle $\phi_{j,k}$ because we know that

$$\left[\frac{C_j(k\nu_o)}{A_j(k\nu_o)}\right]^2 + \left[\frac{S_j(k\nu_o)}{A_j(k\nu_o)}\right]^2 = \frac{[C_j(k\nu_o)]^2 + [S_j(k\nu_o)]^2}{[A_j(k\nu_o)]^2}$$

$$= \frac{[C_j(k\nu_o)]^2 + [S_j(k\nu_o)]^2}{[C_j(k\nu_o)]^2 + [S_j(k\nu_o)]^2}$$

$$= 1.$$

Therefore, $\left(\frac{C_j(k\nu_o)}{A_j(k\nu_o)}, \frac{S_j(k\nu_o)}{A_j(k\nu_o)}\right)$ is a point on the circle of radius 1 centered at the origin. Consequently, there is a phase angle $\phi_{j,k}$ such that

$$\frac{C_j(k\nu_o)}{A_j(k\nu_o)} = \cos(\phi_{j,k}) \quad \text{and} \quad \frac{S_j(k\nu_o)}{A_j(k\nu_o)} = \sin(\phi_{j,k})$$

which give us the equations in (8.3).

Using the two equations in (8.3), along with Equation (8.1), we obtain

$$g_j(t_m) = \mathsf{w}_j(t_m) \cdot \sum_{k=1}^{L} A_j(k\nu_o) \cos(2\pi \cdot k\nu_o t_m - \phi_{j,k}). \tag{8.4}$$

This equation exhibits the essential features of granular synthesis. Although further refinements can be made to it (see Exercise 8.1.3), we shall work with this equation because of its relative simplicity.

Step 2. The sound grains g_j created in Step 1 are combined to create the total waveform. Since the time intervals starting at each τ_j overlap, we add together the sound grains to produce the total waveform g:

$$g(t_m) = \sum_{j=1}^{J} g_j(t_m), \quad \text{for each } t_m. \tag{8.5}$$

This completes the synthesis process.

Programs such as METASYNTH have a rich collection of graphical tools for "painting" spectrograms, i.e., specifying graphically the amplitudes $\{A_j(k\nu_o)\}_{k=1}^{M}$. A video demonstration of METASYNTH can be accessed from this link:

$$\text{MetaSynth slideshow demo} \tag{8.6}$$

at the book's web site (go to Videos/Chapter 8). This demonstration is important for this chapter because it illustrates many of the ideas we describe here: synthesis using phase-shifted cosines (they call it "sine wave synthesis"), phase vocoding (used for the message "Welcome to MetaSynth"), and MIDI synthesis. A number of compositions created with METASYNTH can be found at this link:

MetaSynth composition examples   (8.7)

at the book's web site. Another interesting spectrogram synthesis program is PHOTOSOUNDER, we shall discuss some examples produced with PHOTOSOUNDER later in the chapter.

Specifying the phase angles, $\{\phi_{j,k}\}$, and window functions $\{w_j\}$, is automatically handled by the synthesis procedure of the METASYNTH program. Having the user specify amplitudes, and the program specify phases and windowings, is a sensible approach. As we showed in Chapters 4 and 5, amplitudes have a clear musical interpretation. Phase angles and windowing functions are far more technical in nature, and are better handled behind the scenes by the synthesizing program. It is beyond the scope of this book to discuss the technicalities of choosing phases appropriately (often this information is proprietary). The choice of window functions is also somewhat involved. One aspect that we explore in the exercises is the effect of a windowing function on the timbre of the resulting sound. How windowing relates to timbre is an important consideration in choosing windowing functions. Programs like METASYNTH have palettes of instruments to choose from, and the choice of instrument will affect which windowing functions are employed in the synthesis procedure. In the next section we will describe a beautiful example of musical synthesis, known as *phase vocoding*.

Exercises

8.1.1. One way of demonstrating the role of attack and decay of instrumental notes in sound perception is to play a recording in reverse. Using AUDACITY, load the sound file `piano_clip.wav` (available from the audio files at the book's web site) and select Effect/Reverse to reverse the recording. Play the recording and identify what type of instrument its sound most closely resembles, and explain why this effect occurs in terms of the change of attack and decay of the reversed sound.

8.1.2. In Figure 8.2 we show two waveforms synthesized with the same six harmonics, but using two different window functions. Listen to the sound of these two waveforms, and identify which one seems closer to a piano sound. (Audio files for these waveforms are available at the book's web site.) Explain why the one waveform resembles a piano sound, basing your explanation on the form of the two waveforms. What instrument does the other synthesis sound like?

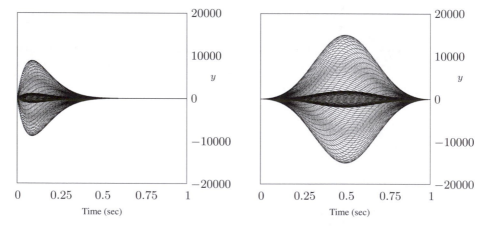

Figure 8.2. Left: Synthesized signal containing six harmonics, with quick onset and more gentle decay. Right: Same harmonics, but multiplied by different window function to have both gentle onset and gentle decay.

8.1.3. The harmonics in trumpet tones can sometimes exhibit a slower onset and faster decay for the higher harmonics, as shown in Figure 8.3. Modify Equation (8.4) to take this effect into account.

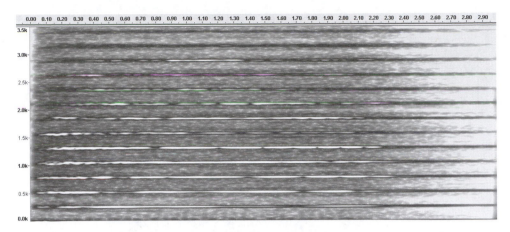

Figure 8.3. Spectrogram of trumpet playing note C_4. Higher frequency harmonics, above 6th harmonic, have slower onsets and slightly faster decay than lower frequency ones.

8.1.4. The effect of *reverberation* (fast multiple echoing) is frequently employed in synthesizing digital sound, in order to simulate effects like recording within a studio or auditorium. One model for creating reverberation is to add together damped time-delayed versions of the original sound. If g is the original sound recording, then explain how the equation

$$f(t_m) = g(t_m) + \sum_{\ell=1}^{M} r(kj)g(t_{m-\ell j}) \qquad (8.8)$$

would provide reverberation (the function r is a damping function).

8.1.5. In the METASYNTH demonstration, at the link given in (8.6), they say that METASYNTH provides "envelope control." Explain how this term, *envelope control*, relates to the synthesis formula in Equation (8.4).

8.1.6. Suppose that reverberation of sound is *frequency-dependent*. In other words, harmonics of different frequencies will echo with different damping functions (see Exercise 8.1.4). Show how this frequency dependent reverberation can be modeled by a modification of the synthesis procedure described in this section.

8.2 Phase Vocoding

In this section we discuss a widely used type of granular synthesis, known as *phase vocoding*. Phase vocoding is the method used by AUTOTUNER. Here we will show how it is used by the composer, Imogen Heap, for her song, *Hide and Seek*. Using phase vocoding, Heap is able to sing multiple note chords, which is generally impossible with the unaided voice.[1] By using phase vocoding, Heap is able to sing the words to her song in the form of three-note and four-note chords. In principle, a chord of any number of notes can be sung using phase vocoding.

Phase vocoding involves some technicalities that we shall bypass in our initial discussion. Later we will describe more of its details. We begin with a basic example.

8.2.1 A Basic Example

A nice outline of phase vocoding, created by the PHOTOSOUNDER program, is provided by the video at this link: Simple Vocoding (Photosounder) at the book's web site. It would be good for the reader to watch this video before reading our discussion.

The essential idea behind phase vocoding is shown in Figure 8.4. An initial waveform, in this case a recording of the phrase "I'm sorry Dave, I'm afraid I can't do that,"[2] has its spectrogram shown on

---

[1]Although there are "throat singers" who can produce several pitches at once, their pitches resemble instrumental tones rather than song lyrics.

[2]Spoken by the rogue computer HAL in the movie, *2001: A Space Odyssey*.

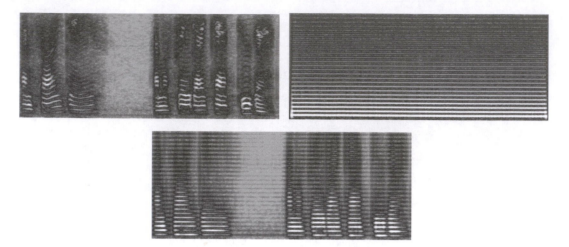

Figure 8.4. Simple illustration of phase vocoding. Top left: Spectrogram of speech recording. Top right: Spectrogram of carrier wave, consisting of several equal amplitude harmonics, having electronic buzzing timbre. Bottom: Spectrogram of product of two spectrograms above. It is spectrogram for phase vocoded signal.

the top left of the figure. The spectrogram on the top right is for a buzzing tone, produced by a collection of several constant tone harmonics of equal amplitude. The values of these spectrograms are simply multiplied, value for value, to produce the spectrogram shown at the bottom of the figure. A synthesis from this bottom spectrogram then produces the phase vocoded sound recording, a buzzing speech that contains characteristics of both the original speech and the buzzing (carrier) wave.

The example just given shows the basic feature of phase vocoding: *Multiplying amplitudes of two signals produces a hybrid signal containing features of both.* The phase vocoding procedure in general, however, is considerably more involved. Phase vocoding uses a more musical carrier signal, but this carrier signal will still have amplitude 1 for each of its constituent frequencies. To be precise, phase vocoding uses the amplitude-phase sound grains g_j, similar to those described by Equation (8.4):

$$g_j(t_m) = \mathsf{w}_j(t_k) \cdot \sum_{k=1}^{L} \mathcal{A}_j(k\nu_o) \cos(2\pi \cdot k\nu_o t_m - \widetilde{\phi}_{j,k}). \tag{8.9}$$

The amplitudes $\mathcal{A}_j(k\nu_o)$ in this equation are computed from a ***modulating signal***, f, for each time segment that begins at time τ_j. Although some technical refinements are employed, one can initially think of an amplitude $\mathcal{A}_j(k\nu_o)$ as equal to a windowed amplitude estimator as described in Chapter 4. For each time segment beginning at τ_j, the windowed amplitude estimator $\mathcal{A}_w(k\nu_o)$ described in Section 4.6 can be used as the amplitude $\mathcal{A}_j(k\nu_o)$ in Formula (8.9). Typically, the modulating signal f is from speech or from singing. The phase angles $\widetilde{\phi}_{j,k}$ come from a ***carrier signal*** $\widetilde{f}$, which usually has a more tonal sound. We now describe how these phase angles $\widetilde{\phi}_{j,k}$ are defined.

For each time segment that begins at time τ_j, compute cosine and sine coefficients, $\{\widetilde{\mathcal{C}}_j(k\nu_o)\}$ and $\{\widetilde{\mathcal{S}}_j(k\nu_o)\}$, of the ***carrier signal*** $\widetilde{f}$. Also compute the amplitudes

$$\widetilde{\mathcal{A}}_j(k\nu_o) = \sqrt{\left[\widetilde{\mathcal{C}}_j(k\nu_o)\right]^2 + \left[\widetilde{\mathcal{S}}_j(k\nu_o)\right]^2}.$$

as with the amplitude calculations for the modulating signal. Although there are various technical refinements involved, these computations can be initially thought of as the windowed cosine, sine, and amplitude estimators we discussed in Section 4.6. Now, compute *normalized cosine and sine coefficients*, $\widetilde{\mathcal{C}}'_j(k\nu_o)$ and $\widetilde{\mathcal{S}}'_j(k\nu_o)$, using these formulas:

$$\widetilde{\mathcal{C}}'_j(k\nu_o) = \frac{\widetilde{\mathcal{C}}_j(k\nu_o)}{\widetilde{\mathcal{A}}_j(k\nu_o)} \quad \text{and} \quad \widetilde{\mathcal{S}}'_j(k\nu_o) = \frac{\widetilde{\mathcal{S}}_j(k\nu_o)}{\widetilde{\mathcal{A}}_j(k\nu_o)}. \tag{8.10}$$

provided $\widetilde{\mathcal{A}}_j(k\nu_o)$ is greater than or equal to a positive threshold value. One possible choice for this threshold value is the amplitude value where frequency $k\nu_o$ would be just audible (to most listeners), or say half of that just audible amplitude. If $\widetilde{\mathcal{A}}_j(k\nu_o)$ is less than the threshold value, then $\widetilde{\mathcal{C}}'_j(k\nu_o)$ and $\widetilde{\mathcal{S}}'_j(k\nu_o)$ are both set equal to 0. (For simplicity, in the discussion that follows, we assume that all of the amplitudes are above threshold values.)

The normalized cosine and sine coefficients defined in (8.10) produce a phase-shifted cosine wave with amplitude 1. To see why, we note that the equations in (8.10) imply that

$$\left[\widetilde{\mathcal{C}}'_j(k\nu_o)\right]^2 + \left[\widetilde{\mathcal{S}}'_j(k\nu_o)\right]^2 = 1. \tag{8.11}$$

This equation tells us that $\left(\widetilde{\mathcal{C}}'_j(k\nu_o), \widetilde{\mathcal{S}}'_j(k\nu_o)\right)$ is a point on the circle of radius 1 centered at the origin. Consequently, there is an angle $\widetilde{\phi}_{j,k}$ for which

$$\widetilde{\mathcal{C}}'_j(k\nu_o) = \cos\widetilde{\phi}_{j,k} \quad \text{and} \quad \widetilde{\mathcal{S}}'_j(k\nu_o) = \sin\widetilde{\phi}_{j,k}.$$

Via the cosine subtraction formula, we then have

$$\widetilde{\mathcal{C}}'_j(k\nu_o)\cos(2\pi \cdot k\nu_o t_m) + \widetilde{\mathcal{S}}'_j(k\nu_o)\sin(2\pi \cdot k\nu_o t_m)$$

$$= \cos\widetilde{\phi}_{j,k}\cos(2\pi \cdot k\nu_o t_m) + \sin\widetilde{\phi}_{j,k}\sin(2\pi \cdot k\nu_o t_m)$$

$$= \cos(2\pi k\nu_o t_m - \widetilde{\phi}_{j,k}).$$

This calculation shows how the phase-shifted cosines in Equation (8.9) are obtained from the carrier signal $\widetilde{f}$.

Adding up the sound grains in Equation (8.9) gives the phase-vocoded signal g. The phase-vocoded signal g is defined by

$$g(t_m) = \sum_{j=1}^{J} g_j(t_m), \quad \text{for each } t_m. \tag{8.12}$$

The full expression for the phase-vocoded signal can also be written down by combining (8.9) and (8.12). We have

$$g(t_m) = \sum_{j=1}^{J}\left[\mathsf{w}_j(t_k) \cdot \sum_{k=1}^{L}\mathcal{A}_j(k\nu_o)\cos(2\pi \cdot k\nu_o t_m - \widetilde{\phi}_{j,k})\right] \tag{8.13}$$

for each time-value t_m. (To be precise, only those terms in the sum over k are used for which a normalized amplitude $\widetilde{\mathcal{A}}_j(k\nu_o)$ is above a threshold.) As we have explained, the amplitudes $\mathcal{A}_j(k\nu_o)$ in Equation (8.13) are obtained from the modulating signal f, while the phase-shifted cosines are obtained from the carrier signal $\widetilde{f}$.

We have described the basic scheme for phase vocoding. There are a couple of more refinements that we shall discuss later. At this point, it is probably better to describe a musical example that illustrates the value of all this mathematics. The song composed by Imogen Heap, *Hide and Seek*, uses a phase vocoding procedure to map her singing voice as a modulator onto the carrier signal provided by her piano playing.

8.2.2 Imogen Heap's *Hide and Seek*

Imogen Heap has produced an artful example of phase vocoding for her song, *Hide and Seek*. In Figure 8.5 we show a spectrogram of the beginning of the song. The constant tone harmonics, which appear as horizontal line segments, are a result of using digital piano tones for the carrier signal. The fluctuation of amplitudes that are apparent within those line segments, are due to the multiplication by voice amplitudes in the phase vocoding process. Notice especially that these modified line segments lie along harmonics that result from chords played by the piano. Heap's voice sounds as if she is singing these chords. The combination of human vocals with piano chords creates a timbre that sounds a bit like organ music. Although we do not have information on all of the exact steps used by Heap to process her sound recording, we can sketch the basic details.

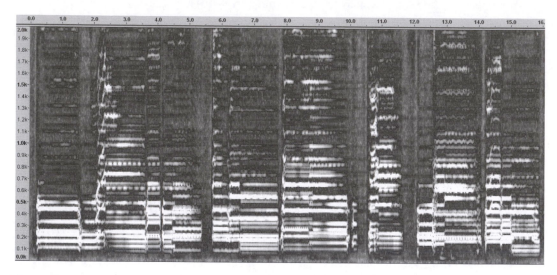

Figure 8.5. Spectrogram from Imogen Heap's song, *Hide and Seek*. Horizontal harmonics are phase-shifted cosines of piano harmonics multiplied by amplitudes from Heap's vocals.

Our previous discussion would seem to imply that the amplitudes for the modulating signal must multiply the carrier signal's phase-shifted cosines for matching frequencies. This is not necessary, however. The modulating signal amplitudes, $\{\mathcal{A}(k\nu_o)\}$, and the carrier signal's phase-shifted cosines *can be treated independently*. In this case, for each windowed time segment, the amplitudes for the harmonics of Heap's voice are identified (from maxima within the set of all amplitudes) and the amplitude for her fundamental is used to multiply the phase-shifted cosines corresponding to the fundamentals of *each of the notes in a piano chord* (those fundamentals also being identified from maxima within the amplitudes for the piano chord). A similar process is performed on each of the harmonics in the piano chord. The amplitudes for each of the overtone harmonics from Heap's voice signal are used as amplitudes for the phase-shifted cosines corresponding to the overtone harmonics from the piano chord. (See Figure 8.6.) The effect is marvelous; the phase-vocoded signal sounds as if Heap is singing chords with 3 notes or more.

Remark 8.2.1. A couple of additional points should be made in connection with phase vocoding. *First*, when the sound grains are added together in Equation (8.12), there should be minimal interference between overlapping cosine functions. A major source of such interference is when two cosine waves of the same frequency are out of phase with each other. By out of phase with each other, we mean that some of their values at times t_m are not aligned, even sometimes of opposite sign. A widely used method for preventing interference of this kind is to slightly adjust the phases in Equation (8.13). The phases for cosines of the same frequency from adjacent windowings are adjusted so that these cosines are exactly in phase, in the sense that their maximum values are occurring at precisely the same time values. This method effectively eliminates any audible interference between overlapping sound grains. *Second*, when pitch shifting is done, as in the song by Imogen Heap, some care must be taken that the formants from human singing are preserved. For instance, the formant amplification factors should generally not be shifted to higher frequencies than normal for human singing. Shifting formants to higher frequencies than normal for human singing produces the "Alvin and the Chip-

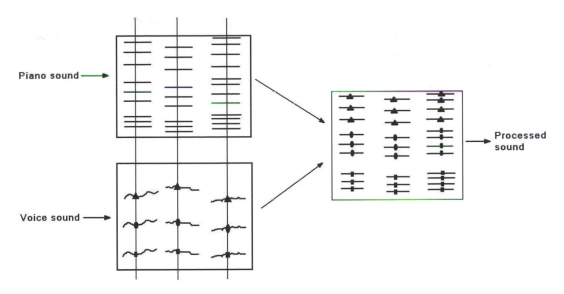

Figure 8.6. Schematic Diagram of Phase Vocoding for Singing Three-Note and Four-Note Chords. Rectangles indicate peak values for voice fundamentals, which are substituted for amplitudes of piano fundamentals. Rounded rectangles indicate peak values for 2nd harmonics of voice, which are substituted for amplitudes of piano's 2nd harmonics. Triangles indicate peak values for 3rd harmonics of voice, which are substituted for amplitudes of piano's 3rd harmonics.

munks effect." Fortunately, it is possible to shift the amplitudes to higher frequencies while keeping the formant amplifying factors unchanged. The technical details for how this is done, however, are rather complex so we shall not delve into them here.[3]

Exercises

8.2.1. Verify that Equation (8.11) holds.

8.2.2. For the point $(0.2, \sqrt{0.96})$ on the circle of radius 1 centered at the origin, find the phase angle ϕ between 0 and $\pi/2$ for which
$$\cos\phi = 0.2, \quad \sin\phi = \sqrt{0.96}.$$

8.2.3. For the point $(\sqrt{0.84}, 0.4)$ on the circle of radius 1 centered at the origin, find the phase angle ϕ between 0 and $\pi/2$ for which
$$\cos\phi = \sqrt{0.84}, \quad \sin\phi = 0.4.$$

8.2.4. What characteristic of organ tones might make them similar in sound to the phase-vocoded vocals in *Hide and Seek*?

8.3 Time Stretching and Time Shrinking

Another application of granular synthesis is time-stretching, or time-shrinking, of recorded sound *without changing pitch.* If a musical recording is simply played twice as long, using half-speed, all the tones within the music are dropped in pitch by an octave. Here we describe a way to prevent that pitch drop. For simplicity, we concentrate on the case of playing the music for twice as long, thereby reducing the tempo by half.

---

[3] A good discussion can be found in W. Sethares, *Rhythm and Transforms*, Springer, 2007.

For simplicity, suppose that the time interval for tempo change contains only the time coordinates τ_1, τ_2, τ_3, τ_4, and τ_5, as beginnings of each time windowing. The basic scheme for doubling the sound length of the recording, without altering the pitch, is shown in Figure 8.7.

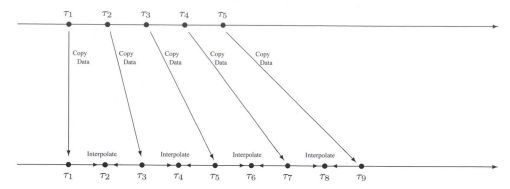

Figure 8.7. Scheme for doubling sound length, without changing pitch. The arrows labeled Copy Data indicate that the cosine and sine coefficients for those time segments are copied to the new time segments. For example, the time segment beginning with τ_2 has its coefficients copied to the time segment beginning with τ_3, and so on. The labels, Interpolate, indicate that the cosine and sine coefficients for those time segments are obtained by interpolation from the cosine and sine coefficients for adjacent time segments. For example, the time segment beginning with τ_2 has its coefficients defined by interpolation of the coefficients from the time segments beginning with τ_1 and τ_3. In the text we describe how interpolation is done.

The cosine and sine coefficients at the time coordinates τ_1, τ_2, ..., τ_5, will be used to create new cosine and sine coefficients for the windowings beginning at the time coordinates τ_1, τ_2, ..., τ_9. Of course, ***to prevent loss of data, all of the cosine and sine coefficients for time segments beyond τ_5 are simply shifted forward in time beyond τ_9***. For τ_1 the cosine and sine coefficients are left as they are:

$$\mathcal{C}_1^{\mathrm{new}}(k\nu_o) = \mathcal{C}_1^{\mathrm{old}}(k\nu_o)$$
$$\mathcal{S}_1^{\mathrm{new}}(k\nu_o) = \mathcal{S}_1^{\mathrm{old}}(k\nu_o). \tag{8.14}$$

At time τ_2, the following formulas are used to assign new values to $\mathcal{C}_2(k\nu_o)$ and $\mathcal{S}_2(k\nu_o)$:

$$\mathcal{C}_2^{\mathrm{new}}(k\nu_o) = \frac{1}{2}\mathcal{C}_1^{\mathrm{old}}(k\nu_o) + \frac{1}{2}\mathcal{C}_2^{\mathrm{old}}(k\nu_o)$$
$$\mathcal{S}_2^{\mathrm{new}}(k\nu_o) = \frac{1}{2}\mathcal{S}_1^{\mathrm{old}}(k\nu_o) + \frac{1}{2}\mathcal{S}_2^{\mathrm{old}}(k\nu_o). \tag{8.15}$$

For τ_3, the cosine and sine coefficients are defined by

$$\mathcal{C}_3^{\mathrm{new}}(k\nu_o) = \mathcal{C}_2^{\mathrm{old}}(k\nu_o)$$
$$\mathcal{S}_3^{\mathrm{new}}(k\nu_o) = \mathcal{S}_2^{\mathrm{old}}(k\nu_o). \tag{8.16}$$

Similar formulas are used for defining $\{\mathcal{C}_j(k\nu_o)\}$ and $\{\mathcal{S}_j(k\nu_o)\}$ for $j = 4, 5, 6, 7, 8$. [Note: The formulas in (8.14) through (8.16) are given as simple examples of interpolation. In practice, more sophisticated interpolation formulas may be employed.]

Once all of the new cosine and sine coefficients are obtained, a granular synthesis is performed. The resulting sound file will extend the frequency data contained in the cosine and sine coefficients over a time interval extending through the windowings from τ_1 to τ_9, which effectively doubles the length of the original time interval. The pitches are unchanged, however, since the pitches are determined by the frequency data encoded by the cosine and sine coefficients.

Example 8.3.1. Time-stretched bird call. One area where time-stretching is of major importance is in studying birdsong. Often birdsong has such a fast tempo that our human ears have trouble catching all of the nuances in the song. Here time-stretching, without pitch change, can be a real aid to us. As an example, we show in Figure 8.8 the spectrograms of an Osprey call and its time-stretching by a factor of 2. The time-stretching preserves the pitch, but also slows down the call's speed enough that it is possible to hear more of the structure of the sound in the call. Clearly that is an advantage in studying birdsong.

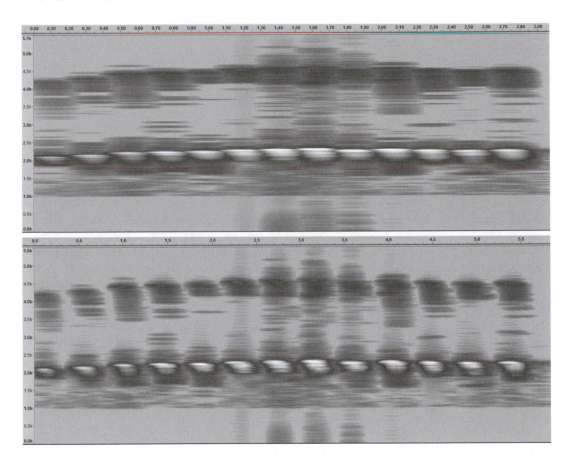

Figure 8.8. Top: Osprey call. Bottom: Time-stretching of the Osprey call by a factor of 2.

Example 8.3.2. Extreme time-stretching. An interesting case of extreme time-stretching is shown in the video at this link:

Extreme Time Stretching (Photosounder)

at the book's web site. This video shows the process of using PHOTOSOUNDER to slow down the phrase, "I'm sorry Dave, I'm afraid I can't do that" that we mentioned earlier, to a tempo that is 200 times slower without changing the pitches. All speech aspect is lost in the resulting sound, instead we hear something like a mass of strings from an orchestra.

We now briefly discuss the case of time-shrinkage. To keep things simple, we examine how to shrink the length of a recording by half. The essential idea is to, in some sense, reverse the scheme shown in Figure 8.7. For simplicity, suppose that the time interval for shrinkage contains only the time coordinates τ_1, τ_2, τ_3, τ_4, and τ_5, as the beginnings of each time windowing. The cosine and sine coefficients for the window starting at τ_1 will be simply left as is. The cosine and sine coefficients for the window starting at τ_2 will be copied to a new windowing that begins at $\tau_{1.5}$, where $\tau_{1.5}$ is defined by

$$\tau_{1.5} = \frac{1}{2}\tau_1 + \frac{1}{2}\tau_2. \tag{8.17}$$

Then the cosine and sine coefficients for the window starting at τ_3 are copied to the window starting at τ_2. The cosine and sine coefficients for the window starting at τ_4 will be copied to a new windowing that begins at $\tau_{2.5}$, where $\tau_{2.5}$ is defined by

$$\tau_{2.5} = \frac{1}{2}\tau_2 + \frac{1}{2}\tau_3. \tag{8.18}$$

Finally, the cosine and sine coefficients for the windowing beginning at τ_5 are copied to the windowing beginning at τ_3, and the coefficients for all subsequent windowings are shifted back in time (from τ_j to τ_{j-2}).

Now that all the coefficients are assigned to windowings, a granular synthesis is performed. The resulting sound file is generated from frequency data contained in the cosine and sine coefficients over a time interval extending through the windowings from τ_1 to τ_3, which effectively halves the length of the original time interval. The pitches are unchanged, however, since the pitches are determined by the frequency data in the cosine and sine coefficients.

Example 8.3.3. Time-morphing. By combining time-shrinkage and time-stretching, we can perform elaborate ***time-morphing*** of digital sound. A fascinating example of time-morphing can be found in the sound synthesis, *One Drop Is Enough,* composed using METASYNTH by Helge Krabye. This sound synthesis was created, in its entirety, by combining time-morphings and pitch shiftings from a single recording of a water drop splashing. To listen to the synthesis, and to read more discussion of its creation, go to the following link:

"One Drop is Enough" details and recording

at the book's web site.

Remark 8.3.1. In Example 8.3.3(a), the time-morphing was done on each individual beat in the music. Those beats were detected automatically from the recorded sound, rather than from a score. Which raises the question: *How does a computer recognize the beats within a digital recording?* This is a very challenging technical problem that is beyond the scope of this book. The book by William Sethares, *Rhythm and Transforms,* discusses this problem in great detail.

Exercises

8.3.1. Describe a scheme for time-stretching digital sound by a factor of 1.5, without changing pitch. Use formulas similar to the interpolation formulas in (8.15).

8.3.2. Describe a scheme for time-stretching digital sound by a factor of 1.25, without changing pitch. Use formulas similar to the interpolation formulas in (8.15).

8.3.3. Describe a scheme for time-stretching digital sound by a factor of $4/3$, without changing pitch. Use formulas similar to the interpolation formulas in (8.15).

8.3.4. If a beat lasts for $1/2$ second, and its first half is time-shrunk by a factor of ρ (where $0 < \rho < 1$), then by what factor must its second half be time-stretched so as to produce a total length of $1/2$ seconds?

8.3.5. Describe a scheme for time-shrinking digital sound by a factor of $2/3$, without changing pitch. Use time-reassignment formulas similar to the ones given in Equations (8.17) and (8.18).

8.3.6. Describe a scheme for time-shrinking digital sound by a factor of $1/3$, without changing pitch. Use time-reassignment formulas similar to the ones given in Equations (8.17) and (8.18).

8.3.7. Describe a scheme for time-shrinking digital sound by a factor of $3/4$, without changing pitch. Use time-reassignment formulas similar to the ones given in Equations (8.17) and (8.18).

The *Strawberry Fields Forever* recording trick

The following three exercises have to do with the Beatles' song, *Strawberry Fields Forever*. Specifically, they all refer to the point where the Beatles' recording engineers spliced together parts from two different versions of the song, recorded in two different keys whose notes all differed by a half step. During this remix, one part was played back at a slower rate in order to drop all its pitches by a quarter step, while the other part was played back at a faster rate in order to raise all its pitches by a quarter step. This remix then had all of its pitches in the same key. The splicing point where this trick was performed occurs at about 1 minute into the recording. Before doing these three exercises, please listen to a recording of *Strawberry Fields Forever*.

8.3.8. Why does the voice of the singer (John Lennon) seem to change slightly after the splice? For this question, remember that the trick adjusts the pitches, so he is singing on key throughout. [Hint: Think about formants.]

8.3.9. Explain why the drums sound different after the splicing.

8.3.10. Explain why the rhythm is altered after the splicing.

8.4 MIDI Synthesis

In this section we briefly describe MIDI synthesis of musical sound. MIDI stands for Musical Instrument Device Interface. It is a popular method of computer synthesis based on musical scores for instruments. A MIDI file is a computer file that specifies pitch values and durations for notes, the instruments that play each note, and the volume level for each note. The MUSESCORE program can read MIDI files and produce a score from them, and it can also save a score in MIDI format. An excellent free player of MIDI files is the WINDOWS program, THE MUSIC ANIMATION MACHINE, available at the link, Music Animation Machine, at the book's web site (go to Software). In Figure 8.9 we show a brief score fragment from a Bach piece and a display of the MIDI file for it, captured from its playback by THE MUSIC ANIMATION MACHINE.

The bottom part of Figure 8.9 shows the MIDI playback display. The rectangles indicate the length of the notes and, according to their vertical position, the pitches of the notes. It is easy to see how these ***note rectangles*** line up with the notes indicated in the score.

THE MUSIC ANIMATION MACHINE has several additional ways for displaying MIDI playback. For example, it has a mode called triads (LATTICE) which displays the *Tonnetz*. It uses numbers like 2 for D and 4♯ for F♯, and the *Tonnetz* is oriented with fifths moving horizontally rather than vertically. Nevertheless, the essential features of the *Tonnetz*, such as fundamental triads appearing as triangles, are retained in the version displayed by THE MUSIC ANIMATION MACHINE.

There are several ways in which MIDI files are used by computers to synthesize sound. Most computers nowadays have some type of sound card which will perform "wavetable synthesis." This synthesis method is analogous to granular synthesis in that a windowed sum of sines and cosines is generated for each note rectangle in the MIDI file. The quality of the sound generated is dependent on the choices of window functions used to generate attack and decay of notes, and the sine and cosine coefficients used to control fundamentals and overtones. This is analogous to granular synthesis, except the note rectangles in MIDI files generally have longer durations than the time intervals used in granular synthesis.

A second way of synthesizing sound is to use recorded notes from actual instruments. The data for these recordings is stored in a data structure called a *soundfont*.[4] Because the soundfont is using

---

[4]A soundfont is a public file format developed by Creative Sound Lab.

Figure 8.9. Top: Two Measures from Bach score (measures 13 and 14 of *Inventio 11, BWV 782*). Bottom: MIDI time durations and pitch indications (upper segments are for the upper staff, lower segments are for the lower staff). Rectangle around last eight notes in upper staff, and lower rectangle around corresponding eight MIDI time durations, are discussed in Exercise 8.4.1.

recordings of actual instrumental notes, there is quite realistic playback. The recorded notes are mapped to the note rectangles to create the synthesized sound. At least that is what is done when a note rectangle matches a pitch and duration of a recorded note stored in the soundfont. Since there is a wide variety of note durations and possible pitches for notes, another method must be used when there is no matching recording. Soundfonts allow for pitch shifting in the following way. A recording of a note in the soundfont at a different pitch can be analyzed to obtain its cosine and sine coefficients. Those coefficients can then be used to multiply sine and cosine functions of the correct frequencies for the desired pitch. The coefficients for the nth harmonic of the recorded sound are used as coefficients for the nth harmonic of the playback sound (which will typically have a different frequency than the frequency of the nth harmonic of the recorded sound). This ability to shift pitches, essentially arbitrarily, also allows for changing time durations of notes. For instance, suppose the note rectangle is $2/3$ the length of the nearest length recorded note in the soundfont. Then the values of the synthesized sound will be specified to play at $3/2$ the speed of the recorded note, and the coefficients from that recorded note will be pitch shifted by an amount corresponding to decreasing the pitch of each harmonic by a factor of $3/2$. The pitch shifting is needed because, without it, a playback at $3/2$ speed would raise the pitches of all harmonics by a factor of $3/2$. For more information on soundfonts, and MIDI playback in general, use the link, Soundfont details, at the book's web site (go to Chapter 8). Overviews of the soundfont format can be found there, as well as a complete technical specification.

A third type of MIDI synthesis is a rather new one. Presently, it is only available for synthesizing piano tones. A commercial venture, called pianoteq, has developed a method for significantly improving the synthesis of MIDI piano tones. One weakness of conventional MIDI piano synthesis is that the full sound properties of an acoustic piano are not modeled by the playback system. For example, in an acoustic piano, when different strings are not damped they can sometimes sympathetically vibrate in response to the vibrations of other strings. This sympathetic vibration is called ***resonance***. A classic demonstration of resonance can be done as follows. Gently press down and hold the C_3 key on the piano keyboard, so that it is not damped. Then quickly strike the G_4 key. As the sound of the G_4 string starts to fade away, you will hear a continuation of the G_4 tone that is emanating from the sympathetic vibration of the undamped C_3 string. This resonance occurs because the fundamental for G_4 is the 3rd harmonic of C_3. The sound vibrations for this G_4 fundamental, resulting from the G_4 string, induce the resonant vibration of the undamped C_3 string, thereby amplifying the G_4 pitch you hear.[5] This effect is important in piano technique. When the score indicates pedaling, the performer will hold down a pedal with his/her foot. Holding down the pedal lifts the dampers off of all the piano strings, allowing them to freely vibrate in resonance with other tones being played. When this pedaling technique is well done, the piano is said to "sing." Conventional MIDI playback does not include this resonance singing for the piano. The pianoteq system, however, uses sophisticated (and proprietary) versions of the synthesis described in Equation (8.4) so as to include resonance effects.

Including resonance is just one improvement that pianoteq achieves over conventional MIDI synthesis. Another improvement is that pianoteq models the reflections of the sound waves off of the piano walls. Including these extremely fast reverberations increases the richness of the timbre of pianoteq's tones.

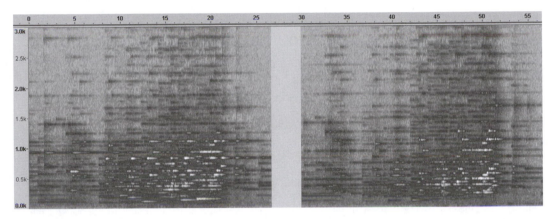

Figure 8.10. Spectrogram of two different instruments playing Debussy's *Clair de Lune*. Left: Rendition using the pianoteq synthesized piano. Right: Rendition on Petrof acoustic grand piano. Petrof rendition shows higher harmonics more prominently.

It is interesting to use spectrograms to investigate the performance of pianoteq. In Figure 8.10 we show a spectrogram of a recording comparing two renditions of a passage from Debussy's *Clair de lune*. The first half of the recording is a pianoteq rendition of the passage, and the second half is a rendition on a Petrof acoustic grand piano. It is clear from this example, that pianoteq is quite comparable to a concert-level acoustic piano. One can also hear, and see it verified on the spectrogram, that the higher harmonics are more prominent for the Petrof piano. Although pianoteq is not precisely reproducing a concert-level acoustic grand piano, such as the Petrof, it is getting close. Furthermore, pianoteq has the advantages that (1) *it never goes out of tune*, and (2) *it is easily transportable*.

Exercises

8.4.1. In Figure 8.9 the last 8 notes in the upper staff of the score are contained in a rectangle, and their MIDI time-durations are shown in another rectangle below them. Explain how these notes exhibit a diatonic scale shift but do ***not*** exhibit a transposition.

8.4.2. In Figure 8.11 we show the score of measures 5 through 10 of a Mozart piece, and below it a portion of the MIDI note rectangles for those measures. Identify which portion of the score corresponds to these rectangles.

---

[5]As a control, you can repeat the experiment with C_3 and F_4. You will not hear any resonance sound from the C_3 string, because F_4 is not a harmonic of C_3.

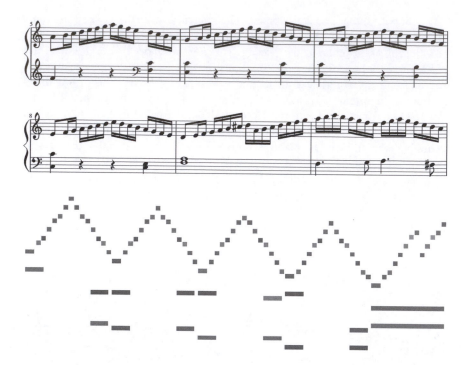

Figure 8.11. Top: Measures 5 through 10 of Mozart score (*Sonata Facile, K. 545*). Bottom: MIDI time-durations and pitch indications for part of those measures.

8.4.3. In Figure 8.11 we show the score of measures 5 through 10 of a Mozart piece, and below it a portion of the MIDI note rectangles for those measures. Explain the relationship between the symmetries in the MIDI note rectangles and the symmetries in the note patterns in the score. (By "symmetries" we mean note transformations of the kind discussed in Sections 3.3 through 3.5.)

Appendix A

Exercise Solutions

This appendix contains solutions to all the odd-numbered exercises.

Chapter 1

Section 1.1

1.1.1 50 Hz.

1.1.3 Harmonics: 110 Hz, 220 Hz, 330 Hz. Fundamental: 110 Hz.

1.1.5 Harmonics: 215 Hz, 430 Hz, 645 Hz, 860 Hz, 1075 Hz, 1290 Hz, 1505 Hz. Fundamental: 215 Hz.

1.1.7 The flute tone sounds more pure because it has only a few significant (large amplitude) harmonics beyond its fundamental. The piano tone has many more significant harmonics, in addition to its fundamental, so it has a richer tone.

1.1.9 Harmonics: 440 Hz, 880 Hz, 1320 Hz, 1760 Hz, 2200 Hz, 2640 Hz, 3080 Hz. Fundamental: 440 Hz. According to Table 1.2, the note being played is A in the 4th octave.

Section 1.2

1.2.1 The two octaves lower G has fundamental $\frac{3}{8}\nu_o$. Its harmonics and the harmonics of D then satisfy:

two octaves lower G:	$\frac{3}{8}\nu_o$	$\frac{3}{4}\nu_o$	$\frac{9}{8}\nu_o$	$\frac{3}{2}\nu_o$	$\frac{15}{8}\nu_o$	$\frac{9}{4}\nu_o$	$\frac{21}{8}\nu_o$	$3\nu_o$	$\frac{27}{8}\nu_o$	$\frac{15}{4}\nu_o$	$\ldots$
D:			$\frac{9}{8}\nu_o$			$\frac{9}{4}\nu_o$			$\frac{27}{8}\nu_o$		$\ldots$

Since $\frac{9}{8}\nu_o = 3 \cdot \left(\frac{3}{8}\nu_o\right)$, it follows that all multiples of the fundamental for D are also multiples of $\frac{3}{8}\nu_o$. Therefore, all the harmonics for D match perfectly with a subset of the harmonics of the lower pitch G.

1.2.3 The two octaves lower F has fundamental $\frac{1}{3}\nu_o$, while the fundamental for A is $\frac{5}{3}\nu_o$. Therefore,

$$\frac{5\nu_o}{3} \cdot k = \frac{\nu_o}{3} \cdot (5k), \quad k = 1, 2, 3, \ldots$$

proving that all the harmonics of A are a subset of the harmonics of this lower pitch F. Likewise, the fundamental for C is $2\nu_o$, and so

$$2\nu_o \cdot k = \frac{\nu_o}{3} \cdot (6k), \quad k = 1, 2, 3, \ldots,$$

which proves that all the harmonics of C are a subset of the harmonics of this lower pitch F.

1.2.5 To obtain an octave equivalent note with fundamental in the range between ν_o and $2\nu_o$, it is necessary to divide $6\nu_o$ by 4. That produces a fundamental of $\frac{3}{2}\nu_o$, the fundamental for the note G, which is already on the scale.

Section 1.3

1.3.1 E_4.

1.3.3 A_4.

1.3.5 Table 1.1: nearest to A_3.   Table 1.2: nearest to A_3, but quite flat.

Section 1.4

1.4.1 E $F^\sharp$ $G^\sharp$ A B $C^\sharp$ $D^\sharp$ E.

1.4.3 **(a)** The F-major scale is

$$F \quad G \quad A \quad B^\flat \quad C \quad D \quad E \quad F.$$

(b) The fifth note down from F in the F-major scale is $B^\flat$. The $B^\flat$-major scale is

$$B^\flat \quad C \quad D \quad E^\flat \quad F \quad G \quad A \quad B^\flat.$$

1.4.5 C D $E^\flat$ F G $A^\flat$ B C.

1.4.7 **(a)** The D-Dorian mode is constructed using these calculations:

$$2 \xrightarrow{+2} 4 \xrightarrow{+1} 5 \xrightarrow{+2} 7 \xrightarrow{+2} 9 \xrightarrow{+2} 11 \xrightarrow[-12 \text{ at top}]{+1} 0 \xrightarrow{+2} 2$$
$$D \qquad E \qquad F \qquad G \qquad A \qquad B \qquad\qquad C \qquad D.$$

(b) The G-Dorian mode is

$$G \quad A \quad B^\flat \quad C \quad D \quad E \quad F \quad G.$$

(c) The C-Dorian mode is

$$C \quad D \quad E^\flat \quad F \quad G \quad A \quad B^\flat \quad C.$$

1.4.9 **(a)** The F-Lydian mode is constructed using these calculations:

$$5 \xrightarrow{+2} 7 \xrightarrow{+2} 9 \xrightarrow{+2} 11 \xrightarrow[-12 \text{ at top}]{+1} 0 \xrightarrow{+2} 2 \xrightarrow{+2} 4 \xrightarrow{+1} 5$$
$$F \qquad G \qquad A \qquad B \qquad\qquad C \qquad D \qquad E \qquad F.$$

(b) The D-Lydian mode is

$$D \quad E \quad F^\sharp \quad G^\sharp \quad A \quad B \quad C^\sharp \quad D.$$

(c) The C-Lydian mode is

$$C \quad D \quad E \quad F^\sharp \quad G \quad A \quad B \quad C.$$

1.4.11 If A_1 has fundamental ν_o, then A_2 has fundamental $2\nu_o = r^{12}\nu_o$, A_3 has fundamental $4\nu_o = r^{24}\nu_o$, $C_4^\sharp$ has fundamental $r^4 \cdot 4\nu_o = r^{28}\nu_o$, and E_4 has fundamental $r^7 \cdot 4\nu_o = r^{31}\nu_o$. We have $r^{12} = 2$ and $r^{24} = 4$. We also have $r^4 \cdot 4 \approx \frac{5}{4} \cdot 4 = 5$, and $r^7 \cdot 4 \approx \frac{3}{2} \cdot 4 = 6$. Since $2, 4, 5$, and 6 are all positive integers, the harmonics of $A_2, A_3, C_4^\sharp$, and E_4 are all subsets of the harmonics of A_1.

1.4.13 The melodic A-minor scale is

$$A \quad B \quad C \quad D \quad E \quad F^\sharp \quad G^\sharp \quad A.$$

Section 1.5

1.5.1 **(a)** $\log_2 32 = 5$ **(b)** $\log_r \left(r^6\right) = 6$ **(c)** $\log 100 = 2$ **(d)** $\log_4 64 = 3$.

1.5.3 **(a)** $\log_2 5 \approx 2.3$ **(b)** $\log_2 10 \approx 3.3$ **(c)** $\log_2 15 \approx 3.9$ **(d)** $\log_2 20 \approx 4.3$.

1.5.5 **(a)** 2.34 **(b)** 1.39 **(c)** 2.21 **(d)** 4.18.

1.5.7 **(a)** 39.2 cents **(b)** 31.19 cents **(c)** 53.27 cents **(d)** -76.96 cents **(e)** -234.47 cents **(f)** -15.67 cents.

1.5.9 For the first pair of frequencies, the difference in cents is 19.71. For the second pair of frequencies, the difference in cents is 14.37. So the second pair of frequencies, $\nu_o = 240$ Hz and $\nu_o' = 242$ Hz, is closer in pitch.

1.5.11 The frequency of the actual fundamental is 215 Hz, and the closest frequency from Table 1.2 is 220 Hz. The difference in cents between these two frequencies is 39.8 cents.

1.5.13 **(a)** The frequency $\frac{16}{9}\nu_o$ is only 3.9 cents lower in frequency than the fundamental for B$^\flat$ on the chromatic scale (which is $\mathbf{r}^{10}\nu_o$). **(b)** The frequency $\frac{16}{9}\nu_o$ is 27.3 cents higher in frequency than $\frac{7}{4}\nu_o$. **(c)** The frequency ratio between C and F in just tuning is $\frac{4}{3}$, and that is also the ratio between F and B$^\flat$ in just tuning when $\frac{16}{9}\nu_o$ is used as fundamental for B$^\flat$. If $\frac{7}{4}\nu_o$ were used as fundamental for B$^\flat$, then the pitch change from F to B$^\flat$ would be heard as significantly less than from C to F.

1.5.15 80 decibels.

1.5.17 120 decibels.

1.5.19 The range of multiples of I_o is from $10^2 I_o$ to $10^{10} I_o$.

Chapter 2

Section 2.1

2.1.1 F, A, G, C, C, A, G, B, E, D, F, C, B, G$^\sharp$, D$^\flat$, B.

2.1.3 Upper Staff: E, G, B, B, G, E, A, G, A, B, C$^\sharp$, D$^\flat$.
Lower Staff: G, F, C, D, F, E, B, E, A, C, B$^\flat$, G$^\sharp$.

2.1.5 A, C, B, E, B, F, D$^\flat$, B$^\flat$, G$^\sharp$, D, F, A, C$^\sharp$, A, C, D.

2.1.7 Treble Clef Staff: G, B, D, G, E, C$^\flat$, A$^\sharp$, F　　Alto Clef Staff: E, C, A, F, G$^\sharp$, B, D, G$^\flat$
Tenor Clef Staff: F, B, G, E, C$^\sharp$, F, A, D　　Bass Clef Staff: A, G$^\flat$, E, A$^\sharp$, C, E, F$^\sharp$, A.

Section 2.2

2.2.1 The equations with fractions for the three measures are

$$\frac{1}{8} + \frac{1}{8} + \frac{1}{4} + \frac{1}{4} + \frac{1}{4} = \frac{4}{4}$$

$$\left(\frac{1}{4} + \frac{1}{8}\right) + \frac{1}{8} + \frac{1}{2} = \frac{4}{4}$$

$$\frac{3}{32} + \frac{1}{32} + \frac{1}{8} + \frac{1}{4} + \frac{1}{2} = \frac{4}{4}.$$

2.2.3 The equations with fractions for the three measures are

$$\frac{1}{8} + \frac{1}{8} + \frac{1}{8} + \frac{1}{8} + \frac{1}{16} + \frac{1}{16} + \frac{1}{8} = \frac{3}{4}$$

$$\frac{1}{4} + \left(\frac{1}{16} + \frac{1}{32}\right) + \frac{1}{32} + \frac{1}{8} + \frac{1}{4} = \frac{3}{4}$$

$$\frac{3}{32} + \frac{1}{32} + \frac{1}{8} + \frac{1}{32} + \frac{1}{32} + \frac{1}{16} + \left(\frac{1}{32} + \frac{1}{64}\right) + \frac{1}{64} + \frac{1}{16} + \frac{1}{4} = \frac{3}{4}.$$

2.2.5 For the first measure, the equation is

$$\left(\frac{1}{4}\right) + \frac{1}{8} + \frac{1}{8} + \frac{2}{8} = \frac{3}{4}.$$

For the triplet, each note is $\frac{1}{12}$ duration.

For the second measure, the equation is

$$\left(\frac{1}{16}\right) + \frac{1}{16} + \frac{1}{8} + \left(\frac{1}{2}\right) = \frac{3}{4}.$$

For the first triplet, each note is $\frac{1}{48}$ duration. For the second triplet, each note is $\frac{1}{6}$ duration.

For the third measure, the equation is

$$\left(\frac{1}{32}\right) + \frac{1}{32} + \frac{1}{32} + \frac{1}{32} + \frac{1}{8} + \frac{1}{2} = \frac{3}{4}.$$

For the triplet, each note is $\frac{1}{96}$ duration.

2.2.7 For the first measure, the equation is

$$\left(\frac{1}{8}\right) + \frac{1}{8} + \left(\frac{1}{4}\right) + \left(\frac{1}{4} + \frac{1}{8}\right) + \frac{1}{8} = \frac{4}{4}.$$

For the first triplet, each note is $\frac{1}{24}$ duration. For the second triplet, each note is $\frac{1}{12}$ duration.

For the second measure, the equation is

$$(1) = \frac{4}{4}.$$

Each note in the triplet is $\frac{1}{3}$ duration.

For the third measure, the equation is

$$(1) = \frac{4}{4}.$$

The first two notes in the triplet are $\frac{1}{3}$ duration. The last two notes are each $\frac{1}{6}$ duration.

2.2.9 The lengths are 1 sec, 0.234375 sec, and 0.5 sec, respectively.

2.2.11 The equations with fractions for the four measures are

$$\frac{2}{8} + \frac{1}{8} + \frac{1}{8} = \frac{2}{4}$$

$$\frac{2}{8} + \frac{1}{4} = \frac{2}{4}$$

$$\frac{2}{8} + \left(\frac{1}{8} + \frac{1}{16} + \frac{1}{32}\right) + \frac{1}{32} = \frac{2}{4}$$

$$\frac{4}{8} = \frac{2}{4}.$$

2.2.13 Adding the $1/32^{\text{nd}}$ note produces the given equation. Solving for x we have

$$x = \frac{3}{4} - \frac{1}{32}$$

$$= \frac{24 - 1}{32} = \frac{23}{32}$$

$$= \frac{16 + 4 + 2 + 1}{32} = \frac{1 + 2 + 4 + 16}{32}$$

$$= \frac{1}{32} + \frac{1}{16} + \frac{1}{8} + \frac{1}{2}.$$

2.2.15 In dancing to music, steps are usually taken on notes receiving emphasis. In rock music, the melody typically has emphasis on the first and third beats, while the back beat has emphasis on the second and fourth beats. Therefore, *every* beat has emphasis from some part of the music, so you cannot take a step on an off beat.

Section 2.3

2.3.1 The 7 sharp key signature is

which is $C^\sharp$-major, or $A^\sharp$-minor. Its notes are $C^\sharp$, $D^\sharp$, $E^\sharp$, $F^\sharp$, $G^\sharp$, $A^\sharp$, $B^\sharp$, $C^\sharp$. All the notes are now sharps, so no further key signatures with sharps appear on the circle of fifths. The notes $E^\sharp$ and $B^\sharp$ are enharmonic with the notes F and C, respectively. This key signature denotes a scale enharmonic with the scale for the $5\flat$ key signature on the circle of fifths.

The 7 flat key signature is

which is $C^\flat$-major, or $A^\flat$-minor. Its notes are $C^\flat$, $D^\flat$, $E^\flat$, $F^\flat$, $G^\flat$, $A^\flat$, $B^\flat$, $C^\flat$. All the notes are now flats, so no further key signatures with flats appear on the circle of fifths. The notes $C^\flat$ and $F^\flat$ are enharmonic with the notes B and E, respectively. This key signature denotes a scale enharmonic with the scale for the $5\sharp$ key signature on the circle of fifths.

2.3.3 The explanation is essentially the same as for major scales. On the left of Figure 2.8, if we view the scale marked by solid dots as beginning at hour 9, then it is the natural A-minor scale. If we also view the scale marked by open circles as starting at hour 4, then it is the natural E-minor scale. The diagram then shows the one note F as being sharped. Since every natural minor scale could be plotted in the same way as the natural A-minor scale, with its starting and ending hour at hour 9, this geometric argument applies to every natural minor scale. Similar reasoning, based on the diagram on the right of Figure 2.8, shows why going down a fifth produces a single flatted note.

2.3.5 We know that $-3 + 7 = +7 - 3$, so the clock hours are the same as well. Adding -3 to all the hours for the notes of a major key produces a natural minor key. Adding $+7$ to all the hours for notes is equivalent to going up a fifth in both major and minor keys. If we first add -3, and then $+7$, this is equivalent to going up a fifth in a natural minor key. Since this equals adding $+7$, and then -3, this is also equivalent to obtaining the natural minor key *from the next key on the circle of fifths going clockwise*. Similar reasoning with -3 and $+5$ applies to going down a fifth.

Chapter 3

Section 3.1

3.1.1 The solution is shown here:

Per. 4th	Unison	Maj. 2nd	Min. 2nd	Unison	Maj. 2nd	Maj. 2nd	Maj. 2nd	Maj. 2nd
+5	0	+2	+1	0	+2	+2	−2	−2

3.1.3 The solution is shown here:

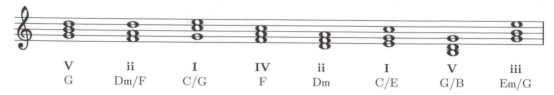

V	ii	I	IV	ii	I	V	iii
G	Dm/F	C/G	F	Dm	C/E	G/B	Em/G

3.1.5 The solution is shown here:

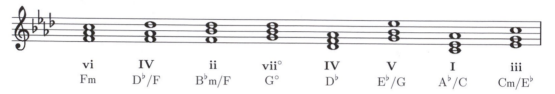

vi	IV	ii	vii°	IV	V	I	iii
Fm	D♭/F	B♭m/F	G°	D♭	E♭/G	A♭/C	Cm/E♭

3.1.7 The G-major scale is

$$G \quad A \quad B \quad C \quad D \quad E \quad F^\sharp \quad G$$

so the minor chord **iii** in G-major starts on B and is created by the half steps shown on the left:

$$
\begin{array}{c}
F^\sharp \\
+4 \uparrow \\
D \\
+3 \uparrow \\
B \\
\\
\text{B-minor}
\end{array}
\qquad\qquad
\begin{array}{c}
F^\sharp \\
+3 \uparrow \\
D^\sharp \\
+4 \uparrow \\
B \\
\\
\text{B-major}
\end{array}
$$

giving the B-minor chord, B-D-F♯. While if we use the same number of half steps as for a major chord, with root B, we get the result shown on the right above, which is the B-major chord.

3.1.9 The solution is shown here:

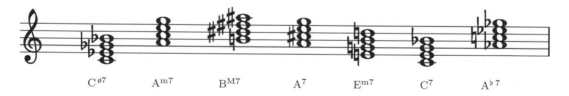

C^{ø7}	A^{m7}	B^{M7}	A^7	E^{m7}	C^7	A^{♭7}

Section 3.2

3.2.1 $\frac{11}{15} = 73.33\%$.

3.2.3 The G^7 chord is G-B-D-F. It contains the triadic chord, G-B-D, which anticipates the C-major chord, C-E-G. The note F contains, as every third harmonic, harmonics of C. Moreover, F is just two hours from G on the chromatic clock. It follows that the presence of F reinforces the anticipation of the C-major chord.

3.2.5 **(a)** Here is the tally:

$$F^\sharp_3 : 2, \quad G_3 : 2, \quad G^\sharp_3 (A^\flat_3) : 2, \quad A_3 : 3,$$

$$A^\sharp_3 : 3, \quad B_3 : 2, \quad C_4 : 2, \quad C^\sharp_4 : 3,$$

$$D_4 : 4, \quad D^\sharp_4 : 2, \quad E_4 : 1, \quad F_4 : 1, \quad F^\sharp_4 : 1.$$

(b) Here is the tally:

Interval, half steps	Number of Times
Tritone, ± 6	2
Minor 7th, ± 10	3
Major 7th, ± 11	3
$\pm$Octave or more	13

The percentage of these intervals, out of the total of 47, is $\frac{21}{47} = 44.68\%$.

3.2.7 **(a)** The key signature has one flat, which could indicate G-major *or* E-minor. The first chord, however, is G-major. This G-major chord is emphasized again at the beginning of measure 1, and at the end of measure 2 tied over into the beginning of measure 3, and then repeated again in measure 3. The key is therefore G-major. **(b)** Six out of seven accidentals in the first line are all chromatic (outside the key of G-major). The exception is the G$^\natural$, which does belong to the key. **(c)** The last four chords are B$^{\phi 7}$, E^7, A^7, and D. They are related by descending fifths for their roots.

3.2.9 The progressions that we are allowing here are

$$\mathbf{I} \rightarrow \mathbf{IV}, \quad \mathbf{I} \rightarrow \mathbf{V},$$

$$\mathbf{IV} \rightarrow \mathbf{I}, \quad \mathbf{IV} \rightarrow \mathbf{V},$$

$$\mathbf{V} \rightarrow \mathbf{I}.$$

For F_{n+1} the number of sequences of progressions fall into two separate categories: those which end with $\mathbf{V}$, or those which end with either $\mathbf{I}$ or $\mathbf{IV}$. Since $\mathbf{V}$ can only follow from $\mathbf{I}$ or $\mathbf{IV}$, and those latter two chords trace back uniquely one more step to all of the ending chords for sequences of length $n - 1$, the number of sequences of length $n + 1$ that end with $\mathbf{V}$ is equal to F_{n-1}. The other category are those sequences of length $n + 1$ that end with either $\mathbf{I}$ or $\mathbf{IV}$. Since each of those ending chords trace back uniquely to all of the ending chords for sequences of length n, the number of sequences of length $n + 1$ that end with $\mathbf{I}$ or $\mathbf{IV}$ is equal to F_n. Therefore, $F_{n+1} = F_n + F_{n-1}$.

3.2.11 **(a)** We have

$$0 \xrightarrow{+2} 2 \xrightarrow{+2} 4 \xrightarrow{+3} 7 \xrightarrow{+2} 9 \xrightarrow{+3} 0$$
$$\text{C} \quad\quad \text{D} \quad\quad \text{E} \quad\quad \text{G} \quad\quad \text{A} \quad\quad \text{C.}$$

(b) The notes are, by measure:

$$A, A, C, D, F, F, D,$$
$$C, C, D, C,$$
$$A, A, C, D, F, F, D,$$
$$C, C, D, C,$$
$$C, C, C, A, C,$$
$$D, D, C,$$
$$A, G, A, C, A, G,$$
$$F, F, G, F.$$

(c) We have

$$5 \xrightarrow{+2} 7 \xrightarrow{+2} 9 \xrightarrow{+3} 0 \xrightarrow{+2} 2 \xrightarrow{+3} 5$$
$$\text{F} \quad\quad \text{G} \quad\quad \text{A} \quad\quad \text{C} \quad\quad \text{D} \quad\quad \text{F}$$

which amounts to adding 5 to each hour for the pentatonic C-major scale. This scale should be called the pentatonic F-major scale.

3.2.13 **(a)** The middle pitch notes for the chords in Figure 3.11 are

$$\text{C D E F G A B C}$$

which is the C-major scale. **(b)** Here are the fundamental triadic chords, in root position, for the natural E-minor scale:

The middle pitch notes for these chords are

G A B C D E F♯ G

which is the scale for the key of G-major. **(c)** The reason that the middle pitch notes of the fundamental chords for each natural minor key consist of the notes for its relative major key is the following. First, since the middle pitch note for the chord **i** is 3 half steps from the root note, that makes it $3 - 3 = 0$ half steps from the tonic note for the relative major key. Therefore, the middle pitch note is the tonic note for the relative major key. Second, since all of the remaining chords are successively shifted up one scale position, it follows that all of the notes of these chords are notes on the scale described by the key signature for the staff. Since that key signature also corresponds to the relative major key, and the middle pitch notes begin with the tonic for that major key, it follows that they form the scale for the relative major key.

3.2.15 Here are the fundamental triadic chords in C-Dorian mode:

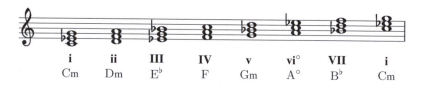

The main reason that C-Dorian mode will sound like a minor key is that its tonic chord **i** is minor. Furthermore, its mediant chord **III** is major, which is also true for a minor key. These two chords are shared by the C-minor key, so music in C-Dorian can tend to sound like music in C-minor.

Section 3.3

3.3.1 The solution is the following:

3.3.3 A solution is shown in Figure A.1.

3.3.5 In the score shown here:

a triplet of three notes is labeled as a in the upper staff. The following succession of diatonic scale shifts is applied to this triplet:

$$\mathcal{S}_0, \mathcal{S}_{-1}, \mathcal{S}_0, \mathcal{S}_{-1}, \mathcal{S}_0, \mathcal{S}_{-1}, \mathcal{S}_0, \mathcal{S}_{-1}, \mathcal{S}_0, \mathcal{S}_{-1}, \mathcal{S}_0, \mathcal{S}_{-1}, \mathcal{S}_0, \mathcal{S}_{-1}, \mathcal{S}_0$$

and that produces all of the notes in the upper staff. The notes in the lower staff are just repetitions of the notes in the upper staff, one octave lower in pitch. The chords from which these triplets are taken, are the following:

E-G-B♭, D-F-A, C-E-G, B♭-D-F, A-C-E, G-B♭-D, F-A-C, E-G-B♭.

Figure A.1. Five diatonic scale shifts in Mozart passage: S_3, S_{-4}, S_3, S_{-4}, and S_1. There is also a succession of three note motives within two boxes marked S, where the first two notes of each motive are diatonically scale shifted (first by S_1, then by S_{-1}, and finally by S_1). Notes in second box S are obtained from diatonic scale shift S_0 applied to notes in first box S (producing exact repetition of notes).

Therefore, in the key of D-minor, we have the chord progression sequence:

$$\mathbf{ii}^\circ \to \mathbf{i} \to \mathbf{VII} \to \mathbf{VI} \to \mathbf{v} \to \mathbf{iv} \to \mathbf{III} \to \mathbf{ii}^\circ.$$

which cycles through all of the fundamental diatonic chords in the key, starting and ending with $\mathbf{ii}^\circ$.

3.3.7 On the score below, the group of four notes on the upper staff marked a is mapped successively to a next group of four notes by the sequence of diatonic scale shifts S'_{-1}, S'_{+1}, and S'_{-1}. The group of four notes on the lower staff marked b is mapped successively to a next group of four notes by the sequence of diatonic scale shifts S'_{+1}, S'_{-1}, and S'_{+1}.

The chords that the notes on the two staves are selected from are: E-G$^\sharp$-B, then D$^\sharp$-F$^\sharp$-A-C, then E-G$^\sharp$-B, then D$^\sharp$-F$^\sharp$-A-C. The chord progression sequence is: $E \to D^{\sharp\,\circ 7} \to E \to D^{\sharp\,\circ 7}$.

3.3.9 There are two ways to solve this problem. **Solution 1:** On the score below, the group of three notes on the upper staff marked a is mapped successively to a next group of three notes by the sequence of diatonic scale shifts S'_0, S'_0 (not the same as the first), S_2, S_{-4}, S'_0, S'_0 (not the same as the previous one), S_2. The group of four notes on the lower staff marked b is mapped to the next four notes by the diatonic scale shift S_{-2}.

The chords that the notes on the two staves are selected from are: B$^\flat$-D-F, then F-A-C, then B$^\flat$-D-F, then part of F-A-C (notes F and A), then G-B$^\flat$-D, then D-F-A, then G-B$^\flat$-D, and then part of D-F-A (notes D and F).

The chord progression sequence is (for the key of $B^\flat$-major):

$$I \to V \to I \to V \to vi \to iii \to vi \to iii. \tag{A.1}$$

Solution 2: There is a second way of describing the tonal sequencing. Treating the first four highest pitch notes, D_5, C_5, D_5, F_5, as a motive, and applying $\mathcal{S}_{-2}$, we get the next set of four highest pitches: $B_4^\flat, A_4, B_4^\flat, D_5$. Similarly, treating the first six middle pitch notes, $F_4, F_4, F_4, F_4, A_4, A_4$, as a motive, and applying $\mathcal{S}_{-2}$, we get the next set of six middle pitches: $D_4, D_4, D_4, D_4, F_4, F_4$. We also have shown above that the lowest four pitches, marked as b on the lower staff, can be treated as a motive, and by again applying $\mathcal{S}_{-2}$, we get the second set of four lowest pitches. This alternative approach treats the music as consisting of the simultaneous playing of three separate melodies. In each case, the melodies are constructed using the same diatonic scale shift $\mathcal{S}_{-2}$. Listening carefully to the score, as it plays in MUSESCORE, you should be able to hear these three voices combining to create the total sound.

Section 3.4

3.4.1 The solution is the following:

3.4.3 The solution is the following:

3.4.5 A modified diatonic scale shift $\mathcal{S}'_0$ is applied to the first three notes in the first measure, to obtain the first three notes in the sixth measure. The diatonic scale shift $\mathcal{S}_1$ is applied to the notes in measures 2 and 3, to obtain the notes in measures 7 and 8. The diatonic scale inversion $\mathcal{IS}_{-3}$ is applied as shown in Figure 3.26.

3.4.7 For the second measure, there are just D, $G^\sharp$, B, and E notes. These are the notes for the E^7 chord, E-$G^\sharp$-B-D. For the third measure, all of the notes are from the A-minor chord, A-C-E.

3.4.9 In the score shown below:

the four notes in the rectangle a are transformed by the sequence $\mathcal{S}_{-2}, \mathcal{IS}_{-2}, \mathcal{S}_2, \mathcal{IS}_2, \mathcal{S}_{-2}$. While the four notes in the rectangle b are transformed by the sequence $\mathcal{S}_2, \mathcal{IS}_2, \mathcal{S}_{-2}, \mathcal{IS}_{-2}, \mathcal{S}_2$. Notice how the signs of the subscripts are the opposite of those in the first sequence.

3.4.11 In the upper staff of the last measure shown in Figure 3.33, the first pair of $1/32^{\text{nd}}$ notes is transformed by the following sequencing of diatonic scale shifts:

$$\mathcal{S}_6, \mathcal{S}_{-1}, \mathcal{S}_{-1}, \mathcal{S}_{-1}, \mathcal{S}_{-1}, \mathcal{S}_{-1}, \mathcal{S}_{-1}.$$

It is interesting to note that, after the final diatonic scale shift, the pair of notes has returned to its original position. That is because $6 - 1 - 1 - 1 - 1 - 1 - 1 = 0$.

On the other hand, there is a longer timescale sequencing of the last three sets of beamed $1/32^{\text{nd}}$ notes. Beginning with the second beaming of four notes (the notes E_6, D_6, D_6, C_6), the diatonic scale shift $\mathcal{S}_{-2}$ is applied twice. There is some musical rationale for this longer timescale view. Because the piece is written in $\frac{4}{4}$ time, these groups of four beamed notes each have the duration of one beat.

Section 3.5

3.5.1 The solution is shown in the following score:

3.5.3 The solution is shown in the following score:

3.5.5 The notes that sound for Horn I are $B^\flat$ (continued from first note), $B^\flat$, $B^\flat$, F, F, F. The notes that sound for Horn II are $B^\flat$ (continued from first note), $B^\flat$, $B^\flat$, F, F, $B^\flat$.

3.5.7 The chromatic inversion is IT_{-2}, applied to the first four notes of the upper staff to get the next four notes.

3.5.9 The chromatic inversion IT_4 converts a G-major chord to a E-minor chord, and converts a G-minor chord to an E-major chord. The calculations are as follows:

$$
\begin{array}{ccccc}
D & & & B \\
+3 \uparrow & & T_4 & -4 \downarrow \\
B & & \nearrow & G \\
+4 \uparrow & & & -3 \downarrow \\
G & & & E \\
\text{(G-major)} & \xrightarrow{\mathsf{IT}_4} & \text{(E-minor)}
\end{array}
\qquad
\begin{array}{ccccc}
D & & & B \\
+4 \uparrow & & T_4 & -3 \downarrow \\
B^\flat & & \nearrow & G^\sharp \\
+3 \uparrow & & & -4 \downarrow \\
G & & & E \\
\text{(G-minor)} & \xrightarrow{\mathsf{IT}_4} & \text{(E-major)}
\end{array}
$$

3.5.11 The chromatic inversion IT_2 converts an F-major chord to a C-minor chord, and converts an F-minor chord to an C-major chord. The calculations are as follows:

$$
\begin{array}{ccccc}
C & & & G \\
+3 \uparrow & & T_2 & -4 \downarrow \\
A & & \nearrow & E^\flat \\
+4 \uparrow & & & -3 \downarrow \\
F & & & C \\
\text{(F-major)} & \xrightarrow{\mathsf{IT}_2} & \text{(C-minor)}
\end{array}
\qquad
\begin{array}{ccccc}
C & & & G \\
+4 \uparrow & & T_2 & -3 \downarrow \\
A^\flat & & \nearrow & E \\
+3 \uparrow & & & -4 \downarrow \\
F & & & C \\
\text{(F-minor)} & \xrightarrow{\mathsf{IT}_2} & \text{(C-major)}
\end{array}
$$

3.5.13 The solution is shown in the following score:

3.5.15 **(a)** The transposition T_4 converts a G^7 chord to a B^7 chord, and the chromatic inversion IT_5 converts a

G^7 chord to a $D^{\emptyset 7}$ chord. The calculations are as follows:

$$
\begin{array}{cccc}
F & A & F & C \\
+3\uparrow & +3\uparrow & +3\uparrow & -4\downarrow \\
D & F^\sharp & D & A^\flat \\
+3\uparrow & +3\uparrow & +3\uparrow \quad T_5 & -3\downarrow \\
B & D^\sharp & B & F \\
+4\uparrow & +4\uparrow & +4\uparrow & -3\downarrow \\
G & \xrightarrow{+4} B & G & D \\
(G^7) & \xrightarrow{T_4} (B^7) & (G^7) \xrightarrow{IT_5} & (D^{\emptyset 7})
\end{array}
$$

(b) Every transposition preserves the half step changes, so a seventh chord will be mapped to a seventh chord. **(c)** Every chromatic inversion will reverse the signs of the half step changes. In this case, from $+4, +3, +3$ to $-4, -3, -3$. These negative half step changes will start from the highest pitch in the chord. From that chord's lowest pitch, the half step changes are then $+3, +3, +4$, so it is a half-diminished seventh chord.

3.5.17 **(a)** The transposition T_5 converts a $B^{\emptyset 7}$ chord to an $E^{\emptyset 7}$ chord, and the chromatic inversion IT_5 converts a $B^{\emptyset 7}$ chord to an $F^{\sharp 7}$ chord. The calculations are as follows:

$$
\begin{array}{cccc}
A & D & A & E \\
+4\uparrow & +4\uparrow & +4\uparrow & -3\downarrow \\
F & B^\flat & F & C^\sharp \\
+3\uparrow & +3\uparrow & +3\uparrow \quad T_5 & -3\downarrow \\
D & G & D & A^\sharp \\
+3\uparrow & +3\uparrow & +3\uparrow & -4\downarrow \\
B & \xrightarrow{+5} E & B & F^\sharp \\
(B^{\emptyset 7}) & \xrightarrow{T_5} (E^{\emptyset 7}) & (B^{\emptyset 7}) \xrightarrow{IT_5} & (F^{\sharp 7})
\end{array}
$$

(b) Every transposition preserves the half step changes, so a half-diminished seventh chord will be mapped to a half-diminished seventh chord. **(c)** Every chromatic inversion will reverse the signs of the half step changes. In this case, from $+3, +3, +4$ to $-3, -3, -4$. These negative half step changes will start from the highest pitch in the chord. From that chord's lowest pitch, the half step changes are then $+4, +3, +3$, so it is a seventh chord.

Chapter 4

Section 4.1

4.1.1 The frequency spread is about 60 Hz, with a rate of 7.5 cycles/sec.

4.1.3 **(a)** The notes for the chord in the first measure are G, B, D, $F^\sharp$. The half steps between these notes on the chromatic clock are $4, 3, 4$, so the chord is classified as a major seventh chord. The note G fades out (as shown in the spectrogram, confirming what our ears tell us), and the minor chord B, D, $F^\sharp$ becomes more prominent. This creates a transition from a major to minor sound, which contributes to the pathos of the music. **(b)** For the second measure, the notes in the chord are D, $F^\sharp$, A, $C^\sharp$. The half steps between these notes on the chromatic clock are $4, 3$, and 4, so the chord is classified as a major seventh chord. Again, there is a transition from major to minor sound as the bass note D_2 fades out.

4.1.5 As the vibrato increases, the tone quality is "softened."

4.1.7 The solution is shown in Figure A.2.

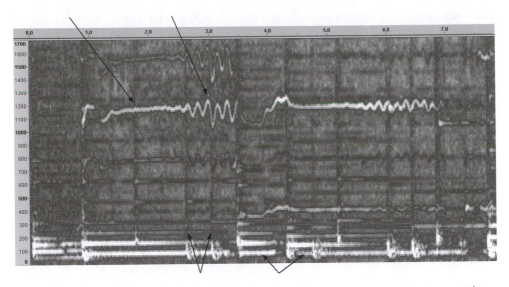

Figure A.2. Spectrogram from recording of *Bohemian Rhapsody*. First arrow on top left points to 3rd harmonic of sung tone, and second arrow points to vibrato. First double arrow on bottom left points to two drum strikes. Second double arrow points to fundamentals for two piano tones.

Section 4.2

4.2.1 **(a)** freq. = 213 Hz, amplitude 32, phase $\phi = 0.3$. **(b)** freq. = 164 Hz, amplitude 240, phase $\phi = -0.2$. **(c)** freq. = 128 Hz, amplitude 120, phase $\phi = 0$. **(d)** freq. = 122 Hz, amplitude 340, phase $\phi = -\pi/2$. **(e)** freq. = 60 Hz, amplitude 240, phase $\phi = \pi$.

4.2.3 **(a)** A_3 **(b)** $D_4^\sharp$ **(c)** C_5 **(d)** B_2 **(e)** $G_4^\sharp$.

Section 4.3

4.3.1 **(a)** The amplitudes, frequencies, phases, and pitches are shown in the following table.

Amplitude	Frequency	Phase	Pitch
300	262	0.3	C_4
600	330	0.3	E_4
100	392	0.29	G_4

(b) The amplitudes, frequencies, phases, and pitches are shown in the following table.

Amplitude	Frequency	Phase	Pitch
400	82	0.5	E_2
200	98	0.5	G_2
100	110	0.5	A_2

(c) The amplitudes, frequencies, phases, and pitches are shown in the following table.

Amplitude	Frequency	Phase	Pitch
1200	220	-0.2	A_3
800	294	-0.2	D_4
100	370	0	$F_4^\sharp$
90	740	-0.3	$F_5^\sharp$

4.3.3 There are 4 lobes shown (half of a lobe is at the left end, and another half is at the right end). This matches exactly with the beat frequency of $454 - 450 = 4$ beats per second described in the text.

4.3.5 If we use $\theta_a = \big((2\pi\nu t - \phi) + (2\pi\omega t - \theta)\big)/2$ and $\phi_d = \big((2\pi\nu t - \phi) - (2\pi\omega t - \theta)\big)/2$, then we obtain

$$\cos(2\pi\nu t - \phi) + \cos(2\pi\omega t - \theta) = \cos(\theta_a + \phi_d) + \cos(\theta_a - \phi_d)$$
$$= 2\cos(\phi_d)\cos(\theta_a)$$
$$= 2\cos\big(2\pi\tfrac{\nu-\omega}{2}\, t - \tfrac{\phi-\theta}{2}\big)\cos\big(2\pi\tfrac{\nu+\omega}{2}\, t - \tfrac{\phi+\theta}{2}\big).$$

Hence, we obtain that

$$y = A\cos(2\pi\nu t - \phi) + A\cos(2\pi\omega t - \theta)$$

satisfies

$$y = 2A\cos\big(2\pi\tfrac{\nu-\omega}{2}\, t - \tfrac{\phi-\theta}{2}\big)\cos\big(2\pi\tfrac{\nu+\omega}{2}\, t - \tfrac{\phi+\theta}{2}\big).$$

The beating phenomenon in this case will be essentially the same as the one described in the text, because the envelope here is $\pm 2A\cos\big(2\pi\tfrac{\nu-\omega}{2}\, t - \tfrac{\phi-\theta}{2}\big)$. This envelope is just a shifting, by the phase constant $\tfrac{\phi-\theta}{2}$ of the envelope for the case discussed in the text. Since the frequency factor, $\tfrac{\nu-\omega}{2}$, is the same for both cases, the beat frequency will be the same for both cases as well. The other factor, $\cos\big(2\pi\tfrac{\nu+\omega}{2}\, t - \tfrac{\phi+\theta}{2}\big)$, is just a shifting by the phase constant $\tfrac{\phi+\theta}{2}$ of the factor $\cos\big(2\pi\tfrac{\nu+\omega}{2}\, t\big)$ discussed in the text. The pitch frequency is $\tfrac{\nu+\omega}{2}$ in both cases.

Section 4.4

4.4.1 $\dfrac{15\text{ beats}}{0.5\text{ sec}} = 30\,\dfrac{\text{beats}}{\text{sec}}$ and the frequency difference is about $760 - 730 = 30$. The results match.

4.4.3 Let the lower pitch have fundamental ν_o and the higher pitch have fundamental ν_o'. Then $\beta = \nu_o' - \nu_o$ is the number of beats per second between the first harmonics. Since the second harmonics are $2\nu_o$ and $2\nu_o'$, we have $2\nu_o' - 2\nu_o = 2(\nu_o' - \nu_o) = 2\beta$ as the number of beats per second between the second harmonics. Likewise, we have $3\nu_o' - 3\nu_o = 3(\nu_o' - \nu_o) = 3\beta$ as the number of beats per second between the third harmonics.

4.4.5 The repeated notes for the lower voice are E_3 and F_3, which have fundamentals of about 165 Hz and 175 Hz, respectively. They produce a beating tone of fundamental $\frac{1}{2}(165 + 175) = 170$ Hz. The repeated notes for the upper voice are B_3 and C_4, which have fundamentals of about 247 Hz and 262 Hz, respectively. They produce a beating tone of fundamental $\frac{1}{2}(247 + 262) = 254.5$ Hz. If the amplitudes are equal, then these two subsidiary tones with fundamentals of 170 Hz and 254.5 Hz will be heard as beating tones. If the amplitudes are not equal, then there are two other tones for the higher amplitude frequencies, creating a total of four tones.

Section 4.5

4.5.1 **(a)** The frequencies are about 112 Hz and 224 Hz, which are approximately fundamentals for the notes A_2 and A_3, respectively. **(b)** The amplitudes are about 200 for frequency 112 Hz, and 100 for frequency 224 Hz.

4.5.3 **(a)** The frequencies are about 256 Hz and 320 Hz, which are approximately fundamentals for the notes C_4 and E_4 on the just tuned scale (see Table 1.1). **(b)** The amplitudes are about 100 for frequency 256 Hz, and 25 for frequency 320 Hz.

4.5.5 Here is a table of approximate frequencies and their approximate pitches (your estimates may differ slightly from ours):

Frequency	Pitch
330	E_4
660	E_5
990	B_5
1320	E_6
1650	$G_6^\sharp$
1980	B_6
2310	D_7

The note being played is E_4.

4.5.7 Yes, they do match the notes E_4 and F_4 listed in Figure 4.18.

Section 4.6

4.6.1 **(a)** The frequencies are about 144 Hz and 168 Hz, which are closest to the fundamentals for the notes D_3 and E_3, respectively. **(b)** The amplitudes are about 200 for frequency 144 Hz, and 100 for frequency 168 Hz.

4.6.3 **(a)** The frequencies are about 345 Hz and 384 Hz, which are closest to the fundamentals for the notes F_4 and G_4, respectively. **(b)** The amplitudes are about 800 for 384 Hz, and 100 for 345 Hz.

4.6.5 Here is a table of approximate frequencies and their pitches (your estimates may differ slightly from ours):

Frequency	Pitch
349	F_4
698	F_5
1047	C_6
1396	F_6
1745	A_6

The note being played is F_4.

Section 4.7

4.7.1 For $k = 1, 2, 3, \ldots, N/2 - 1$, we find that:

$$\mathcal{S}(k\nu_o) = \frac{2}{N} \sum_{m=1}^{N} f(t_m) \sin(2\pi k \nu_o t_m)$$

$$= \beta_j \frac{2}{N} \sum_{m=1}^{N} \sin(2\pi j \nu_0 t) \sin(2\pi k \nu_o t_m)$$

$$= \beta_j \frac{1}{N} \sum_{m=1}^{N} \left\{ \cos\left(2\pi[j - k]\nu_0 t_m\right) - \cos\left(2\pi[j + k]\nu_0 t_m\right) \right\}.$$

We then have

$$\mathcal{S}(k\nu_o) = \beta_j \frac{1}{N} \sum_{m=1}^{N} \cos\left(2\pi[j - k]\nu_0 t_m\right) - \beta_j \frac{1}{N} \sum_{m=1}^{N} \cos\left(2\pi[j + k]\nu_0 t_m\right). \tag{A.2}$$

Using Theorem 4.7.1, alternately for $\ell = j - k$ and $\ell = j + k$, we find that both of the sums in (A.2) are 0 when $k \neq j$. However, when $k = j$, the first sum equals $\beta_j \cdot 1 = \beta_j$. Thus, we have found that

$$\mathcal{S}(k\nu_o) = \begin{cases} \beta_j, & \text{when } k = j \\ 0, & \text{when } k \neq j. \end{cases} \tag{A.3}$$

4.7.3 For $k = 1, 2, 3, \ldots, N/2 - 1$, we find that:

$$\mathcal{C}(k\nu_o) = \frac{2}{N} \sum_{m=1}^{N} f(t_m) \cos(2\pi k\nu_o t_m)$$

$$= \beta_j \frac{2}{N} \sum_{m=1}^{N} \sin(2\pi j\nu_o t) \cos(2\pi k\nu_o t_m)$$

$$= \beta_j \frac{1}{N} \sum_{m=1}^{N} \left\{ \sin\left(2\pi[j+k]\nu_o t_m\right) - \sin\left(2\pi[j-k]\nu_o t_m\right) \right\}.$$

We then have

$$\mathcal{C}(k\nu_o) = \beta_j \frac{1}{N} \sum_{m=1}^{N} \sin\left(2\pi[j+k]\nu_o t_m\right) - \beta_j \frac{1}{N} \sum_{m=1}^{N} \sin\left(2\pi[j-k]\nu_o t_m\right). \tag{A.4}$$

Using Theorem 4.7.1, alternately for $\ell = j - k$ and $\ell = j + k$, we find that both of the sums in (A.4) are 0. Thus, we have found that

$$\mathcal{C}(k\nu_o) = 0, \quad \text{for each } k. \tag{A.5}$$

4.7.5 The completed table is the following:

ν	$k\nu_o$	A_k	$\mathcal{A}_w(\nu - k\nu_o)$	$A_k \mathcal{A}_w(\nu - k\nu_o)$	$\mathcal{A}_w(k\nu_o)$
260	256	100	0.88	88	88
260	264	100	0.88	88	88
326	320	50	0.75	37.5	37.5
326	328	50	0.97	48.5	48.4

There are two equal heights for the peaks at 256 Hz and 264 Hz because there are two equal values of 0.88 for $\mathcal{A}_w(\nu - k\nu_o)$ at $\nu - k\nu_o = \pm 4$.

Chapter 5

Section 5.1

5.1.1 The spectrogram reveals her powerful breaths at about 8.0 sec and 12.5 sec. These breaths appear as long vertical bands, a kind of percussive effect. There is a prominent chiaroscuro timbre from about 10.0 sec to 11.6 sec. There is a change from absence of chiaroscuro timbre, during the time interval 1.0 sec to 2.2 sec, to presence of chiaroscuro timbre, during the time interval 2.2 sec to 5.2 sec. Finally, we can see the close match of her 2nd harmonic with a string harmonic at about 1400 Hz during the time interval 0.5 sec to 5.0 sec.

5.1.3 **(a)** Her *passaggio* is shown within a box drawn on the spectrogram in Figure A.3. Notice that her fundamental within this box covers an approximate range of 622 Hz to 740 Hz, i.e., between E-flat and F-sharp at the top of the staff. Within this range, a slight smudging of her tone occurs, confirming her difficulty in negotiating the *passaggio*. **(b)** When the frequencies for notes fall between formants, this could cause trouble with voice amplification. For example, Fleming refers to "tones between E-flat and F-sharp at the top of the staff." That is a frequency range from about 622 Hz to 740 Hz, and it covers a range of smaller amplification lying between two formant peaks (see the graph on the left of Figure 5.2).

5.1.5 The harmonics for the vocals are contained in the frequency range of 500 Hz to 1500 Hz. Because of the spacing between harmonics, it appears that the fundamentals do not show up clearly in the spectrogram (they must be below 500 Hz). John Lennon, via multi-tracking, is singing a chordal progression, with good *legato* and very little *vibrato*. Some of the closer harmonics are showing a beating effect, although this does not create a dissonance in the sound.

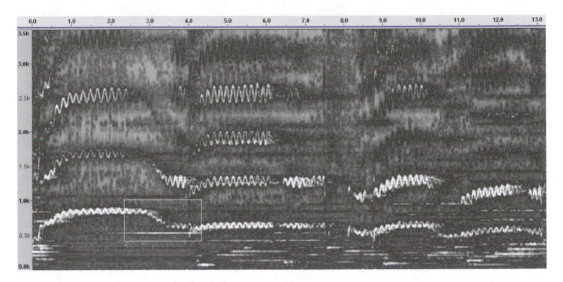

Figure A.3. Spectrogram from recording of *O Mio Babbino Caro* with Renée Fleming. Her *passaggio* is shown within box at lower left.

5.1.7 Using $A_k = c$, Equation (5.1) has the form:

$$\widetilde{A}_k = F(k\nu_1) \cdot c \tag{A.6}$$

for $k = 1, 2, 3, \ldots$. Equation (A.6) shows that the set of voiced amplitudes $\{\widetilde{A}_k\}$ will follow the profile of the graph of cF when ν_1 is a relatively low frequency, since cF is assumed to be continuous and a small value for ν_1 will give many closely spaced points on the graph of cF. For example, in Figure 5.3, we used $c = 1$ and a relatively small frequency ν_1, so the graph of $\{\widetilde{A}_k\}$ closely follows the profile of the formant F in that figure.

Section 5.2

5.2.1 The cup mute dampens the lower frequencies, so the muted trumpet sounds "brighter" due to the increased relative loudness of the higher overtones. This shows clearly in the increased brightness of the overtones above 1200 Hz.

5.2.3 The piano spectrogram shows much more uniform brightness through a greater frequency range than the orchestration. This is audible in the recordings as well; the piano chords are much more dynamic than the orchestral chords due to their fuller spectral content.

5.2.5 **(a)** E_4 **(b)** D_4.

5.2.7 The spectrogram for *Variations On A Theme By Erik Satie* shows much more brightness for higher frequency harmonics than the spectrogram for the piano rendition of *Gymnopedie I*. The *Variations* piece uses different instruments, a guitar and flute, which increases the relative loudness of higher frequency pitches and overtones. The instrumentation, in relation to Satie's score, appears to be based mostly on splitting the notes from the upper and lower staves and having them played by these different instruments. The upper staff notes are played by a flute, and the lower staff notes by a guitar. In the original, a sustained note $F_4^\sharp$ is held by the piano throughout four measures. It quickly fades out to inaudibility. In the variation, however, the flute holds this note quite audibly through all four measures.

5.2.9 A transposition corresponds to a multiplication by a frequency factor, $\mathbf{r}^k$ for some integer k. This has the effect of shifting the frequency positions of harmonics from musical tones. For the two measures of the score, the transposition applied to the first measure is T_{-5}, which corresponds to multiplying the harmonics of each note by the factor $\mathbf{r}^{-5}$. If the instrument used to play the measures were a piano, then the spectrogram for the first measure would appear as a set of four line segments of equal length (since all the notes are quarter notes) for each set of harmonics. Each of these groups of four harmonics would be separated by frequency factors of

$\mathbf{r}^4$, $\mathbf{r}^3$, and $\mathbf{r}^{-3}$. For the second measure, the spectrogram would appear similar, just shifted down in frequency due to multiplication of the harmonics from the first measure by $\mathbf{r}^{-5}$.

5.2.11 **(a)** In playing a violin or guitar, a vibrato can be created by using one's finger to rapidly move the string back and forth at the position being pressed down. This causes a rapid alternation of increasing/decreasing tension, and thus a rapid alternation of increasing/decreasing pitch. For an electric guitar, there is a second technique involving the "whammy bar." When the wammy bar is pulled down, it rapidly alters the tension on the string, inducing a correspondingly rapid alteration of pitch. **(b)** Assuming that increasing humidity will increase the density of a string, there will be a decrease in pitch because of Formula (5.2). Conversely, if decreasing humidity causes the string's density to decrease, then there will be an increase in pitch. This is why the humidity around concert pianos must be carefully controlled to prevent the pianos from going out of tune.

5.2.13 The player gently touches the string at one of two nodes, either $\frac{1}{4}L$ or $\frac{3}{4}L$, where L is the length that determines the fundamental ν_o. This causes a relative dampening of all the harmonics that are **not** integer multiples of $4\nu_o$. (Note: The player does **not** touch the node $\frac{1}{2}L$. Touching that node would not suppress the 2nd harmonic, $2\nu_o$, nor would it suppress any of the odd multiples of $2\nu_o$, such as $6\nu_o$, $10\nu_o$, $14\nu_o$, and so on.)

5.2.15 The tones are D_5, $F_7^\sharp$, and A_6. They form a D-major chord.

Section 5.3

5.3.1 It is an elaboration of the pattern described in Section 5.3.6. There are increasingly rapidly occurring divisions into strata, marked off by pounding tympani notes. Within the strata at the start there are glissandos plus the repeating of the theme (including successive transpositions). After 25 seconds, the tempo of the tympani rapidly accelerates, with the strata sectioning the theme into parts (rather than entire repetitions between the tympani notes). Trills and tremolo[1] are used throughout the piece in the strings. The subtle dissonance created by the trilling appears in the beating interference between closely spaced harmonics at around 1.5 kHz in the first 10 seconds. The final 15 seconds consists of a slower tempo resolving melody, ending with a final long chord played with tremolo by the strings (the tremolo appearing as an alternation of closely spaced bright and dark bands within the constant line segments for the string harmonics).

5.3.3 **(a)** There is a blue note between about 0.4 and 0.7 seconds, created by striking E_4 and F_4 simultaneously. The blue note frequency is about 339 Hz. There is a second blue note between about 2.2 and 2.5 seconds, created by striking F_4 and $F_4^\sharp$ simultaneously. The blue note frequency is about 360 Hz. There is a third blue note between about 6.0 and 6.2 seconds, created by striking B_3 and C_4 simultaneously. The blue note frequency is about 254 Hz. **(b)** There is a bass line melody played throughout, in accompaniment to Ellington's melody. The orchestra provides rhythmic bursts of notes.

Chapter 6

Section 6.1

6.1.1 Here are the graphs for the solution ($\mathbf{F}^\sharp$ can also be labeled $\mathbf{G}^\flat$, and $\mathbf{D}^\sharp$ as $\mathbf{E}^\flat$):

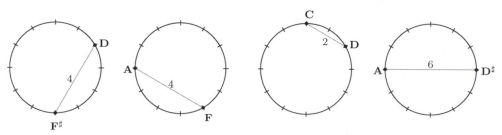

---

[1] A **trill** is a rapid, repeating, alternation between two adjacent notes. **Tremolo** is an extremely rapid repetition of a note, so that the sound is perceived as a wavering in volume of a single note.

6.1.3 **(a)** $\frac{8\pi}{12} = \frac{2\pi}{3}$ radians, **(b)** $\frac{-6\pi}{12} = \frac{-\pi}{2}$ radians, **(c)** $\frac{-10\pi}{12} = \frac{-5\pi}{6}$ radians, **(d)** $\frac{18\pi}{12} = \frac{3\pi}{2}$ radians.

6.1.5 Here are the graphs for the solution:

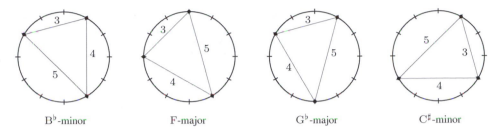

    $B^\flat$-minor   F-major $G^\flat$-major $C^\sharp$-minor

6.1.7 **(a)** $44 \equiv 2 \bmod 7$, **(b)** $83 \equiv 11 \bmod 12$, **(c)** $122 \equiv 2 \bmod 24$, **(d)** $-63 \equiv 7 \bmod 10$, **(e)** $38 \equiv 6 \bmod 16$ **(f)** $-12 \equiv 2 \bmod 7$.

6.1.9 **(a)** The best way to label the hours of the clock for displaying the C-major scale is shown on the left of Figure A.4. **(b)** The best way to label the hours of the clock for displaying the G-major scale is shown on the right of Figure A.4. **(c)** A diatonic scale shift $\mathcal{S}_k$, for the C-major scale, is a rotation of the clock on the left of Figure A.4 by an angle of $k \cdot 2\pi/7$ about its center. If k is positive, then the rotation is clockwise. If k is negative, then the rotation is counter-clockwise. A formula for $\mathcal{S}_k$ is the following mapping for each hour m:

$$\mathcal{S}_k : \ m \to k + m \ \bmod 7.$$

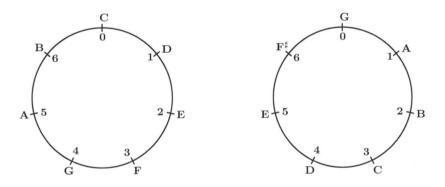

Figure A.4. Left: 7-hour clock for C-major scale. Right: 7-hour clock for G-major scale.

Section 6.2

6.2.1 The graphs for the solution are shown here:

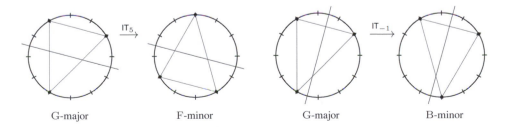

     G-major   F-minor   G-major   B-minor

6.2.3 The graphs for the solution are shown here:

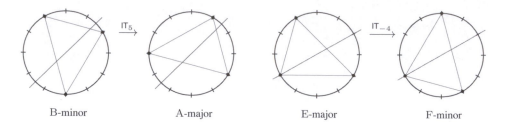

| B-minor | A-major | E-major | F-minor |

6.2.5 (a) $\mathcal{I}_3$, (b) $\mathcal{I}_2$, (c) $\mathcal{I}_5$, (d) $\mathcal{I}_8$. For the smaller of the two hours, multiply by 2 to get the value of k for $\mathcal{I}_k$. So, if the hours are j and $6 + j$, then the pitch class chromatic inversion is $\mathcal{I}_{2j}$.

6.2.7 Exercise 6.2.1: $\mathcal{I}_7$ and $\mathcal{I}_1$. Exercise 6.2.3: $\mathcal{I}_3$ and $\mathcal{I}_4$.

6.2.9 The geometric interpretation of diatonic scale inversion is that each such inversion corresponds to a reflection through a mirror line drawn through one of the following pairs of hours: $0, 3.5$ or $0.5, 4$ or $1, 4.5$ or $1.5, 5$, or $2, 5.5$ or $2.5, 6$ or $3, 6.5$. The generalization of pitch class chromatic inversion to this diatonic case is that each diatonic scale inversion is equal to one of the mappings $\mathcal{I}_k$ given by

$$\mathcal{I}_k : \quad m \to k - m \mod 7$$

for $k = 0, 1, 2, \ldots, 6$.

Section 6.3

6.3.1 (a) 16-hour clock. (b) 18-hour clock. (c) 12-hour clock. (d) 16-hour clock.

6.3.3 The clock is shown on the left of Figure A.5. It has note onsets marked at hours $0, 2, 4, 8, 9, 12$.

6.3.5 The clock is shown in the middle of Figure A.5. It has note onsets marked at hours $0, 2, 4, 8, 10, 11$.

6.3.7 The clock with note onsets for the first rhythmic passage is shown on the right of Figure A.5. The first passage has 7 quarter note chords preceding the half note chord. The second passage has only 5 quarter note chords preceding the half note chord, so it begins at hour 4, as indicated by the arrow in the figure. This is a jumping ahead by 4 hours. The third and fourth passages have the same note onsets as the first, so they begin at hour 0 on the clock shown in the figure.

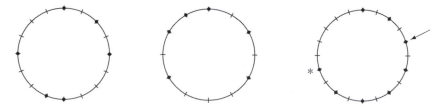

Figure A.5. Graphs for solutions to exercises. Left: Clock with note onsets for Exercise 6.3.3. Middle: Clock with note onsets for Exercise 6.3.5. Right: Clock with note onsets for Exercise 6.3.7. The note onset for each tied half note and eighth note chord is indicated by the asterisk at hour 14, and the note onset for the beginning of the second rhythmic passage is indicated by the arrow pointing to hour 4.

6.3.9 The solution for parts **(a)** and **(b)** is shown here:

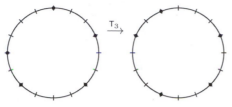

Original rhythm   Time-shifted rhythm   Different starting times

The formula for part **(c)** is

$$\mathsf{T}_3: \quad m \rightarrow 3 + m \mod 16.$$

6.3.11 The solution for parts **(a)** and **(b)** is shown here:

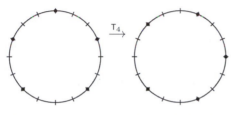

Original rhythm   Time-shifted rhythm   Different starting times

The formula for part **(c)** is

$$\mathsf{T}_4: \quad m \rightarrow 4 + m \mod 16.$$

6.3.13 The clock diagrams are as follows:

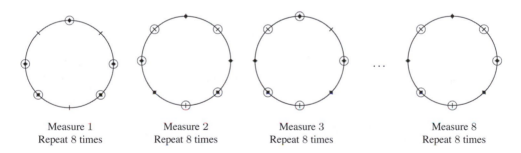

Measure 1   Measure 2   Measure 3   Measure 8
Repeat 8 times   Repeat 8 times   Repeat 8 times   Repeat 8 times

One rhythm repeats unchanged in each measure, while the second rhythm is time-shifted by T_1 from one measure to the next. On an 8-hour clock, the two rhythms return to perfect synchrony in the 9th measure.

6.3.15 The completed tables are the following:

(a)

Measure:	1	2	3	4	5	6	7	8	9
Rhythmic Distance:	0	6	4	2	6	2	4	6	0

(b)

Measure:	1	2	3	4	5	6	7	8	9
Rhythmic Distance:	0	2	4	6	6	6	4	2	0

(c)

Measure:	1	2	3	4	5	6	7	8	9	10	11	12	13
Rhythmic Distance:	0	8	6	4	6	4	8	4	6	4	6	8	0

6.3.17 There is no unique answer to this exercise. The pitch classes for the notes given are **G, B, G, A, A, F**. Applying the transposition T_5 to those pitch classes gives the following pitch classes:

$$\mathbf{C}, \mathbf{E}, \mathbf{C}, \mathbf{D}, \mathbf{D}, \mathbf{A}^{\sharp}. \tag{A.7}$$

The notes for the last two measures will be chosen from these pitch classes. The note onsets in the first two measures are $0, 2, 4, 8, 9, 12$. Applying the time-shift T_2 to those onsets gives the following note onsets:

$$2, 4, 6, 10, 11, 14. \tag{A.8}$$

The following score shows one way of choosing notes from the pitch classes in (A.7), with the note onsets in (A.8), to form the third and fourth measures:

6.3.19 There is no unique answer to this exercise. The pitch classes for the notes given are **G, B, G, A, A, F**. Applying the transposition T_{-5} to those pitch classes gives the following pitch classes:

$$\mathbf{D}, \mathbf{F}^{\sharp}, \mathbf{D}, \mathbf{E}, \mathbf{E}, \mathbf{C}. \tag{A.9}$$

The notes for the next two measures will be chosen from these pitch classes. The note onsets in the first two measures are $0, 2, 4, 8, 9, 12$. Applying the time-shift T_{-3} to those onsets, using $\mod 16$ arithmetic and rewriting to standard order, gives the following note onsets:

$$1, 5, 6, 9, 13, 15. \tag{A.10}$$

These will be used as the note onsets for the next two measures.

For the last two measures, we apply the transposition T_5 to the pitch classes in Equation (A.9) and get the following pitch classes:

$$\mathbf{G}, \mathbf{B}, \mathbf{G}, \mathbf{A}, \mathbf{A}, \mathbf{F}. \tag{A.11}$$

The notes for the last two measures will be chosen from these pitch classes. We apply the time-shift T_2 to the note onsets in Equation (A.10), and write the resulting note onsets in standard order. This gives the following note onsets:

$$1, 3, 7, 8, 11, 15. \tag{A.12}$$

The following score shows one way of choosing notes from the pitch classes in (A.9), with the note onsets in (A.10), to form the third and fourth measures, and notes from the pitch classes in (A.11), with the note onsets in (A.12), to form the fifth and sixth measures:

Section 6.4

6.4.1 For $\mathsf{T}_2 \mathcal{I}_3$, we have each hour m mapped as follows:

$$m \xrightarrow{\mathsf{T}_2} 2 + m \mod 16 \xrightarrow{\mathcal{I}_3} 3 - (2 + m) \mod 16 = 1 - m \mod 16.$$

So hour m is mapped to hour $1 - m$, which is exactly what $\mathcal{I}_1$ does. For $\mathcal{I}_8 \mathsf{T}_{10}$, we have each hour m mapped as follows

$$m \xrightarrow{\mathcal{I}_8} 8 - m \mod 16 \xrightarrow{\mathsf{T}_{10}} 10 + (8 - m) \mod 16 \equiv 2 - m \mod 16.$$

So hour m is mapped to hour $2 - m$, which is exactly what $\mathcal{I}_2$ does. For $\mathcal{I}_{10} \mathcal{I}_6$, we have each hour m mapped as follows:

$$m \xrightarrow{\mathcal{I}_{10}} 10 - m \mod 16 \xrightarrow{\mathcal{I}_6} 6 - (10 - m) \mod 16$$

$$= -4 + m \mod 16 \equiv 12 + m \mod 16.$$

So hour m is mapped to hour $12 + m$, which is exactly what T_{12} does. For $\mathcal{I}_{11}\mathsf{T}_{10}$, we have each hour m mapped as follows:

$$m \xrightarrow{\mathcal{I}_{11}} 11 - m \mod 16 \xrightarrow{\mathsf{T}_{10}} 10 + (11 - m) \mod 16 \equiv 5 - m \mod 16.$$

So hour m is mapped to hour $5 - m$, which is exactly what $\mathcal{I}_5$ does.

6.4.3 From Exercise 6.4.2, we know that $\mathsf{T}_6\mathsf{R} = \mathsf{R}\,\mathsf{T}_6$ and $\mathsf{R}\,\mathsf{R} = \mathsf{T}_0$ (the identity transformation). Therefore, $\mathsf{R}\,\mathsf{T}_6\,\mathsf{R}\,\mathcal{I}_8 = \mathsf{T}_6\mathcal{I}_8$, and for $\mathsf{T}_6\mathcal{I}_8$, we have each hour m mapped as follows:

$$m \xrightarrow{\mathsf{T}_6} 6 + m \mod 12 \xrightarrow{\mathcal{I}_8} 8 - (6 + m) \mod 12 = 2 - m \mod 12.$$

So hour m is mapped to hour $2 - m$, which is exactly what $\mathcal{I}_2$ does. Likewise, we have $\mathsf{R}\,\mathcal{I}_9\,\mathsf{R}\,\mathsf{T}_6 = \mathcal{I}_9\mathsf{T}_6$ and for $\mathcal{I}_9\mathsf{T}_6$, we have each hour m mapped as follows:

$$m \xrightarrow{\mathcal{I}_9} 9 - m \mod 12 \xrightarrow{\mathsf{T}_6} 6 + (9 - m) \mod 12 = 15 - m \mod 12 \equiv 3 - m \mod 12.$$

So hour m is mapped to hour $3 - m$, which is exactly what $\mathcal{I}_3$ does. We also have $\mathsf{R}\,\mathsf{T}_{10}\,\mathsf{R}\,\mathsf{T}_9 = \mathsf{T}_{10}\mathsf{T}_9$ and for $\mathsf{T}_{10}\mathsf{T}_9$, we have each hour m mapped as follows:

$$m \xrightarrow{\mathsf{T}_{10}} 10 + m \mod 12 \xrightarrow{\mathsf{T}_9} 9 + (10 + m) \mod 12 \equiv 7 + m \mod 12.$$

So hour m is mapped to hour $7 + m$, which is exactly what T_7 does. Finally, we have $\mathsf{R}\,\mathcal{I}_8\,\mathsf{R}\,\mathcal{I}_{11} = \mathcal{I}_8\mathcal{I}_{11}$ and for $\mathcal{I}_8\mathcal{I}_{11}$, we have each hour m mapped as follows:

$$m \xrightarrow{\mathcal{I}_8} 8 - m \mod 12 \xrightarrow{\mathcal{I}_{11}} 11 - (8 - m) \mod 12 = 3 + m \mod 12.$$

So hour m is mapped to hour $3 + m$, which is exactly what T_3 does.

6.4.5 $\mathcal{I}_7, \mathsf{R}\mathcal{I}_5, \mathsf{T}_3, \mathsf{R}\mathcal{I}_1$.

6.4.7 The solution is shown here:

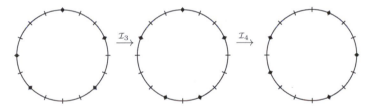

6.4.9 The given two measures have note onsets described by the hours $0, 3, 6, 9, 12$ on a 16-hour clock. Applying $\mathcal{I}_3$ to these hours gives the hours $3, 0, 13, 10, 7$. In standard form, these hours are $0, 3, 7, 10, 13$. There is no unique score with these note onsets, but here is one possibility:

6.4.11 The pitch classes for the notes in the given 3 measures are $\mathbf{A}, \mathbf{F}, \mathbf{A}, \mathbf{F}, \mathbf{F}, \mathbf{A}, \mathbf{F}, \mathbf{A}, \mathbf{A}, \mathbf{A}, \mathbf{F}, \mathbf{A}$. Applying $\mathsf{R}\mathcal{I}_1$ to these pitches, gives the pitches:

$$\mathbf{E}, \mathbf{G}^\sharp, \mathbf{E}, \mathbf{E}, \mathbf{E}, \mathbf{G}^\sharp, \mathbf{E}, \mathbf{G}^\sharp, \mathbf{G}^\sharp, \mathbf{E}, \mathbf{G}^\sharp, \mathbf{E}. \tag{A.13}$$

The note onsets, on an 18-hour clock, for the given 3 measures are

$$0, 2, 4, 6, 7, 8, 10, 11, 12, 13, 14, 16.$$

Applying T_2 to these note onsets, we get these note onsets (after using mod 18 arithmetic and writing in standard order):

$$0, 2, 4, 6, 8, 9, 10, 12, 13, 14, 15, 16. \tag{A.14}$$

There is no unique score with the pitch classes in (A.13) and the note onsets in (A.14), but here is one possibility:

Section 6.5

6.5.1 The Euclidean algorithm to find the greatest common divisor (g.c.d.) for each case goes as follows:

(a)	**(b)**	**(c)**
$84 = 24 + 60$	$120 = 36 + 84$	$440 = 200 + 240$
$84 = 2 \cdot 24 + 36$	$120 = 2 \cdot 36 + 48$	$440 = 2 \cdot 200 + 40$
$84 = 3 \cdot 24 + 12$	$120 = 3 \cdot 36 + 12$	$200 = 40 + 160$
$24 = 12 + 12$	$36 = 12 + 24$	$200 = 2 \cdot 40 + 120$
$24 = 2 \cdot 12 + 0$	$36 = 2 \cdot 12 + 12$	$440 = 3 \cdot 40 + 80$
g.c.d. is 12	$36 = 3 \cdot 12 + 0$	$200 = 4 \cdot 40 + 40$
	g.c.d. is 12	$200 = 5 \cdot 40 + 0$
		g.c.d. is 40

6.5.3 The procedure goes as follows:

$$1\,1\,1\,1\,1\,1\,1\,1,\ 0\,0\,0\,0 \qquad (12 = 8 + 4)$$

$$1\,0\ 1\,0\ 1\,0\ 1\,0,\ 1\,1\,1\,1 \qquad (8 = 4 + 4)$$

$$1\,0\,1\ 1\,0\,1\ 1\,0\,1\ 1\,0\,1, \qquad (8 = 2 \cdot 4 + 0).$$

So the final sequence is 101101101101 which, on the chromatic clock, marks off the hours $0, 2, 3, 5, 6, 8, 9, 11$. These are the hours for

$$\text{B} \quad \text{C} \quad \text{D} \quad \text{E}^\flat \quad \text{F} \quad \text{G}^\flat \quad \text{G}^\sharp \quad \text{A} \quad \text{B}.$$

if we take the tonic note to be B.

6.5.5 The procedure goes as follows:

$$1\,1\,1\,1,\ 0\,0\,0\,0\,0\,0\,0\,0 \qquad (12 = 4 + 8)$$

$$1\,0\ 1\,0\ 1\,0\ 1\,0,\ 0\,0\,0\,0 \qquad (12 = 2 \cdot 4 + 4)$$

$$1\,0\,0\ 1\,0\,0\ 1\,0\,0\ 1\,0\,0, \qquad (12 = 3 \cdot 4 + 0)$$

and that gives the sequence 1 0 0 1 0 0 1 0 0 1 0 0.

6.5.7 The procedure goes as follows:

$$1\,1\,1,\ 0\,0\,0\,0\,0 \qquad (8 = 3 + 5)$$

$$1\,0\ 1\,0\ 1\,0,\ 0\,0 \qquad (8 = 2 \cdot 3 + 2)$$

$$1\,0\,0\ 1\,0\,0,\ 1\,0, \qquad (3 = 2 + 1)$$

and that gives the sequence 1 0 0 1 0 0 1 0.

6.5.9 The hours of the pitch classes for S are $0, 4, 8$. If T_0 is applied to these, we get $0, 4, 8$ so S is unchanged. If T_4 is applied, we get $4, 8, 0$, and that gives the same set of pitch classes for S (different order, but same set). If T_8 is applied, we get $8, 0, 4$, and that also gives the same set of pitch classes for S.

6.5.11 (a) Lowering $\mathbf{G}^\sharp$ by one half step gives pitch classes $\mathbf{C}, \mathbf{E}, \mathbf{G}$. These are pitch classes for the C-major chord. Raising $\mathbf{G}^\sharp$ by one half step gives pitch classes $\mathbf{C}, \mathbf{E}, \mathbf{A}$. These are pitch classes for the A-minor chord. (b) Lowering $\mathbf{E}$ by one half step gives pitch classes $\mathbf{C}, \mathbf{E}^\flat, \mathbf{G}^\sharp$. These are pitch classes for the $\mathbf{A}^\flat$-major chord ($\mathbf{G}^\sharp$ being enharmonic with $\mathbf{A}^\flat$). Raising $\mathbf{E}$ by one half step gives pitch classes $\mathbf{C}, \mathbf{F}, \mathbf{G}^\sharp$. These are pitch classes for the F-minor chord ($\mathbf{G}^\sharp$ being enharmonic with $\mathbf{A}^\flat$). (c) Lowering $\mathbf{C}$ by one half step gives pitch classes $\mathbf{B}, \mathbf{E}, \mathbf{G}^\sharp$. These are pitch classes for the E-major chord. Raising $\mathbf{C}$ by one half step gives pitch classes $\mathbf{C}^\sharp, \mathbf{E}, \mathbf{G}^\sharp$. These are pitch classes for the $\mathbf{C}^\sharp$-minor chord.

6.5.13 The procedure goes as follows:

$$1\,1\,1\,1\,1\,1\,1,\ 0\,0\,0\,0\,0\,0\,0\,0\,0 \quad (16 = 7 + 9)$$

$$10\ 10\ 10\ 10\ 10\ 10\ 10,\ 0\,0 \quad (16 = 2 \cdot 7 + 2)$$

$$1\,0\,0\ 1\,0\,0,\ 10\ 10\ 10\ 10\ 10 \quad (7 = 2 + 5)$$

$$1\,0\,0\,1\,0\ 1\,0\,0\,1\,0,\ 10\ 10\ 10 \quad (7 = 2 \cdot 2 + 3)$$

$$1\,0\,0\,1\,0\,1\,0\ 1\,0\,0\,1\,0\,1\,0,\ 10 \quad (7 = 3 \cdot 2 + 1)$$

and that gives the sequence $1\,0\,0\,1\,0\,1\,0\,1\,0\,0\,1\,0\,1\,0\,1\,0$.

6.5.15 The procedure goes as follows:

$$1\,1\,1\,1\,1\,1\,1\,1\,1\,1,\ 0\,0\,0\,0\,0\,0 \quad (16 = 10 + 6)$$

$$10\ 10\ 10\ 10\ 10\ 10,\ 1\,1\,1\,1 \quad (10 = 6 + 4)$$

$$1\,0\,1\ 1\,0\,1\ 1\,0\,1\ 1\,0\,1,\ 10\ 10 \quad (6 = 4 + 2)$$

$$1\,0\,1\,1\,0\ 1\,0\,1\,1\,0,\ 1\,0\,1\ 1\,0\,1 \quad (4 = 2 + 2)$$

and that gives the sequence $1\,0\,1\,1\,0\,1\,0\,1\,1\,0\,1\,0\,1\,1\,0\,1$. On a 16-hour clock, this sequence marks off hours

$$0, 2, 3, 5, 7, 8, 10, 12, 13, 15$$

which are the note onsets for the rhythm played by the *quijada* in Figure 6.11.

Section 6.6

6.6.1 The probabilities are

$$1: \frac{12}{66} \qquad 2: \frac{12}{66} \qquad 3: \frac{12}{66} \qquad 4: \frac{12}{66} \qquad 5: \frac{12}{66} \qquad 6: \frac{6}{66}$$

or, in simplified form

$$1: \frac{2}{11} \qquad 2: \frac{2}{11} \qquad 3: \frac{2}{11} \qquad 4: \frac{2}{11} \qquad 5: \frac{2}{11} \qquad 6: \frac{1}{11}.$$

The graph of these probabilities, and the dissonance score, are shown in Figure A.6.

6.6.3 The probabilities are

$$1: 0 \qquad 2: \frac{6}{15} \qquad 3: 0 \qquad 4: \frac{6}{15} \qquad 5: 0 \qquad 6: \frac{3}{15}$$

or, in simplified form

$$1: 0 \qquad 2: \frac{2}{5} \qquad 3: 0 \qquad 4: \frac{2}{5} \qquad 5: 0 \qquad 6: \frac{1}{5}$$

The graph of these probabilities, and the dissonance score, are shown in Figure A.6.

6.6.5 The probabilities are

$$1: \frac{3}{21} \qquad 2: \frac{3}{21} \qquad 3: \frac{5}{21} \qquad 4: \frac{4}{21} \qquad 5: \frac{4}{21} \qquad 6: \frac{2}{21}.$$

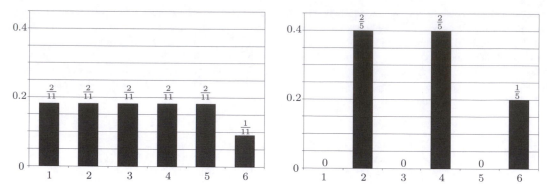

Figure A.6. Left: Interval probabilities for chromatic scale (Exercise 6.6.1). Dissonance score $3.636363\ldots$. Right: Interval probabilities for whole tone scale (Exercise 6.6.3). Dissonance score 3.6.

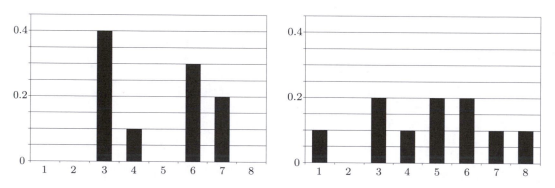

Figure A.7. Note onsets and interval frequencies. Left: *Bossa Nova* rhythm. Right: *Soukous* rhythm.

The dissonance score is $\frac{70}{21} = 3.333\ldots$.

6.6.7 **(a)** For both scales, the numbers of occurrences of intervals for distinct pitch classes are

$$1\text{: } 2 \qquad 2\text{: } 5 \qquad 3\text{: } 4 \qquad 4\text{: } 3 \qquad 5\text{: } 6 \qquad 6\text{: } 1.$$

(b) The combined counts are, after throwing out 6's:

$$1\text{: } 4 \qquad 2\text{: } 10 \qquad 3\text{: } 8 \qquad 4\text{: } 6 \qquad 5\text{: } 12 \qquad 6\text{: } 0.$$

(c) The probabilities are

$$1\text{: } \frac{1}{10} \qquad 2\text{: } \frac{1}{4} \qquad 3\text{: } \frac{1}{5} \qquad 4\text{: } \frac{3}{20} \qquad 5\text{: } \frac{3}{10} \qquad 6\text{: } 0.$$

The dissonance score is 2.8.

6.6.9 The frequencies for the *Bossa Nova* rhythm are graphed on the left of Figure A.7.

6.6.11 The frequencies for the *Soukous* rhythm are graphed on the right of Figure A.7.

6.6.13 **(a)** For a major or minor chord, the set of chromatic distances is $\{3, 4, 5\}$. Since major or minor chords have the same set of chromatic distances, they are all Z-related. **(b)** Since a transposition corresponds to a rotation of the chromatic clock, it will preserve all possible chromatic distances between pitch classes in a given set. Therefore, that set and its transposition are Z-related. The same reasoning applies to a chromatic inversion. **(c)** The intervals between successive pitch classes $\mathbf{C}, \mathbf{C}^\sharp, \mathbf{E}, \mathbf{F}^\sharp$ on the chromatic clock are $\{1, 3, 2, 6\}$, while the intervals between successive pitch classes $\mathbf{C}, \mathbf{C}^\sharp, \mathbf{D}^\sharp, \mathbf{G}$ on the chromatic clock are $\{1, 2, 4, 5\}$. Since those sets of intervals are not the same, the two sets of pitch classes cannot be transpositions or chromatic inversions of each other.

6.6.15 The two sets of pitch classes both have the following number of occurrences of chromatic distances:

$$1:\ 2 \qquad 2:\ 1 \qquad 3:\ 2 \qquad 4:\ 2 \qquad 5:\ 2 \qquad 6:\ 1$$

so they are Z-related. The intervals between successive pitch classes $\mathbf{E}, \mathbf{F}, \mathbf{G}^\sharp, \mathbf{A}, \mathbf{B}$ on the chromatic clock are $\{1, 3, 1, 2, 5\}$, while the intervals between successive pitch classes $\mathbf{D}^\sharp, \mathbf{F}^\sharp, \mathbf{A}, \mathbf{B}^\flat, \mathbf{B}$ on the chromatic clock are $\{3, 3, 1, 1, 4\}$. Since those collections of intervals are not the same, the two sets of pitch classes cannot be transpositions or chromatic inversions of each other.

Section 6.7

6.7.1 The musical matrix is shown here:

	I_0	I_9	I_3	I_6	
T_0	6	3	9	0	RT_0
T_3	9	6	0	3	RT_3
T_9	3	0	6	9	RT_9
T_6	0	9	3	6	RT_6
	$\mathsf{R}\mathsf{I}_0$	$\mathsf{R}\mathsf{I}_9$	$\mathsf{R}\mathsf{I}_3$	$\mathsf{R}\mathsf{I}_6$	

6.7.3 The musical matrix is shown here:

	I_0	I_3	I_6	I_9	
T_0	9	0	3	6	RT_0
T_9	6	9	0	3	RT_9
T_6	3	6	9	0	RT_6
T_3	0	3	6	9	RT_3
	$\mathsf{R}\mathsf{I}_0$	$\mathsf{R}\mathsf{I}_3$	$\mathsf{R}\mathsf{I}_6$	$\mathsf{R}\mathsf{I}_9$	

6.7.5 The pitch classes are from the set $\{1, 4, 7, 10\}$ which are separated clockwise on the chromatic clock by 3, and there are exactly four pitch classes in the set. Therefore, the transpositions and chromatic inversions for this set are exactly the ones in the group $\mathfrak{C}_S$. Consequently, the MUSICAL MATRIX ALGORITHM applies to any tone row from this set of pitch classes, provided the transformations are taken from the group $\mathfrak{C}_S$.

6.7.7 The musical matrix is shown here:

	I_0	I_{10}	I_6	I_2	I_8	I_4	
T_0	2	0	8	4	10	6	RT_0
T_2	4	2	10	6	0	8	RT_2
T_6	8	6	2	10	4	0	RT_6
T_{10}	0	10	6	2	8	4	RT_{10}
T_4	6	4	0	8	2	10	RT_4
T_8	10	8	4	0	6	2	RT_8
	$\mathsf{R}\mathsf{I}_0$	$\mathsf{R}\mathsf{I}_{10}$	$\mathsf{R}\mathsf{I}_6$	$\mathsf{R}\mathsf{I}_2$	$\mathsf{R}\mathsf{I}_8$	$\mathsf{R}\mathsf{I}_4$	

6.7.9 The musical matrix is shown here:

	I_0	I_8	I_6	I_4	I_{10}	I_2	
T_0	8	4	2	0	6	10	RT_0
T_4	0	8	6	4	10	2	RT_4
T_6	2	10	8	6	0	4	RT_6
T_8	4	0	10	8	2	6	RT_8
T_2	10	6	4	2	8	0	RT_2
T_{10}	6	2	0	10	4	8	RT_{10}
	RI_0	RI_8	RI_6	RI_4	RI_{10}	RI_2	

6.7.11 The musical matrix is shown in Figure A.8.

	I_0	I_{10}	I_6	I_4	I_5	I_7	I_9	I_8	I_{11}	I_1	I_2	I_3	
T_0	10	8	4	2	3	5	7	6	9	11	0	1	RT_0
T_2	0	10	6	4	5	7	9	8	11	1	2	3	RT_2
T_6	4	2	10	8	9	11	1	0	3	5	6	7	RT_6
T_8	6	4	0	10	11	1	3	2	5	7	8	9	RT_8
T_7	5	3	11	9	10	0	2	1	4	6	7	8	RT_7
T_5	3	1	9	7	8	10	0	11	2	4	5	6	RT_5
T_3	1	11	7	5	6	8	10	9	0	2	3	4	RT_3
T_4	2	0	8	6	7	9	11	10	1	3	4	5	RT_4
T_1	11	9	5	3	4	6	8	7	10	0	1	2	RT_1
T_{11}	9	7	3	1	2	4	6	5	8	10	11	0	RT_{11}
T_{10}	8	6	2	0	1	3	5	4	7	9	10	11	RT_{10}
T_9	7	5	1	11	0	2	4	3	6	8	9	10	RT_9
	RI_0	RI_{10}	RI_6	RI_4	RI_5	RI_7	RI_9	RI_8	RI_{11}	RI_1	RI_2	RI_3	

Figure A.8. Musical matrix for Exercise 6.7.11.

6.7.13 Suppose that the first hour in the initial tone row for T_0 is the hour m. Now consider any other hour in that initial tone row, say hour j. The first hour m in the initial tone row is also the first hour in the column labeled by I_0. To go from hour m to hour j requires adding $j - m$. Hence the column that begins with hour j is labeled by I_{j-m}. Since $I_0 = \mathcal{I}_{2m}$, it will produce from that hour j the hour $2m - j$ at the corresponding position in the first column. Adding $j - m$ to $2m - j$, in order to go to the column headed by j, we obtain $2m - j + j - m = m$. Thus, we have hour m at the diagonal position corresponding to the column headed by hour j in the initial tone row. This argument shows that the same hour m lies along all of the diagonal positions, which is what we needed to show.

6.7.15 There are $12 \cdot 11 \cdot 10 \cdots 2 \cdot 1 = 479,001,600$ musical matrices.

Chapter 7

Section 7.1

7.1.1 **(b)** $f \xrightarrow{\mathcal{R}} A^\flat$ (or $G^\sharp$)   **(c)** $B \xrightarrow{\mathcal{L}} e^\flat$ (or $d^\sharp$)   **(d)** $a \xrightarrow{\mathcal{P}} A$ **(e)** $d^\flat \xrightarrow{\mathcal{R}} E$ (or $F^\flat$)   **(f)** $A^\flat \xrightarrow{\mathcal{L}} c.$

7.1.3 **(a)** $c^\flat \xrightarrow{\mathcal{P}} E^\flat$ **(b)** $a \xrightarrow{\mathcal{R}} C$ **(c)** $b \xrightarrow{\mathcal{L}} G$ **(d)** $B \xrightarrow{\mathcal{P}} b$ **(e)** $c \xrightarrow{\mathcal{R}} E^\flat$ (or $D^\sharp$)   **(f)** $D \xrightarrow{\mathcal{L}} g^\flat$ (or $f^\sharp$).

7.1.5   The note for the $C_3^\flat$ tone appears in the score for the lyric "von," as shown here:

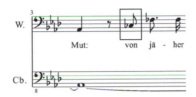

Because $C_3^\flat$ is enharmonic with B_2, its 2nd harmonic will equal the fundamental of B_3, which is approximately 247 Hz (using Table 1.2 on page 13). This B_3 tone can be regarded as flatted from C_4, which has a fundamental of approximately 262 Hz. That frequency is close to the 5th harmonic of $A_1^\flat$, which is approximately $5 \cdot 52 = 260$ Hz. This illustrates how the harmonics in the music are related to the mapping $A^\flat \xrightarrow{\mathcal{P}} a^\flat$ (i.e., $\mathbf{A^\flat}$-C-$E^\flat \xrightarrow{\mathcal{P}} \mathbf{A^\flat}$-$C^\flat$-$E^\flat$).

7.1.7   We have $C \xrightarrow{T_1} D^\flat \xrightarrow{\mathcal{R}} b^\flat$ and $C \xrightarrow{\mathcal{R}} a \xrightarrow{T_1} b^\flat$, so both $T_1\mathcal{R}$ and $\mathcal{R}T_1$ map C to $b^\flat$. Furthermore, $c \xrightarrow{T_1} d^\flat \xrightarrow{\mathcal{R}} E$ and $c \xrightarrow{\mathcal{R}} E^\flat \xrightarrow{T_1} E$, so both $T_1\mathcal{R}$ and $\mathcal{R}T_1$ map c to E. The geometric interpretations of $\mathcal{R}$ and T_1 are now used. We have that T_1 is a rotation of the chromatic clock by one hour clockwise, and $\mathcal{R}$ is a reflection about the mirror line through the side of a chordal triangle of chromatic distance 4 for either a major or minor chord. Therefore, it follows that $T_1\mathcal{R}$ and $\mathcal{R}T_1$ will produce the same chord for either a major or minor triangle by simply rotating the diagrams that describe the result for C and c. For example, if our starting chord were E, we could plot the pitch classes for the chromatic scale on the chromatic clock, starting with **E** at hour 0. Then we would get that $\mathcal{R}T_1$ and $T_1\mathcal{R}$ map E to the same chord, because we have shown they do so for C. A similar argument applies for e, and so for any major or minor chords.

7.1.9   We have $C \xrightarrow{T_1} D^\flat \xrightarrow{\mathcal{L}} f$ and $C \xrightarrow{\mathcal{L}} e \xrightarrow{T_1} f$, so both $T_1\mathcal{L}$ and $\mathcal{L}T_1$ map C to f. Furthermore, $c \xrightarrow{T_1} d^\flat \xrightarrow{\mathcal{L}} A$ and $c \xrightarrow{\mathcal{L}} A^\flat \xrightarrow{T_1} A$, so both $T_1\mathcal{L}$ and $\mathcal{L}T_1$ map c to A. The geometric interpretations of $\mathcal{L}$ and T_1 are now used. We have that T_1 is a rotation of the chromatic clock by one hour clockwise, and $\mathcal{L}$ is a reflection about the mirror line through the side of a chordal triangle of chromatic distance 3 for either a major or minor chord. Therefore, it follows that $T_1\mathcal{L}$ and $\mathcal{L}T_1$ will produce the same chord for either a major or minor triangle by simply rotating the diagrams that describe the result for C and c. For example, if our starting chord were A, we could plot the pitch classes for the chromatic scale on the chromatic clock, starting with **A** at hour 0. Then we would get that $\mathcal{L}T_1$ and $T_1\mathcal{L}$ map A to the same chord, because we have shown they do so for C. A similar argument applies to the chord a, and so for any major or minor chords.

7.1.11   To show that $T_k\mathcal{P} = \mathcal{P}T_k$, we observe that $\mathcal{P}$ simply lowers the middle pitch of a major chord by a half step, and raises the middle pitch of a minor chord by a half step. That operation is clearly independent of the rotation by T_k. Therefore, we can either apply T_k before applying $\mathcal{P}$ or after applying T_k, the result will be the same either way. To show that $T_k\mathcal{L} = \mathcal{L}T_k$, we first observe that $T_1\mathcal{L} = \mathcal{L}T_1$ was shown for Exercise 7.1.9. But then we have

$$
\begin{aligned}
T_2\mathcal{L} &= T_1 T_1 \mathcal{L} \\
&= T_1 \mathcal{L} T_1 \\
&= \mathcal{L} T_1 T_1 \\
&= \mathcal{L} T_2
\end{aligned}
$$

which proves that $T_2\mathcal{L} = \mathcal{L}T_2$. A similar chain of steps can be done to prove that $T_k\mathcal{L} = \mathcal{L}T_k$ for all $k = 3, 4, \ldots$. Since $T_{-j} = T_{12-j}$ for transpositions acting on pitch classes, we will also have $T_k\mathcal{L} = \mathcal{L}T_k$ for all negative integers k. And since T_0 makes no changes in any pitch classes, we clearly have $T_0\mathcal{L} = \mathcal{L}T_0$. Consequently, we have shown that $T_k\mathcal{L} = \mathcal{L}T_k$ for all integers k.

Section 7.2

7.2.1 The notes for the first three chords are taken from the pitch classes for the chord c. (See Figure A.9.) The notes for the next two chords are taken from the pitch classes for the chord $A^\flat$. The notes for the next four chords are taken from the pitch classes for the chord f. The notes for the last chord are taken from the pitch classes for the chord F. The chord progressions for these five measures are related to $\mathcal{P}, \mathcal{R}, \mathcal{L}$ as follows:

$$c \xrightarrow{\mathcal{L}} A^\flat \xrightarrow{\mathcal{R}} f \xrightarrow{\mathcal{P}} F.$$

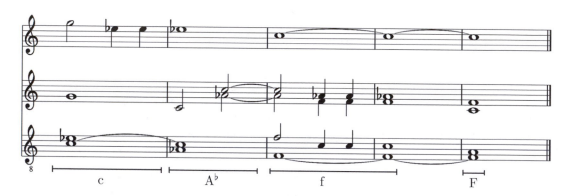

Figure A.9. Location of chords for the ending of the Gesualdo madrigal, "Luci serene e chiare."

7.2.3 The notes for the first measure, and the very beginning of the second measure, are largely taken from the pitch classes for the chord b. The notes in the remainder of the second measure are largely taken from the pitch classes for the chord D, which occurs at the end of the second viola crescendo, followed by the chord A. The chord progressions for these two measures are related to $\mathcal{P}, \mathcal{R}, \mathcal{L}$ as follows:

$$b \xrightarrow{\mathcal{R}} D \xrightarrow{\mathcal{L}\mathcal{R}} A.$$

7.2.5 (a) The sequence of mappings is

$$C \xrightarrow{\mathcal{P}} c \xrightarrow{\mathcal{R}} E^\flat \xrightarrow{\mathcal{P}} e^\flat \xrightarrow{\mathcal{R}} G^\flat \xrightarrow{\mathcal{P}} g^\flat \xrightarrow{\mathcal{R}} A \xrightarrow{\mathcal{P}} a \xrightarrow{\mathcal{R}} C$$

or

$$C \xrightarrow{\mathcal{R}} a \xrightarrow{\mathcal{P}} A \xrightarrow{\mathcal{R}} g^\flat \xrightarrow{\mathcal{P}} G^\flat \xrightarrow{\mathcal{R}} e^\flat \xrightarrow{\mathcal{P}} E^\flat \xrightarrow{\mathcal{R}} c \xrightarrow{\mathcal{P}} C.$$

Either way, only 8 distinct chords are used. **(b)** The sequence of mappings is

$$C \xrightarrow{\mathcal{P}} c \xrightarrow{\mathcal{L}} A^\flat \xrightarrow{\mathcal{P}} a^\flat \xrightarrow{\mathcal{L}} E \xrightarrow{\mathcal{P}} e \xrightarrow{\mathcal{L}} C$$

or

$$C \xrightarrow{\mathcal{L}} e \xrightarrow{\mathcal{P}} E \xrightarrow{\mathcal{L}} a^\flat \xrightarrow{\mathcal{P}} A^\flat \xrightarrow{\mathcal{L}} c \xrightarrow{\mathcal{P}} C.$$

Either way, only 6 distinct chords are used.

Section 7.3

7.3.1 The chordal spine for the G-major scale is $D - b - G - e - C - a$. (See Figure A.10.)

7.3.3 The chordal spine for the B-major scale is $G^b - e^b - B - a^b - E - d^b$. (See Figure A.10.)

7.3.5 The chordal spine for the A^b-major scale is $E^b - c - A^b - f - D^b - b^b$. (See Figure A.10.)

7.3.7 **(a)** The regions of activation on the *Tonnetz* for the whole tone scale are shown in Figure A.10. **(b)** There is no chordal spine for the whole tone scale. One cannot form any major or minor triadic chords from its pitches. **(c)** Two geometric properties distinguishing this whole tone scale from all other scales we have considered are: (1) it has no chordal spine; (2) the hexagons for its pitch classes on the *Tonnetz* do not form a single connected region.

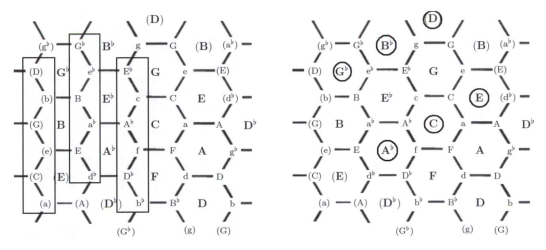

Figure A.10. Left: three chordal spines on *Tonnetz*. Leftmost rectangle encloses chordal spine for G-major scale. Middle rectangle encloses chordal spine for B-major scale. Rightmost rectangle encloses chordal spine for A^b-major scale. Right: Six circles mark off hexagons activated by pitch classes of whole tone scale.

Section 7.4

7.4.1 Six mirror lines can be drawn through the middle of opposite sides, or through opposite corners of each regular hexagon on the *Tonnetz*. For example, see Figure A.11, where reflections through these mirrors, and the *Tonnetz* transformations that they give, are shown for the regular hexagon **G**. It is easily checked that every possible *Tonnetz* transformation for moving across this hexagon, or along one of its edges, is described by one of the six reflections shown there. Since all of the hexagons in the *Tonnetz* were constructed in the same way, it follows that every *Tonnetz* transformation can be realized by one of these six types of reflection. Since a reflection is self-inversive, it follows that all *Tonnetz* transformations are self-inversive.

7.4.3 The graph of the moves for the *Tonnetz* transformations is shown on the left of Figure A.13.

7.4.5 The *Tonnetz* transformation is $E \xrightarrow{T} a$, corresponding to a move across the hexagon **E**.

7.4.7 The chords for the passage are shown in Figure A.12. The *Tonnetz* transformations are

$$F \xrightarrow{T} C \xrightarrow{T} F \xrightarrow{T} B^b \xrightarrow{T} g \xrightarrow{T} D \xrightarrow{T} G \xrightarrow{T} B$$

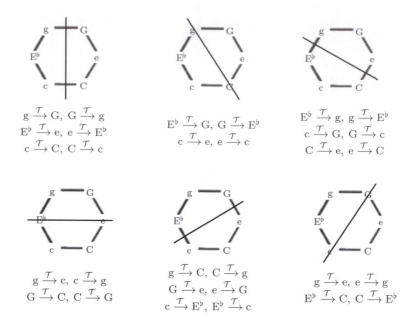

$$g \xrightarrow{T} G, \; G \xrightarrow{T} g$$
$$E^\flat \xrightarrow{T} e, \; e \xrightarrow{T} E^\flat$$
$$c \xrightarrow{T} C, \; C \xrightarrow{T} c$$

$$E^\flat \xrightarrow{T} G, \; G \xrightarrow{T} E^\flat$$
$$c \xrightarrow{T} e, \; e \xrightarrow{T} c$$

$$E^\flat \xrightarrow{T} g, \; g \xrightarrow{T} E^\flat$$
$$c \xrightarrow{T} G, \; G \xrightarrow{T} c$$
$$C \xrightarrow{T} e, \; e \xrightarrow{T} C$$

$$g \xrightarrow{T} c, \; c \xrightarrow{T} g$$
$$G \xrightarrow{T} C, \; C \xrightarrow{T} G$$

$$g \xrightarrow{T} C, \; C \xrightarrow{T} g$$
$$G \xrightarrow{T} e, \; e \xrightarrow{T} G$$
$$c \xrightarrow{T} E^\flat, \; E^\flat \xrightarrow{T} c$$

$$g \xrightarrow{T} e, \; e \xrightarrow{T} g$$
$$E^\flat \xrightarrow{T} C, \; C \xrightarrow{T} E^\flat$$

Figure A.11. Six reflectional symmetries for hexagon **G** in *Tonnetz*, and *Tonnetz* transformations they describe.

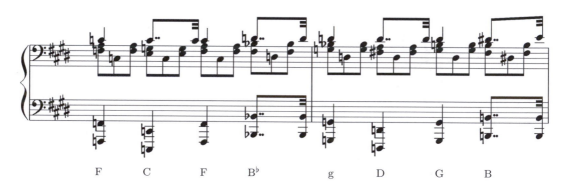

Figure A.12. Chordal analysis of measures 10 and 11 of Chopin's *E Major Prelude, Op. 28 No. 9*.

The graph of the moves for the *Tonnetz* transformations are shown on the right of Figure A.13.

7.4.9 In Tymoczko's model for chord progressions in a major key, the progressions generated by one *Tonnetz* transformation are

$$\mathbf{I} \xrightarrow{T} \mathbf{vi}, \quad \mathbf{I} \xrightarrow{T} \mathbf{IV}, \quad \mathbf{I} \xrightarrow{T} \mathbf{V}, \quad \mathbf{ii} \xrightarrow{T} \mathbf{V}, \quad \mathbf{IV} \xrightarrow{T} \mathbf{ii},$$

$$\mathbf{IV} \xrightarrow{T} \mathbf{I}, \quad \mathbf{V} \xrightarrow{T} \mathbf{I}, \quad \mathbf{vi} \xrightarrow{T} \mathbf{IV}, \quad \mathbf{vi} \xrightarrow{T} \mathbf{ii}, \quad \mathbf{vi} \xrightarrow{T} \mathbf{I},$$

and the progressions generated by two *Tonnetz* transformations are

$$\mathbf{I} \xrightarrow{TT} \mathbf{ii}, \quad \mathbf{IV} \xrightarrow{TT} \mathbf{V}, \quad \mathbf{V} \xrightarrow{TT} \mathbf{vi}, \quad \mathbf{V} \xrightarrow{TT} \mathbf{IV}, \quad \mathbf{vi} \xrightarrow{TT} \mathbf{V}.$$

7.4.11 In the chord progressions for a major key given in Figure 3.10, the progressions generated by one *Tonnetz*

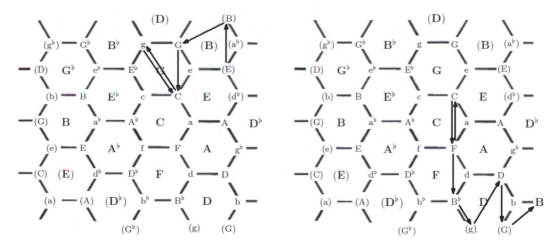

Figure A.13. Left: Moves on *Tonnetz* for Chopin passage in Figure 7.18 on p. 235. Right: Moves on *Tonnetz* for Chopin passage in Figure A.12.

transformation are

$$I \xrightarrow{\mathcal{T}} IV, \quad I \xrightarrow{\mathcal{T}} iii, \quad I \xrightarrow{\mathcal{T}} vi, \quad I \xrightarrow{\mathcal{T}} V,$$

$$ii \xrightarrow{\mathcal{T}} V, \quad iii \xrightarrow{\mathcal{T}} vi, \quad IV \xrightarrow{\mathcal{T}} ii, \quad IV \xrightarrow{\mathcal{T}} I,$$

$$V \xrightarrow{\mathcal{T}} I, \quad vi \xrightarrow{\mathcal{T}} ii, \quad vi \xrightarrow{\mathcal{T}} IV,$$

$$(\text{after } V \to vi): \quad vi \xrightarrow{\mathcal{T}} ii, \quad vi \xrightarrow{\mathcal{T}} IV, \quad vi \xrightarrow{\mathcal{T}} iii, \quad vi \xrightarrow{\mathcal{T}} I,$$

and the progressions generated by two *Tonnetz* transformations are

$$I \xrightarrow{\mathcal{TT}} ii, \quad iii \xrightarrow{\mathcal{TT}} IV, \quad IV \xrightarrow{\mathcal{TT}} V, \quad V \xrightarrow{\mathcal{TT}} vi,$$

$$(\text{after } V \to vi): \quad vi \xrightarrow{\mathcal{TT}} V.$$

7.4.13 See Table A.1.

Table A.1 CHORDS CONNECTED TO TONIC IN SIX MODES OF C-MAJOR SCALE.

Mode	Tonic chord	$\mathcal{T}$ changes type	$\mathcal{T}$ preserves type	Use $\mathcal{TT}$
Major (Ionian)	C	a, e	F, G	d
Dorian	d	F, G	a	C, e
Mixolydian	G	e, d	C	F, a
Phrygian	e	G, C	a	F, d
Lydian	F	a, d	C	G, e
Aeolian	a	C, F	d, e	G

7.4.15 All of the chord progressions correspond to single *Tonnetz* transformations, except for $D^7 \to E^7$ which corresponds to two *Tonnetz* transformations. In Figure A.14 we show the movement on the *Tonnetz* corresponding to these chord progressions. The modulations are between the keys of G-major and A-major.

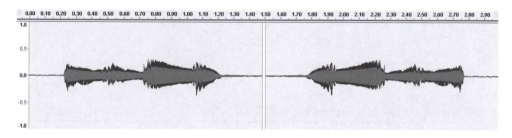

Figure A.14. Moves on *Tonnetz* for *Blue Light* chord progressions in (7.13) on p. 240. The thicker arrows correspond to moves that are done multiple times.

Chapter 8

Section 8.1

8.1.1 In the following figure we have graphed the waveform from `piano_clip.wav` on the left, and its time-reversal on the right:

The original shows quick onsets for the piano tones, due to the percussive effects of the hammer striking the piano strings. The piano tones end with a gentle decay. For the time-reversed tones, the onset and decay occur in reverse order, so the time-reversed tones have a gentle onset. This gentle onset creates a violin-like timbre. The rapid decay of the time-reversed tones somewhat alters the effect, however, so these tones are not perfect simulations of violin tones.

8.1.3 Equation (8.4) is modified to have this form:

$$g_j(t_m) = \sum_{k=1}^{L} \mathsf{w}_{j,k}(t_m)\mathcal{A}_j(k\nu_o)\cos(2\pi \cdot k\nu_o t_m - \phi_{j,k})$$

where the functions $\mathsf{w}_{j,k}(t_m)$ provide frequency-dependent amplitude control. Equivalently, this equation could be expressed as

$$g_j(t_m) = \sum_{k=1}^{L} \mathsf{A}_{j,k}(t_m)\cos(2\pi \cdot k\nu_o t_m - \phi_{j,k}) \tag{A.15}$$

by subsuming the amplitude constants $\mathcal{A}_j(k\nu_o)$ as part of single amplitude functions $A_{j,k}(t_m)$ for each k. ***Note:*** If the windowings in Equation (8.4) are of very short duration, then there is not a great difference between Equations (8.4) and (A.15). When the windowings are of longer duration, however, then Equation (A.15) is better able to model tones like the trumpet tones shown in Figure 8.3.

8.1.5 Envelope control refers to the window functions $w_j(t_m)$ in Equation (8.4).

Section 8.2

8.2.1 The calculation goes as follows:

$$\left[\widetilde{\mathcal{C}}_j'(k\nu_o)\right]^2 + \left[\widetilde{\mathcal{S}}_j'(k\nu_o)\right]^2$$

$$= \frac{\left[\widetilde{\mathcal{C}}_j(k\nu_o)\right]^2}{\left[\widetilde{\mathcal{A}}_j(k\nu_o)\right]^2} + \frac{\left[\widetilde{\mathcal{S}}_j(k\nu_o)\right]^2}{\left[\widetilde{\mathcal{A}}_j(k\nu_o)\right]^2}$$

$$= \frac{\left[\widetilde{\mathcal{C}}_j(k\nu_o)\right]^2 + \left[\widetilde{\mathcal{S}}_j(k\nu_o)\right]^2}{\left[\widetilde{\mathcal{A}}_j(k\nu_o)\right]^2}$$

$$= \frac{\left[\widetilde{\mathcal{C}}_j(k\nu_o)\right]^2 + \left[\widetilde{\mathcal{S}}_j(k\nu_o)\right]^2}{\left[\widetilde{\mathcal{C}}_j(k\nu_o)\right]^2 + \left[\widetilde{\mathcal{S}}_j(k\nu_o)\right]^2}$$

$$= 1.$$

8.2.3 Since $\tan \phi = \dfrac{\sin \phi}{\cos \phi}$, we have

$$\tan \phi = \frac{0.4}{\sqrt{0.84}}$$

Therefore,

$$\phi = \arctan\left(\frac{0.4}{\sqrt{0.84}}\right)$$

$$= 0.4115168\ldots$$

Section 8.3

8.3.1 The coefficient data at time τ_1 is copied to time τ_1 without change. At time τ_2, the following formulas are used to assign new values to $\mathcal{C}_2(k\nu_o)$ and $\mathcal{S}_2(k\nu_o)$:

$$\mathcal{C}_2^{\text{new}}(k\nu_o) = \frac{1}{3}\mathcal{C}_1^{\text{old}}(k\nu_o) + \frac{2}{3}\mathcal{C}_2^{\text{old}}(k\nu_o)$$

$$\mathcal{S}_2^{\text{new}}(k\nu_o) = \frac{1}{3}\mathcal{S}_1^{\text{old}}(k\nu_o) + \frac{2}{3}\mathcal{S}_2^{\text{old}}(k\nu_o). \tag{A.16}$$

At time τ_3, the following formulas are used to assign new values to $\mathcal{C}_3(k\nu_o)$ and $\mathcal{S}_3(k\nu_o)$:

$$\mathcal{C}_3^{\text{new}}(k\nu_o) = \frac{2}{3}\mathcal{C}_2^{\text{old}}(k\nu_o) + \frac{1}{3}\mathcal{C}_3^{\text{old}}(k\nu_o)$$

$$\mathcal{S}_3^{\text{new}}(k\nu_o) = \frac{2}{3}\mathcal{S}_2^{\text{old}}(k\nu_o) + \frac{1}{3}\mathcal{S}_3^{\text{old}}(k\nu_o). \tag{A.17}$$

At time τ_4, the *old* coefficient values of $\mathcal{C}_3(k\nu_o)$ and $\mathcal{S}_3(k\nu_o)$ are copied without change to give the new coefficient values for $\mathcal{C}_4(k\nu_o)$ and $\mathcal{S}_4(k\nu_o)$. This step at τ_4 is the beginning of a new cycle of interpolations of coefficients, similar to what we just described. This cycling is repeated as long as necessary to achieve the time-stretching.

8.3.3 The coefficient data at time τ_1 is copied to time τ_1 without change. At time τ_2, the following formulas are used to assign new values to $\mathcal{C}_2(k\nu_o)$ and $\mathcal{S}_2(k\nu_o)$:

$$\mathcal{C}_2^{\mathrm{new}}(k\nu_o) = \frac{1}{4}\,\mathcal{C}_1^{\mathrm{old}}(k\nu_o) + \frac{3}{4}\,\mathcal{C}_2^{\mathrm{old}}(k\nu_o)$$

$$\mathcal{S}_2^{\mathrm{new}}(k\nu_o) = \frac{1}{4}\,\mathcal{S}_1^{\mathrm{old}}(k\nu_o) + \frac{3}{4}\,\mathcal{S}_2^{\mathrm{old}}(k\nu_o). \tag{A.18}$$

At time τ_3, the following formulas are used to assign new values to $\mathcal{C}_3(k\nu_o)$ and $\mathcal{S}_3(k\nu_o)$:

$$\mathcal{C}_3^{\mathrm{new}}(k\nu_o) = \frac{1}{2}\,\mathcal{C}_2^{\mathrm{old}}(k\nu_o) + \frac{1}{2}\,\mathcal{C}_3^{\mathrm{old}}(k\nu_o)$$

$$\mathcal{S}_3^{\mathrm{new}}(k\nu_o) = \frac{1}{2}\,\mathcal{S}_2^{\mathrm{old}}(k\nu_o) + \frac{1}{2}\,\mathcal{S}_3^{\mathrm{old}}(k\nu_o). \tag{A.19}$$

At time τ_4, the following formulas are used to assign new values to $\mathcal{C}_4(k\nu_o)$ and $\mathcal{S}_4(k\nu_o)$:

$$\mathcal{C}_4^{\mathrm{new}}(k\nu_o) = \frac{3}{4}\,\mathcal{C}_3^{\mathrm{old}}(k\nu_o) + \frac{1}{4}\,\mathcal{C}_4^{\mathrm{old}}(k\nu_o)$$

$$\mathcal{S}_3^{\mathrm{new}}(k\nu_o) = \frac{3}{4}\,\mathcal{S}_3^{\mathrm{old}}(k\nu_o) + \frac{1}{4}\,\mathcal{S}_4^{\mathrm{old}}(k\nu_o). \tag{A.20}$$

At time τ_5, the *old* coefficient values of $\mathcal{C}_4(k\nu_o)$ and $\mathcal{S}_4(k\nu_o)$ are copied without change to give the new coefficient values for $\mathcal{C}_5(k\nu_o)$ and $\mathcal{S}_5(k\nu_o)$. This step at τ_5 is the beginning of a new cycle of interpolations of coefficients, similar to what we just described. This cycling is repeated for as long as necessary to achieve the time-stretching.

8.3.5 The coefficient data at time τ_1 is copied to time τ_1 without change. The coefficient data for the windowing beginning at τ_2 is copied to a windowing beginning at the time $\tau_{5/3}$ defined by

$$\tau_{5/3} = \frac{1}{3}\,\tau_1 + \frac{2}{3}\,\tau_2.$$

The coefficient data for the windowing beginning at τ_3 is copied to a windowing beginning at the time $\tau_{7/3}$ defined by

$$\tau_{7/3} = \frac{2}{3}\,\tau_2 + \frac{1}{3}\,\tau_3.$$

The coefficient data for the windowing beginning at τ_4 is copied to the windowing beginning at τ_3. This step at τ_4 is the beginning of a new cycle of time-reassignments, similar to what we just described. This cycling is repeated for as long as necessary to achieve the time-shrinkage.

8.3.7 The coefficient data at time τ_1 is copied to time τ_1 without change. The coefficient data for the windowing beginning at τ_2 is copied to a windowing beginning at the time $\tau_{7/4}$ defined by

$$\tau_{7/4} = \frac{1}{4}\,\tau_1 + \frac{3}{4}\,\tau_2.$$

The coefficient data for the windowing beginning at τ_3 is copied to a windowing beginning at the time $\tau_{5/2}$ defined by

$$\tau_{5/2} = \frac{1}{2}\,\tau_2 + \frac{1}{2}\,\tau_3.$$

The coefficient data for the windowing beginning at τ_4 is copied to a windowing beginning at the time $\tau_{13/4}$ defined by

$$\tau_{13/4} = \frac{3}{4}\,\tau_3 + \frac{1}{4}\,\tau_4.$$

The coefficient data for the windowing beginning at τ_5 is copied to the windowing beginning at τ_4. This step at τ_5 is the beginning of a new cycle of time-reassignments, similar to what we just described. This cycling is repeated for as long as necessary to achieve the time-shrinkage.

8.3.9 The drum tonalities are not related to either of the keys involved. Moreover, their percussive onsets contain high amplitudes from the entire range of audible frequencies. Therefore, when the music is lowered in pitch for one part, and raised in pitch for the other part, the difference in pitch within the drum beats is clearly audible.

Section 8.4

8.4.1 The first four rectangles for the MIDI time durations have wider vertical spacing between the second and third notes, and between the third and fourth notes, than the vertical spacing between the rectangles for the remaining four notes. This difference in spacing represents a difference in the frequency ratios for the fundamentals of the notes that these rectangles correspond to in the score. A transposition would keep these frequency ratios, which correspond to the number of half steps between the notes, unchanged. So a transposition is **not** being used. On the other hand, the four notes in the score, corresponding to the first four MIDI rectangles, are moved down as a group by one position on the score to create the second group of four notes. Shifting down by one score position is what the diatonic scale shift S_{-1} does to those first four notes.

8.4.3 A succession of mirror symmetric structures is evident in the upper half of the MIDI time-durations and in the upper staff of the score. This succession of mirror symmetric structures corresponds to a succession of palindromes in the score.

Appendix B

Music Software

There are two computer programs that we have used extensively. AUDACITY is a free audio processor with reasonably good capabilities (especially considering its zero cost). MUSESCORE is a good music scoring program. It is also free.

B.1 AUDACITY

The installation program for AUDACITY can be found at the Software link at the book's web site. An Internet search on the keyword "Audacity" will also get you to it. After AUDACITY is installed, then it must be configured for analyzing music.

B.1.1 Configuring AUDACITY

To configure AUDACITY for analyzing music, select the Edit menu and choose Preferences. (On a MAC, you select Preferences from the Audacity menu.) From the list of choices shown on the left of the preferences window, select Spectrograms. Then make the following three changes:

1. Change the FFT size, which is listed as 256, to 4096.

2. Change the Window type, which is listed as Hanning, to Blackman.

3. Change the Maximum frequency, which is listed as 8000, to 3000.

After making these changes, click the OK button. AUDACITY is now configured for analyzing music. It will remember these settings in the future, so you won't have to change them again. The only exception is the setting for the Maximum frequency value, which you might want to change for different types of music.[1]

B.1.2 Loading and Displaying a Music File

To load a music file, select the File menu and choose Open. (Or, you can also choose Import and then Audio.) AUDACITY recognizes most common formats for music files, such as .wav and .mp3 files. When you load a music file, you will see a blue-colored graph displayed on the screen. It is a graph of the electrical current oscillations in the recording of the music, the music's waveform. If you want a spectrogram displayed, then you must do the following. Click on a small box on the left of the screen, adjacent to the graph displayed, containing the name of the music file. (See Figure B.1.) A menu will appear, and you select Spectrogram from this menu. (Do not select Spectrogram log(f), we will not be using that option.) After you have selected Spectrogram, AUDACITY will display a spectrogram of the music.

---

[1] There are a limited number of situations where precise frequency identification is crucial. In such cases, an FFT size greater than 4096 points can be specified. For example, an FFT size of 32768 will guarantee a frequency accuracy within ± 1 Hz for recorded music.

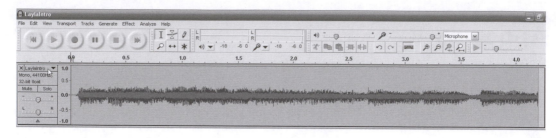

Figure B.1. Producing spectrogram in AUDACITY. Mouse pointer shown on left, at box where you click to display menu for plotting spectrogram.

To get a good view of the spectrogram, it is best to grab the lower border of its graph and pull it down to expand the view. Also, if the music you selected is a stereo recording, then two spectrograms are displayed. Often it is best to just have one spectrogram. To plot just one spectrogram for stereo music, select the Tracks menu and choose Stereo to mono.

B.1.3   Music Files from CDs

One thing that AUDACITY cannot do is directly load music files from CDs. On the AUDACITY web site they recommend using iTunes or CDEX to extract music files from CDs and save them in a form that AUDACITY can read. You can find further information by going to the AUDACITY web site and doing a search using *Importing from CDs*. Another program that converts CD music files is FREE STUDIO MANAGER. It can be downloaded from the Software link at the book's web site. FREE STUDIO MANAGER also provides a great number of quite helpful audio and video utilities.

B.2   MUSESCORE

The installation program for MUSESCORE can be found at the Software link at the book's web site. An Internet search on the keyword "MuseScore" will also get you to it. Once the program is installed, then it is ready to play scores, or create new scores. The scores discussed in this book can all be downloaded from the book's web site. Once you have those scores on your computer, then you select the File menu and choose Open, and then select a score. To play a score, you click on the rightward pointing arrowhead at the top of the screen (and also click on it to stop playback). More options for playback are described in the help file for MUSESCORE.

B.2.1   Different Soundfonts for MUSESCORE

MUSESCORE uses a free, and very small size, soundfont to create the playback of scores. As described in Section 8.4, a soundfont is a collection of pre-recorded sounds of various instruments that are patched into the durations for individual notes within a musical score. The better the quality of the recordings, and the wider the range of instruments recorded, the better the playback will sound. This constraint generally requires a rather large size soundfont. The soundfont that comes with the initial installation of MUSESCORE is rather small, and the playback suffers accordingly. Fortunately, there is a wide variety of soundfonts available on the Internet. Some are quite expensive. A good free one is available from the Software link at the book's web site. Once you have downloaded this soundfont and placed it in a folder on your computer, then you can tell MUSESCORE to use it. You do that by selecting the Display menu and choosing Synthesizer. You then find the item called SoundFont at the bottom of the window that has opened up, and click on the button to the right of the file name shown there. (This button has an icon on it that is supposed to look like a file folder.) When you click this button, it will open up a file search window for you to find and select the soundfont you saved.

MUSESCORE will then use this new soundfont for playing scores.

MUSESCORE has a variety of options for playing scores, for adjusting volume, for changing instruments, for saving audio recordings of scores, and many other features. Clear descriptions of how to use these options can be found in the MUSESCORE help. This help can be obtained by selecting the Help menu and choosing either Local Handbook or Online Handbook. As an example of using the help, if you search on the word "Mixer," you can find out how to change instruments, add reverberation to the playback, adjust the volume of the playback, and so on. The help also describes how to use MUSESCORE to create your own musical scores. All of the musical scores in this book were created using MUSESCORE.

Appendix C

Amplitude and Frequency Results

In this Appendix we prove two results that were stated in Section 4.7. They are Theorem 4.7.1, and the exact estimations of frequency and amplitude for certain waveforms.

C.1   Proof of Theorem 4.7.1

Here is Theorem 4.7.1, as stated in Section 4.7, followed by a mathematically rigorous proof.

Theorem 4.7.1. *Let $\nu_o = \frac{1}{\Omega}$. For $\ell\nu_o = \frac{\ell}{\Omega}$, with $\ell = 0, \pm 1, \pm 2, \ldots, \pm(N-1)$, and points $t_m = \frac{(m-1)\Omega}{N}$ defined in (4.15), we have*

$$\frac{1}{N} \sum_{m=1}^{N} \sin(2\pi\ell\nu_o t_m) = 0 \tag{C.1}$$

and

$$\frac{1}{N} \sum_{m=1}^{N} \cos(2\pi\ell\nu_o t_m) = \begin{cases} 0, & \text{when } \ell \neq 0 \\ 1, & \text{when } \ell = 0. \end{cases} \tag{C.2}$$

Proof. To prove (C.1), we first rewrite the sum on the left side as

$$\sum_{m=1}^{N} \sin(2\pi\ell\nu_o t_m) = \sum_{m=1}^{N} \sin\left(2\pi\ell \, \frac{m-1}{N}\right). \tag{C.3}$$

We make use of the trigonometric identity

$$2\sin(\theta)\cos(\theta) = \sin(\theta + \phi) - \sin(\theta - \phi),$$

after multiplying both sides of (C.3) by $2\cos\left(\pi\frac{\ell}{N}\right)$:

$$\sum_{m=1}^{N} \sin(2\pi\ell\nu_o t_m) \, 2\cos\left(\pi\frac{\ell}{N}\right) = \sum_{m=1}^{N} \sin\left(2\pi\ell \, \frac{m-1}{N}\right) 2\cos\left(\pi\,\frac{\ell}{N}\right)$$

$$= \sum_{m=1}^{N} \left[\sin\left(2\pi\ell \, \frac{m-1/2}{N}\right) - \sin\left(2\pi\ell \, \frac{m-3/2}{N}\right) \right]$$

$$= \sum_{m=1}^{N} \sin\left(2\pi\ell \, \frac{m-1/2}{N}\right) - \sum_{m=1}^{N} \sin\left(2\pi\ell \, \frac{m-3/2}{N}\right).$$

For the last two sums, we peel off the $m = N$ term from the first sum and peel off the $m = 1$ term from the second sum. By a substitution of $m' = m - 1$ for the index, we obtain

$$\sum_{m=1}^{N} \sin(2\pi\ell\nu_o t_m)\, 2\cos\left(\pi\tfrac{\ell}{N}\right) = \sum_{m=1}^{N} \sin\left(2\pi\ell\,\tfrac{m-1/2}{N}\right) - \sum_{m=1}^{N} \sin\left(2\pi\ell\,\tfrac{m-3/2}{N}\right)$$

$$= \sum_{m=1}^{N-1} \sin\left(2\pi\ell\,\tfrac{m-1/2}{N}\right) + \sin\left(2\pi\ell\,\tfrac{N-1/2}{N}\right)$$

$$- \sum_{m=2}^{N} \sin\left(2\pi\ell\,\tfrac{m-3/2}{N}\right) - \sin\left(2\pi\ell\,\tfrac{-1/2}{N}\right)$$

$$= \sum_{m=1}^{N-1} \sin\left(2\pi\ell\,\tfrac{m-1/2}{N}\right) + \sin\left(2\pi\ell\,\tfrac{N-1/2}{N}\right)$$

$$- \sum_{m'=1}^{N-1} \sin\left(2\pi\ell\,\tfrac{m'-1/2}{N}\right) - \sin\left(2\pi\ell\,\tfrac{-1/2}{N}\right).$$

Since the last two sums on the right have exactly the same terms, they cancel each other out. We then use periodicity of the sine function to obtain:

$$\sum_{m=1}^{N} \sin(2\pi\ell\nu_o t_m)\, 2\cos\left(\pi\tfrac{\ell}{N}\right) = \sin\left(2\pi\ell\,\tfrac{N-1/2}{N}\right) - \sin\left(2\pi\ell\,\tfrac{-1/2}{N}\right)$$

$$= \sin\left(2\pi\ell + 2\pi\ell\,\tfrac{-1/2}{N}\right) - \sin\left(2\pi\ell\,\tfrac{-1/2}{N}\right)$$

$$= \sin\left(2\pi\ell\,\tfrac{-1/2}{N}\right) - \sin\left(2\pi\ell\,\tfrac{-1/2}{N}\right)$$

$$= 0.$$

Thus, we have

$$\sum_{m=1}^{N} \sin(2\pi\ell\nu_o t_m)\, 2\cos\left(\pi\tfrac{\ell}{N}\right) = 0.$$

Now, suppose that $\ell \neq \pm N/2$. For the values of ℓ we are considering, we then have $2\cos\left(\pi\tfrac{\ell}{N}\right) \neq 0$, and so we can divide it out. Multiplying by $1/N$, we then obtain (C.1). For $\ell = \pm N/2$, we have

$$\sum_{m=1}^{N} \sin(2\pi\ell\nu_o t_m) = \sum_{m=1}^{N} \sin\left(\pm\pi \cdot (m-1)\right)$$

$$= 0$$

because each term $\sin\left(\pm\pi(m-1)\right)$ equals 0. Therefore, (C.1) holds for all $\ell = 0, \pm 1, \pm 2, \ldots, \pm(N-1)$. When $\ell = \pm 1, \pm 2, \ldots, \pm(N-1)$, the identity

$$\frac{1}{N}\sum_{m=1}^{N} \cos(2\pi\ell\nu_o t_m) = 0$$

is proved in a similar way, except one begins by multiplying by $2\sin\left(\pi\tfrac{\ell}{N}\right)$. The case of $\ell = 0$ was proved in Section 4.7, and so our proof is complete.   □

C.2   Proof of Exact Amplitude and Frequency Estimates

We now show that our estimators, $\mathcal{C}(k\nu_o)$ and $\mathcal{S}(k\nu_o)$, give exact estimates for waveforms that are finite sums of harmonic oscillators, all having frequencies less than $N/2$ and belonging to our set of test frequencies. Such a waveform can be expressed as

$$f(t) = \sum_{j=1}^{M} \left[\alpha_j \cos(2\pi j \nu_o t) + \beta_j \sin(2\pi j \nu_o t) \right] \tag{C.4}$$

where $M < N/2$. Our cosine estimator $\mathcal{C}(k\nu_o)$ is then

$$\mathcal{C}(k\nu_o) = \frac{2}{N} \sum_{m=1}^{N} \left\{ \sum_{j=1}^{M} \left[\alpha_j \cos(2\pi j \nu_o t_m) + \beta_j \sin(2\pi j \nu_o t_m) \right] \cos(2\pi k \nu_o t_m) \right\}.$$

Using the Distributive Law, we can rewrite the right side of this equation as

$$\mathcal{C}(k\nu_o) = \frac{2}{N} \sum_{m=1}^{N} \left\{ \sum_{j=1}^{M} \left[\alpha_j \cos(2\pi j \nu_o t_m) \cos(2\pi k \nu_o t_m) \right. \right.$$
$$\left. \left. + \beta_j \sin(2\pi j \nu_o t_m) \cos(2\pi k \nu_o t_m) \right] \right\}. \tag{C.5}$$

Since the two sums in Equation (C.5) have a finite number of terms, we can rearrange these terms in any order. Therefore, we can rewrite Equation (C.5) as follows:

$$\mathcal{C}(k\nu_o) = \sum_{j=1}^{M} \frac{2}{N} \left\{ \sum_{m=1}^{N} \left[\alpha_j \cos(2\pi j \nu_o t_m) \cos(2\pi k \nu_o t_m) \right. \right.$$
$$\left. \left. + \beta_j \sin(2\pi j \nu_o t_m) \cos(2\pi k \nu_o t_m) \right] \right\}.$$

Using the Distributive Law, we have

$$\mathcal{C}(k\nu_o) = \sum_{j=1}^{M} \left\{ \alpha_j \left[\frac{2}{N} \sum_{m=1}^{N} \cos(2\pi j \nu_o t_m) \cos(2\pi k \nu_o t_m) \right] \right.$$
$$\left. + \beta_j \left[\frac{2}{N} \sum_{m=1}^{N} \sin(2\pi j \nu_o t_m) \cos(2\pi k \nu_o t_m) \right] \right\}. \tag{C.6}$$

From our results for single harmonic oscillators, we know that

$$\frac{2}{N} \sum_{m=1}^{N} \cos(2\pi j \nu_o t_m) \cos(2\pi k \nu_o t_m) = \begin{cases} 1, & \text{when } k = j \\ 0, & \text{when } k \neq j \end{cases}$$

and

$$\frac{2}{N} \sum_{m=1}^{N} \sin(2\pi j \nu_o t_m) \cos(2\pi k \nu_o t_m) = 0, \text{ for all } k.$$

Applying these identities in Equation (C.6), we obtain

$$\mathcal{C}(k\nu_o) = \alpha_k. \tag{C.7}$$

A similar argument shows that

$$\mathcal{S}(k\nu_o) = \beta_k. \tag{C.8}$$

Equations (C.7) and (C.8) show that our estimators give exact values for every waveform that is a finite sum of harmonic oscillators, all having frequencies less than $N/2$ and belonging to our set of test frequencies. In particular, we have shown why our estimators gave exact values for amplitudes and frequencies in Example 4.5.1.

Appendix D

Glossary

We provide here a very brief glossary of musical terms used in the book. If you cannot find the term that you are looking for here, or in the book's index, then you could consult the following link:

Online Music Dictionary

at the book's web site (see Links). This dictionary defines a multitude of musical terms. It is an excellent resource.

a cappella   Singing without instrumental accompaniment.

accidental   A note that does not belong to the key for the music.

arpeggio or **arpeggiated chord**   Chords with their notes played out in a sequence, rather than simultaneously.

basilar membrane   Membrane in the inner ear, composed of a sequence of tiny hairs (cilia) that respond to frequencies in a sound wave.

blue note   Slight variation in pitch of a note, usually flatter. Often blue notes are the third, fifth, and seventh notes of a major scale played a quarter step flatter. They are used in both blues and jazz music.

brass instruments   A family of musical instruments which includes horns, trumpets, cornets, and trombones.

fundamental   1st harmonic of a musical tone, which determines its pitch.

graphic equalizer   An electronic device that allows selective amplification (or reduction) of volume of different ranges of frequency.

legato   Tones played with smooth flowing transitions. (Compare **staccato**.)

ostinato   Continual repetition of a note, or short sequence of notes, over an extended period of time. A short rhythmic, or harmonic, pattern can also be repeated as an ostinato.

overtones   2nd, 3rd, 4th, ..., harmonics of a musical tone. (Compare **fundamental**.)

phoneme   An indivisible unit of sound in speech. Speech is constructed from combinations of phonemes.

riff   An extended melodic structure, often used as a repeated motif. Riffs are frequently employed in jazz music, and also in rock and blues.

staccato   Tones played with sharp transitions. (Compare **legato**.)

staves   The plural of staff. In some countries, a staff is called a stave.

tremolo   Rapid fluctuation in volume of a tone.

vibrato   A periodic fluctuation (oscillation) of pitch of a tone.

Appendix E

Permissions

We gratefully acknowledge the following for permission to reprint copyrighted material:

- Carrie Magin, excerpts from her composition, *Hegira* © Carrie Magin.

- Dmitri Tymoczko, excerpt from his composition, *Piano Games* © Dmitri Tymoczko.

- *Notices of the American Mathematical Society,* figures adapted from

    G.W. Don, K.K. Muir, G.B. Volk, J.S. Walker. Music: Broken Symmetry, Geometry, and Complexity.

    Notices of the American Mathematical Society, **51** (2010), 30–49. © American Mathematical Society.

Every effort has been made to identify copyright holders of previously published materials included in this book. The publisher apologies for any oversights that may have occurred; any errors that may have been made will be corrected in subsequent printings upon notification to the publisher.

Bibliography

J.F. Alm and J.S. Walker. Time-frequency analysis of musical instruments. *SIAM Review,* **44** (2002), 457–476. Available at the link, References, on book's web site.

W. Anku. Circles and time: A theory of structural organization of rhythm in African music. *Music Theory Online,* **6**, Jan. 2000.

G. Assayag, H. Feichtinger, and J.-F. Rodrigues, (Eds.). *Mathematics and Music: a Diderot Mathematical Forum.* Springer, New York, NY, 2002.

P. Ball. *The Music Instinct: How Music Works and Why We Can't Do Without It.* Oxford Univ. Press, Oxford, 2010.

J.W. Beauchamp (Ed.). *Analysis, Synthesis, and Perception of Musical Sounds: The Sound of Music.* Springer, New York, NY, 2007.

D. Barenboim. *Music Quickens Time.* Verso, London, 2008.

L. Bernstein. *The Joy of Music.* Amadeus Press, Pompton Plains, NJ, 2004.

X. Cheng, J.V. Hart, and J.S. Walker. Time-frequency analysis of musical rhythm. *Notices of the American Mathematical Society,* **56** (2009), 344–360. Available at the link, References, on book's web site.

R. Cogan. *New Images of Musical Sound.* Harvard University Press, Cambridge, MA, 1984.

D. Cooke. *The Language of Music.* Oxford Univ. Press, London, 1959.

G. Cooper and L. Meyer. *The Rhythmic Structure of Music.* Univ. of Chicago, Chicago, IL, 1960.

A. Copland. *What to Listen for in Music.* McGraw-Hill, New York, NY, 1957.

A.S. Crans, T.M. Fiore, and R. Satyendra. Musical Actions of Dihedral Groups. *American Mathematical Monthly,* **116** (2009), 479–495.

S. Dance. *The World of Duke Ellington.* Scribner, New York, NY, 1970.

G.W. Don. Brilliant colors provocatively mixed: Overtone structures in the music of Debussy. *Music Theory Spectrum,* **23** (2001), 61–73.

G.W. Don, K.K. Muir, G.B. Volk, and J.S. Walker. Music: Broken Symmetry, Geometry, and Complexity. *Notices of the American Mathematical Society,* **51** (2010), 30–49. Available at the link, References, on book's web site.

W. Everett. *The Beatles As Musicians: Revolver through the Anthology.* Oxford Univ. Press, New York, NY, 1999.

J. Fauvel, R. Flood, and R. Wilson, Eds. *Music and Mathematics: From Pythagoras to Fractals.* Oxford Univ. Press, New York, NY, 2003.

R. Fleming. *The Inner Voice: The Making of a Singer.* Viking, New York, NY, 2004.

S. Flinn, Shepard's Tones web page. Available at Shepard Tones Demo at book's web site (see Videos/Chapter 5).

A. Forte. *The Structure of Atonal Music.* Yale Univ. Press, New Haven, CT, 1973.

H. Goodall. *20th Century Greats — The Beatles.* Channel Four Television, 2004.

H. Helmholtz. *On the Sensations of Tone.* Dover, New York, NY, 1954.

A. Hodeir. *Jazz: Its Evolution and Essence*. Da Capo Press, New York, NY, 1975.

D. Huron. *Sweet Anticipation: Music and the Psychology of Expectation*. MIT Press, Cambridge, MA, 2006.

A.M. Jones. *Studies in African Music*. Oxford Univ. Press, London, 1959.

S. Kostka and D. Payne. *Tonal Harmony, with an Introduction to Twentieth Century Music, Sixth Edition*. McGraw-Hill, New York, NY, 2009.

D. Locke. *Drum Gahu: An Introduction to African Rhythm*. White Cliffs Media, Tempe, AZ, 1998.

METASYNTH music software. Available at the link, Software, on book's web site.

S. McCoy. *Your Voice: An Inside View. Multimedia Voice Science and Pedagogy*. Inside View Press, Princeton, NJ, 2004.

R. Miller. *On the Art of Singing*. Oxford Univ. Press, New York, NY, 1996.

H.F. Olson. *Music, Physics and Engineering*. Dover Publications, New York, NY, 1967.

W. Piston and M. Devoto. *Harmony, 4th edition*. Norton, New York, NY, 1978.

R. Plomp and W.J.M. Levelt, Tonal consonance and critical bandwidth. *J. Acoust. Soc. Am.*, **38** (1965), 548–560.

J. Powell. *How Music Works*. Little, Brown and Company, New York, NY, 2010.

G. Read. *Music Notation, 2nd Edition*. Taplinger Publishing, New York, NY, 1979.

J. Roederer. *The Physics and Psychophysics of Music, 4th Edition*. Springer, New York, NY, 2008.

C. Rosen. *Beethoven's Piano Sonatas: A Short Companion*. Yale Univ. Press, New Haven, CT, 2002.

C. Rosen. *The Classical Style: Haydn, Mozart, Beethoven*. Norton, New York, NY, 1997.

C. Rosen. *The Romantic Generation*. Harvard Univ. Press, Cambridge, MA, 1995.

G. Schuller. *The Swing Era: The Development of Jazz. 1930–1945*. Oxford Univ. Press, New York, NY, 1989.

W. Sethares. *Rhythm and Transforms*. Springer, New York, NY, 2007.

W. Sethares. *Tuning, Timbre, Spectrum, Scale, Second Edition*. Springer, London, 2005.

R.N. Shepard. Circularity in judgments of relative pitch. *J. of the Acoustical Soc. of America*, **36** (1964), 2346–53.

E. Stern. Comments on Louis Armstrong. From *Pandolous* web site, www.pandalous.com, with search topic: Louis Armstrong.

I. Stewart. Faggot's fretful fiasco. Chapter 4, pp. 61–76, of J. Fauvel, R. Flood, and R. Wilson, Eds. (2003).

G. Toussaint. Computational geometric aspects of rhythm, melody, and voice leading. *Computational Geometry* (2010), **43**, 2–22.

G. Toussaint. The rhythm that conquered the world: what makes a "good" rhythm good? *Percussive Notes*, Nov. 2011, 52–59.

G. Toussaint. The *Euclidean* algorithm generates traditional musical rhythms. *Proc. of BRIDGES: Mathematical Connections in Art, Music and Science*. Banff, 2005, 47–56

D. Tymoczko. (2010). Lecture Notes for Music 105 course. Available at Tymoczko Lecture Notes at book's web site (see Links).

D. Tymoczko. (2011a). Lecture Notes for Music 106 course. Available at Tymoczko Lecture Notes at book's web site (see Links).

D. Tymoczko. *A Geometry of Music*. Oxford Univ. Press, New York, NY, 2011b.

J.S. Walker. *Fourier Analysis*. Oxford Univ. Press, New York, NY, 1988.

J.S. Walker. *A Primer on Wavelets and their Scientific Applications, Second Edition*. Chapman & Hall/CRC Press, Boca Raton, FL, 2008.

P. Weiss and R. Taruskin, Eds. *Music in the Western World: A History in Documents*, Schirmer Books, 1984.

Index